Defending Life

Defending Life

The Nature of Host-Parasite Relations

Elling Ulvestad
Haukeland University Hospital, Bergen, Norway
The Gade Institute, University of Bergen, Bergen, Norway

Springer

A C.I.P. Catalogue record for this book is available from the Library of Congress.

ISBN 978-1-4020-5675-8 (HB)
ISBN 978-1-4020-5676-5 (e-book)

Published by Springer,
P.O. Box 17, 3300 AA Dordrecht, The Netherlands.

www.springer.com

Printed on acid-free paper

Sisyphus was condemned by the Gods to ceaselessly roll a rock to the top of a mountain whence it would roll down by it's own weight.
Greek mythology

The struggle itself towards the heights is enough to fill a man's heart.
One must imagine Sisyphus happy.
Albert Camus

Scientific endeavour is academic rock'n rolling.
Anonymous

It's only rock'n roll but I like it.
The Rolling Stones

Table of Contents

Introduction **xi**

Preface **xv**

1 Tracks of thought **1**

1.1. Organismal maintenance **2**

Protoimmunology 2

A Sort of Struggle Between the Parasite and Its Prey 5

Towards a Unified View 8

The side-chain theory 8

Instructive theories 9

The clonal selection theory 10

Generation of diversity 10

The hapten-carrier puzzle 11

Adjuvancy 13

1.2. The war metaphor **14**

1.3. Metaimmunological musings **18**

Speculative Thinking 18

Two opinions 18

Self and other 19

Metaimmunology 19

Mediating dualism 21

Participant and spectator 24

Abduction 26

Evaluation of Hypotheses 29

The thought collective 29

The frame of reference 31

A likelihood rationale 34

The Mishap of Theory 36

Theory in immunological practice 36

Explanation and prediction 37

Predictive accuracy 40

Simplicity versus complexity 40

Parsimony and fit to reality 41

Instrumental realism 42

2 Immunobiology **45**

2.1. The received view **46**

Patterns of Interaction 46

The innate and adaptive subsystems 46

	Stratified security	47
	Design principles	48
	Equipping the Adaptive Toolbox	51
	Lymphopoiesis	51
	Selection within	52
	Selection by antigen	53
	Exemption Explained	54
	Memory	54
	Specificity	59
	Tolerance	63
2.2.	**The integrated view**	**65**
	Revising the View	65
	Immunocompetence	67
	The One and the Many	71
2.3.	**The lens metaphor**	**76**
3	***Adaptive plasticity***	**81**
3.1.	**Being-in-the-world**	**83**
	The Knower and the Known	83
	Worldliness	86
	The Self	88
	The tripartite self	88
	The endangered self	90
	The immune self	92
	Subject, object – both or neither?	92
	The emerging self	93
	A relation which relates to itself	94
3.2.	**Signalling behaviour**	**97**
	Signs Matter	97
	Laws and rules	97
	Biosemiotics	101
	Sign-phenomenology	102
	Immunosemiosis	105
	The Reliability of Signs	108
3.3.	**Signal, decision, response**	**109**
	Tuning the Threshold	109
	The Signal Detection Framework	110
	Discrimination	113
4	***Natura naturans***	**117**
4.1.	**Situating life**	**118**
	Life's Nebulous Nature	118
	The demarcation problem	118
	Unearthing the common ancestor	119

Tracking life 121
Transcendence 121
Process 122
Pattern 123
The Composite Organism 125
Conflict vs. integration 125
Development 126
Modularity 127
Self-Maintenance 129
Embodied drive 129
Autopoiesis 132
Apoptosis 134
4.2. Social evolution **136**
Co-Operation and Conflict 136
Social behaviour 136
Altruism and spitefulness 138
Multi-level selection 140
Shaping the Multicellular Organism 140
Communal unicellular organisms 140
Transitions in individuality 142
The rules they are achanging 142
Vulnerability to cheating 143
Conflict modification 144
Evolvability 145
Constraints on Social Evolution 146
The self-other divide 146
The costs of self-defence 147
Evolution of defence mechanisms 150
4.3. Coevolutionary dynamics **151**
Endosymbiosis 151
Symbiogenesis 151
The mitochondrion 153
Transposable elements 154
Attack and Defence 155
The works of Sisyphus 155
Virulence 156
Multi-level dynamics 160
The "Lucky Split" Hypothesis and the Evolution of Antigen Binding Receptors 162
The immunological big bang 162
RAG-insertion 164
The received repertoire 165
The birth and death model 165

Selection for variability 167
Coping with autoreactive receptors 169

5 Disabled defences **173**
5.1. Failure to perform 173
Goal-Directedness 173
Malfunctioning 175
The Disposable Organism 177
5.2. The re-enacting of ancient conflicts 179
The Evolution and Disruption of Individuality 179
Intraorganismal Parasitism 181
Infection 181
Chimerism 182
Mosaicism 184
Autoimmunity – A Conflict Amongst Evolutionary Individuals 186
Associations to be explained 186
Proximate explanations 188
An ultimate explanation 189
Contesting the explanandum 189
Organismal integrity – a precarious undertaking 191
Stocktaking 193
5.3. Environmental challenges 198
Resources Count, Decisions Amount 198
Allergy – The Spatio-Temporal Mismatch Hypothesis 199
Crossing Barriers 202

Appendix **207**
Innate immunity 207
Adaptive immunity 209

References **211**

Index **235**

Introduction

Defending life is of symbiotic interactions between hosts and parasites. More generally, it is of natural processes and patterns, and of conceptual clarifications and explanatory hypotheses. Or stated differently, the book is about relations – between hosts and parasites, between the scientific mind and the world, and of relations that relate to the self. For the latter, the alluded self may be thought of as inhering within the scientific mind, the immune system, or both.

A major contention of the book is that the immune system depends ontologically on the ecosystem in which it is embedded; it would not have the features it has if it was not related in one way or other to parasitic agents and to the host's own cells and tissues. And as of the parasites; their variegated life styles, which are exquisitely adapted to the challenges encountered in their respective niches, would likewise not be what they are if they evolved in non-hostile environments. Thus, both hosts and parasites evolve to defend themselves against challenges and threats from without. Biological meaningfulness, or what makes a response right or wrong, therefore depends on the existence of structures and practices that have evolved and developed in proximity with significant others.

The importance of a relational perspective can be exemplified by considering interactions that take place between the host's antibodies and the parasite's antigens. The two are what they are in terms of each other; antibodies are host products that bind to foreign bodies, and antigens are foreign bodies that induce the formation of antibodies. But although the two have a structural basis that can be defined in molecular terms, their unique properties as antigen and antibody emerge in action. One can not tell the functional activity of an antibody given its molecular structure, and conversely, there is no way to tell the structure of an antibody from its functional profile. The same argument goes for the antigens. Since the functional properties of the molecules may be multiply realised, analysis at the molecular and relational levels reveal different aspects of reality.

It is a commonly held assumption that molecular data are more fundamental than relational data. This belief probably stems from the important role that molecular knowledge plays in experimental biology. Nonetheless, molecular data would be devoid of meaning if they failed to provide information about higher level functions, for example of host-parasite relations. Explanations at the two levels are therefore neither interchangeable nor redundant, and there is, accordingly, an asymmetrical relation between their explanatory scopes. Whichever is seen as most fundamental depends upon the perspective taken as well as on the pragmatic interests of the investigators.

While relations between organisms are mostly believed to take place between independent entities, the relational view takes a peculiar turn when it comes to understanding immunity. In that case the relational view has to account, not only for the relation between host and parasite, but also for the idea that the immune system relates to itself. The immune system is a mediator between the host organism and the parasite; it is the organism's therapeutic agent so to say. The immune system therefore has to relate both to the host organism in which it dwells and to the parasite that it fights. But by relating to the physiological status of the host organism, the immune system inevitably has to relate to itself in the context of the organism. A complete theory about the immune system should thus account for the manners in which the system relates to itself in the context of other-reactivity.

The tight connections between hosts and parasites, which are of both biological and conceptual sorts, ensures that one can be readily recognized by the other – hosts are entities that harbour parasites while parasites are entities that manage to proliferate and cause malfunctioning of the vertebrate host. Nevertheless, simplicity belies a complex world; the reciprocal interchanges and their outcomes, which have wide-ranging consequences for all life on Earth, are as intricate and multifarious as their conceptualisations are demanding. While much is already known about the processes involved, more empirical work is needed. And since the field is rich on detail but poorer on theory, much conceptual work remains to be done as well.

Defending life is of conceptual more than empirical issues. Still, the treatise is absorbed with empirical detail. The rationale for this is, to use an unfashionable term, dialectical – good concepts serve to validate and highlight empirical detail, and reliable facts are needed to corroborate and expand explanatory concepts. This is well exemplified by the research process; investigations of even the most trivial immune response imposes scientists to be highly selective about which events to include and which to exclude from their description. To reduce the arbitrariness of the selective procedure they therefore embed their investigations within a conceptual framework. And to avoid the risk of merely recasting one-sided and unjustified conceptual labels into new but equally biased labels, they evaluate and harmonise concepts with data. Thus, conceptual and empirical investigations are non-arbitrarily connected and serve to reveal interdependent aspects of the same reality.

The explanatory course taken throughout *Defending life* is dynamic and explorative, and life is investigated at all layers – from molecules up through cells, organisms and ecosystems. Alongside the inverse course, which goes from ecological contingencies down to gene-expression profiles, the approach will hopefully facilitate an advanced understanding of immunocompetence as well as its converse, immunoincompetence. The emphasis on analytical abstractions, coherent patterns and generative mechanisms should make possible the distinction between genuine causality and coincidental associations, and thus increase the understanding of why we observe what we observe.

As indicated by the book's title, immunity is an active process that serves to preserve life. But whether the life being preserved is self or non-self, or something in between, is left open for further analysis in the chapters that follow. These chapters further explicate how the immune systems of vertebrate organisms are tied up to ecological contingencies, and how they, as a result of naturally selected control mechanisms, manage to give flexible responses to changing environments.

Chapter 1 provides a recapitulation of the history of immunology as well as an introduction to thought styles in immunology. Chapter 2 gives an overview of the immune system and the standing of immunobiology. The ecology of organismal immunity is discussed further in chapter 3, which also gives an overview of the peculiarities that the immune system as well as the scientific mind face up to as they dwell in their respective worlds. Chapter 4 lays out the evolutionary interactions and selective influences that generated the adaptive immune system. And finally, the knowledge obtained in the previous chapters – especially the idea that the term parasite should denote diverse agents in addition to infectious microbes, including malignant cells, transplanted cells and autoreactive lymphocytes – is utilised in chapter 5 to address aspects of the malfunctioning immune system and to present a unified hypothesis that purports to explain inflammation, immunosenescence, autoimmunity and allergy.

The invited reader of the exposition is any individual interested in the validity of explanatory arguments. The book contains detailed descriptions of the immune system as well as conceptual clarifications, and should be of interest to biologists, medical scientists, physicians and philosophers of science. I believe that many of the discussions are relevant also for students, both at the graduate and PhD levels, and that the transactions will bring forth a new understanding of the field as it relates to life in general.

I recognize that being an expert on all the material covered in a book of this breadth would be an insurmountable task. Since my own fields of expertise are within immunology and microbiology, my treatment of other areas of biology may be biased. And so may my treatment of biophilosophical topics; coming from the sciences, it was especially hard to get a firm grasp on arguments from the phenomenological and existentialist traditions. Fortunately, upon finishing the book I have been surrounded by a corpus of competitive collaborators who unabashedly invaded my arguments. They include Karl-Henning Kalland, Jarl Giske, Per Jakobsen, and Torbjørn Hansen at the University of Bergen, Nils Roll-Hansen at the University of Oslo, and Alfred I. Tauber at the Boston University. Their criticisms were invaluable in resetting the course when positioned in the wrong direction. Their frivolous attacks on my arguments were often rewarded with success. Nevertheless, because I have some defences of my own they were not able to breach all barriers. Any arguments or facts that are conceived of in a mistaken manner, and which have survived their attacks, are therefore entirely of my own making. Surely, the coevolutionary project that we designate science will reveal the arguments' fitness in due time.

Preface

My first encounters with immunology impressed me with a sense of fascination and bewilderment. I was captivated by the scientific challenges involved in understanding the complex demands and tasks that nature had bestowed upon the immune system, while the lack of explanatory coherence that immunologists provided for their favourite biosystem was disenchanting. The textbooks and research papers of the 1980s were preoccupied with a molecular understanding of immune system activation, and conspicuously lacked in-depth discussions of organising principles around which the immune system functioned. At the time immunology was treated largely as a medical speciality, and little regard was given to integrating it with other biological fields or of laying out the conundrums of the science. There were loads of facts, but less conceptual accounting for the facts. Since then, the stack of facts has increased exponentially, and more than one hundred scientific journals are currently available for their publication.

Like most other biomedical sciences, immunology has profited greatly from major advances in analytical technology. The development of microarrays for analysis of gene expression and of flow cytometry for classification and functional analysis of cells have, for example, allowed researchers to gather and pile huge stocks of observational and experimental data. Along with advances in computer technology and statistical methodology these new technologies have already served to reveal intricate biological networks and to allow sophisticated diagnosing of medical maladies. The degree to which the increasing amounts of facts have led to a comparable increase in functional understanding is, nevertheless, contentious.

While there is a definitive risk that scientists may succumb to facts-overload, there is also a risk that technology may conceal rather than reveal aspects of reality, especially if scientists fail to appreciate the shortcomings of the technology utilised. Cells analysed by flow cytometry and genes analysed by microarrays are for instance marked by fluorescent tags to allow their detection. Untagged cells are not revealed and thus go unnoticed, as are genes not included in the microarrays. While concealment of "noisy" data may be an advantage in many experiments, concealment is definitively a nuisance if it leaves important assets of the cell population or gene network out of reach for the investigator.

Whether observational data contribute to revelation or concealment of reality can not be settled from the data alone; it is the data's role as evidence in favour of one or another hypothesis that serves this function. In the 1920s the German philosopher Martin Heidegger acknowledged that science proceeds as much by conceptual and contextual clarification as by factual gatherings, and thus made a critical remark to the research performed by contemporary biologists. He saw them as being preoccupied with short-sighted collecting of facts, and noted that

> even if they are guided by *unclarified* and *inadequate* theories, all factual investigations of this kind will always and inevitably produce something. Results are results. Certainly, if we are simply concerned with getting a successful result – as is often the case today – then science has performed its task. But the question at issue is this: what do these results imply for our understanding of the relevant subject-area as such, and what do they contribute to our knowledge of the elementary simplicity of the *substantive essence* of animals, plants and matter? (Heidegger 1995, p. 217).

Nearly eighty years later Charles Orosz (2000) similarly complained that contemporary immunology has turned into a facts factory with little regards for interpretation and conceptual clarification. The facts are now so overwhelming that not only are scientists challenged with mining new information from nature; they also need to mine the fact depositories already uncovered by others. He claimed that this preoccupation with data is owing to the industrial interest for applicable facts, and regretfully stated that this has been a hindrance to

> philosophical musings on why these facts are, or how they fit together, i.e., the philosophic side of science. Yet, such philosophic musings provide new perspectives on reality, and open new tracts of the unknown for scientific mining.... More to the point, how can Immunologists fill the void between their satisfying accumulation of immunologic facts and their unsatisfactory understanding of immune function? (Orosz 2000, p. 339).

True enough; immunologists make use of hypotheses and theories to direct their experimental work. But these are more often than not evaluated according to whether or not they are instrumental in controlling the environment, of making successful predictions for further experimentation; whether or not they contribute to a deeper and more realistic understanding of immunity is less debated. This facts-gathering strategy is methodologically challenging because experimentation, which is characterised by simplification and control, is converse to the life it purports to understand; a complex and seemingly spontaneous undertaking. The biochemist Alfred Gilman summarized this succinctly when claiming that "I could draw you a map of all the components in a cell and put all the proper arrows connecting them", but for even the simplest single-celled micro-organism, "I or anybody else would look at that map and have absolutely no ability to predict anything" (Gibbs 2001).

By tackling natural complexity with experimental simplicity, which is a precarious albeit necessary undertaking, any immunologist with an instrumental leaning runs a risk of misapprehending the data's biological value. It is this risk that needs to be controlled, and which, according to Heidegger and Orosz, should be achieved by giving the data a realistic interpretation according to how well they fit into a coherent biological framework. What appears to be a dynamic and ever-changing organisational panoply at the scale of the interacting agents that comprise it may often look to be a single functional entity from a higher scale; higher level explanations are thus often preferable to lower level explanations when it comes to prediction.

Since evolutionary theory permeates all of biology, the data generated at the biochemical level should thus be measured against how well they fit into the framework provided by the higher level theory of natural selection. This strategy has permeated the rich literature on the philosophy of biology, written by biologists and philosophers alike, which has contributed significantly to our understanding of life in general. Still, little has yet been written on the philosophy of immunology. True enough, one of the first treatises attempting to bridge the gap between facts and philosophy, *Genesis and development of a scientific fact*, was written by an immunologist. But even though Ludwik Fleck (1935) in that treatise explored the relation between immunological facts and their theoretical interpretation, the investigation was merely intended to nail down some general philosophical and sociological accounts of the origin of scientific facts; it was not intended as a preamble to a philosophy of immunology.

Neither so were several book length treatises that touched upon historical and philosophical aspects of immunology, including Arthur Silverstein's (1989) *A history of immunology*, Pauline Mazumdar's (1995) *Species and specificity*, Leslie Brent's (1997) *A history of transplantation immunology*, and Edward Golub's (1997) *The limits of medicine*. The treatises provide outstanding descriptions of the development of immunological thought, but do not properly address Orosz's query. A decisive step in that direction was taken with Alfred Tauber's (1994) book *The immune self*, in which the philosophical and scientific tradition that gave rise to the concept of the immune self was analysed. The book explored how the immune self relates to immune system functioning, but did not aspire to give a full conceptual account of the immune system. Neither so did Paul Ewald's (1994) book *Evolution of infectious disease* nor Steven Frank's (2002) book *Immunology and evolution of infectious disease*. While both books gave thorough reviews of mechanisms by which infectious agents evolve during contact with the immune system, their orientations were more towards understanding parasite behaviour than at clarifying the internal workings of the immune system. The first attempt at that task was commenced with Irun Cohen's (2000) book *Tending Adam's garden*, in which he envisioned the immune system as being a cognitive system operating more or less according to the principles utilised by the mind.

Both *The immune self* and *Tending Adam's garden* deliver bold statements that, if proven true, would have wide-ranging consequences for immunological science. Not only would two basic tenets of immunology, the immune self and the clonal selection theory, both originally introduced by Macfarlane Burnet in the middle of the 20th century, be at stake, but so would the theoretical foundation of immunology altogether. Even though Tauber and Cohen present compelling arguments in favour of their views, I still find reason to question some of their arguments and conclusions. In the following pages I will thus argue that especially Cohen pays too little attention to ecological and evolutionary aspects of immune functioning; that they both fail to appreciate the importance of sign operations, what I term immunosemiosis; and that although they acknowledge the shortcomings of the subject-object metaphor that has characterised Western science since Descartes and Kant, they fail to transcend its bearings in their accounts of the immune system.

I strongly oppose the belief that immune activity involves anything like rational deliberations. Nevertheless, I still sustain Cohen's belief that the immune system may be investigated by metaphors and models originally developed to explore mentalistic operations. This is owing to the fact that the immune and central nervous systems face the same type of problem; how to give flexible responses to an ever changing environment. To substantiate my claims I lean heavily on multilevel selection theory, on the American philosopher Charles Sanders Peirce's theoretical discussions of signs and their meaning, and on ecological insights made by Martin Heidegger in his *Being and time*. Multi-level selection theory offers a new way of framing the intricate relations instantiated between evolution and development, and Peirce and Heidegger supplant the insights with a novel formulation of the relations between animals and their lifeworld. They both rejected the Cartesian distinction between the knower and the known, between the mind and the world, by claiming that living entities exhibit an engaged way of relating to their environment; living entities are not cognitive but actional agents that respond to signs in context. They thus replaced the detached observer with the pragmatic and existential actor, an outlook that fits nicely into the framework of multi-level selection theory.

Peirce and Heidegger were primarily occupied with the specific human way of being, but were well aware that animals could be characterised in a similar manner. In the present book I call attention to the fact that also the immune system becomes shaped by significant entities, which may be self or non-self, and that these entities make their impact during both evolutionary and developmental time. Since the response characteristics of the vertebrate immune system are, to a large degree, guided by these interactions, I further assert that the functional as well as the malfunctional immune responses can be explained within the framework.

Throughout the book much space is devoted to historical recapitulations, the rationale being that immunology is a historical science in at least three respects; immunology studies structures and functions that have been shaped by evolution and individual development, the science itself has a history, and thinking about these items is tightly connected to the philosophical thoughts prevailing at each historical epoch. While philosophical theories differ from scientific theories by their greater abstractness and generality, they aim, no less than scientific theories, to reveal the truth. By investigating how the early history of immunology intersects with related concepts in the fields of pathology, allergy, physiology, medicine and biology, and by integrating immunology with evolutionary biology, ecology and the history and philosophy of science, I hope that a coherent picture of immune reactivity will emerge. Details will be adhered to, but an effort will be made to integrate them with explicit explanatory biological models. A fundamental task will be to acquire insight into the relational structure between the immune system and its world of antigens, and to answer the question of how organisms come to be equipped with defence systems. The questions are not easily answered by facts, contemporary science cannot even address the questions experimentally, but answers are of fundamental importance for anyone interested in a deeper understanding of the functional and malfunctional immune system.

CHAPTER 1

Tracks of thought

Up through the centuries the Latin word *immunitas*, a legal term used by the Romans to describe an individual that was exempt from burdens, duty or control, was used only sporadically as a medical term. Hence, when science began to investigate organismal defence in the late nineteenth century, the term immunity was relatively free from conceptual and historical bindings. The connotatively appropriate term was thus available for use by the newly established discipline that set out to investigate the foundations of organismal exemption from infectious disease (Silverstein 1989).

Starting out as a derivative of vaccinology and bacteriology, immunology relatively soon transformed into a major field of its own. Large amounts of data were collected, and owing to the relatively late appearance of the science, the rich literature generated is still accessible. Contemporary researchers can therefore, at least in principle, survey all previous observational and experimental data within their particular field of interest. Still, this would be an insurmountable task, partly owing to the amount of work involved in scrutinizing such an extensive literature, but also because of interpretative difficulties.

While extensive and cumulative generation of data is an important marker of scientific success, the same data are of little use if they do not contribute to a deeper understanding of the field under investigation. The way data are presented is therefore of uttermost importance. Immunological data are often presented by enlisting properties of cells and molecules in minute detail, while less attendance is paid to the immune system as a device that orients the organism in its dealing with the environment. It is a major postulate of this book that only by integrating immunological data with biological models of organismal dealings with the environment can immune system functioning be properly understood. Such integrated explanatory models will hopefully provide a secure platform from where researchers can dive into the flow of facts and, with anticipation, expect to resurface with renewed knowledge without having drowned in the details.

Since there are few if any biological laws, researchers cannot anchor their vessel to a substratum of spatiotemporally unrestricted facts. Biological facts become adjusted to ecological contingencies and epistemological constraints throughout time, and it is therefore necessary to know the history of organismal adaptations as well as the thoughts that have been elaborated to explain these adaptations to fully understand their functions. This premise is often overlooked by immunologists who disregard history, both the history of thoughts and life. They thereby risk to loose sight of the compromises and trade-offs that have shaped contemporary organisms and theories, and thus to obtain a myopic vision of organismal dealings with internal and external environments. Knowledge about history is important, not only for avoiding repetition of the achievements and mistakes of previous researchers, but also for the sake of proper understanding. Getting the history right will also enable researchers to adjust their expectations to scientific achievements in the future. Thus, the increasing gap between the room of experience and the horizon of expectation, which has been created by accelerating increases in empirical knowledge, may, at least to some degree, be reconciled through historiographical knowledge.

1.1. Organismal maintenance

PROTOIMMUNOLOGY

The earliest members of the species *Homo sapiens* that emerged out of Africa 130 000 years ago lived in small bands as hunters and gatherers. The most common infectious diseases that burdened them were caused by microbes that normally complete their life cycles in non-human hosts, so-called zoonotic diseases, and chronic infections caused by microbes that are capable of surviving for long times in infected individuals. While both types of infectious agents could be reliably transmitted, the transmission rates were low and the diseases remained localised to the small communities. Still, the endemic diseases played an important selective part in the evolution of the human species, and those individuals that descended from the survivors eventually learned to live with and to some degree resist the infectious agents (Lederberg 2000).

When husbandry and the organisation of societies began to replace hunting and gathering about 7000 BC a new opportunity for the spread of infectious disease emerged. Living in cities and travelling to foreign places caused mingling and spread of endogenous infections between previously isolated communities, and people came to experience epidemic diseases with a death rate quite unlike anything encountered previously. Many of the pathogens could survive only in large populations in which new hosts were constantly produced through birth and immigration. Hence, the adoption of a civilised life style tended to increase rather than decrease the burden of infectious disease (Cohen 1989).

In Europe, community living was first manifested in the Greek world. Greece was situated at the crossroads of Asia, Europe and Africa, and while it was a good place from where to disseminate civilisation, it was also a good place from which diseases could be spread. The first catastrophic epidemic in the West whose symptoms and signs are well described broke out there in 430 BC, and in the ensuing years deadly epidemics and pandemics beset cities and nations, with great economic, political and social consequences.

The new epidemic diseases were phenotypically distinct and gave rise to acute and reproducible symptoms in a large amount of people. In addition they occurred periodically, had a relatively short incubation time, and induced protective immunity. These characteristics made possible the discerning of some common features of the diseases, and allowed individuals in the afflicted societies to speculate upon causal explanations. In the second book of *The Peloponnesian war* the Greek historian Thucydides (460–404 BC), who was in Athens during the outbreak of plague, described a peculiar phenomenon in those individuals that survived: "Yet it was with those who had recovered from the disease that the sick and the dying found most compassion. These knew what it was from experience, and had now no fear for themselves; for the same man was never attacked twice – never at least fatally". At the time disease was looked upon as a kind of punishment for sinful activities, and exemption from disease would, accordingly, signify that the individual had led a clean and pious life. Persons that had experienced the exemption phenomenon therefore "half entertained the vain hope that they were for the future safe from any disease whatsoever."

The emancipation of medicine from superstition had to await the new attitude of mind that the Ionian philosopher Thales (550 BC) and his followers had been the first to apply to the world about them (Copleston 1962). Their ideas of natural causation spread rapidly, and the followers of Hippocrates, a physician born on the Greek island of Cos around 460 BC, were quick to apply the naturalistic worldview to medicine. Based on records of signs and symptoms of their patients, combined with knowledge of various environmental conditions such as climate and food, the Hippocratic doctors searched for general naturalistic theories that could account for their observations. This approach to medicine contrasted sharply with the religious views that preceded it, and an entirely rational outlook towards disease, in which the causes and symptoms of disease were accounted for in purely natural terms, was presented for the first time (Longrigg 1993).

The Hippocratic doctors were heavily influenced by Empedocles (495–435 BC) who espoused a theory in which matter was considered to consist of an interrelationship between the four elements air, water, fire and earth. Empedocles claimed that objects come into being through the mingling of elements and cease to be through their separation. He also claimed that parts of animals, no less than physical objects, were generated by the same elements. In his poem *On Nature* he thus wrote, "From these arose blood and the various forms of flesh." (Kirk et al. 1983, p. 302).

It is uncertain when the doctrine of the four elements transformed into the four humours, but the humoral theory is clearly set out in the Hippocratic treatise *The nature of man*, written about 400 BC. In it the author, who was probably Hippocrates' son-in-law Polybus, stated that man can not be composed of only one substance, because one substance alone cannot generate another substance. Rather, the body contains several elements and, consequently, disease has a plurality of forms and a plurality of cures.

> The human body contains blood, phlegm, yellow bile and black bile. These are the things that make up its constitution and cause its pains and health. Health is primarily that state in which these constituent substances are in the correct proportion to each other, both in strength and quantity, and are well mixed. Pain occurs when one of the substances presents either a deficiency or an excess, or is separated in the body and not mixed with the others. (Lloyd 1978, p. 262).

While these general pathological doctrines have been criticised for being more or less speculative and arbitrary conjectures, heavily influenced by social ideologies and culture (Galdston 1981), the criticism does not exclude the possibility that the humoral hypothesis was grounded on observations made during bloodletting and performance of autopsies (Fåhræus 1924). Blood that is allowed to flow rapidly into a tall container during therapeutic bloodletting is at first a uniform red fluid. A change then slowly takes place. The upper part clears and becomes a transparent yellow fluid (yellow bile), while the bottom of the vessel accumulates a dark red (black bile). At the surface there is a thin red layer (blood), and in many diseases, for example pneumonia, there will also separate out a layer that is pale white (phlegm). The upper part corresponds to what is nowadays called serum, the bright red part to oxygenated erythrocytes, and the black fluid at the bottom consists of deoxygenated erythrocytes. The white layer consists of fibrin, which increases during severe infectious diseases like pneumonia.

Early Greek medicine clearly recognised the effect of external causes on the body, and a comprehensive theory concerning the elements of the body was elaborated. While the Hippocratic doctors had little knowledge of the causal factors relating to the conditions they encountered, their ideas about epidemic disease were quite elaborate. In *The nature of man* the author stated that "When a large number of people all catch the same disease at the same time, the cause must be ascribed to something common to all and which they all use". And furthermore, "When an epidemic of one particular disease is established, it is evident that it is not the regimen but the air breathed which is responsible. Plainly, the air must be harmful because of some morbid secretion which it contains." (Lloyd 1978, pp. 266–67).

The idea that the body actively responds to external causes was not stated in the Hippocratic writings. Rather, the body was looked upon as a more or less passive receptacle to external influences. The treatment that followed from this theory consisted in restoring the balance of the four humours, and an extensive therapeutic use of bloodletting, cupping and leeches was initiated. The humoral theory was highly successful as measured by the duration of its influence, and way into the 19th century disease was still thought of as an imbalance of humours. As late as in 1833 over 40 million leeches were imported to France for bloodletting (Golub 1997).

That organisms actively defend themselves against infectious agents, a rather novel idea, presupposes knowledge of infectious agents and the existence of a causal relation between the infectious agent and the host's disease. But neither the discoveries of disease inducing microbes nor of specific protection by vaccination managed to change the view of the passive host. So entrenched was the envisioning of the passive organism that Louis Pasteur, a French chemist and vaccinologist, as late as in 1880 assumed that microorganisms, during infection or following vaccination, exhausted an essential nutrient from the host. He explained protection against reinfection by claiming that the depletion of nutrients was so effective during the first encounter with the microbe that the body could not support renewed growth during secondary infection. Specific protection was hypothesised to be owing to highly specialised nutritional requirements for each microbe (Silverstein 1989).

The conceptualisation of the organism as active would seem to require an additional radical break with the biological tradition and the way living entities interact. This break, which was accomplished during the latter half of the 19th century with the publication of Charles Darwin's book *The origin of species*, effectively served to write off the explanatory theories of supernatural causation and humours in balance by providing a novel explanation for life's dynamics.

A SORT OF STRUGGLE BETWEEN THE PARASITE AND ITS PREY

The observation that organisms are exceptionally well adapted to their environments long predates scientific biology. Such adaptations lacked proper scientific explanations in the pre-Darwinian world, and it is thus little wonder that they were subjected to religious speculation. Indeed, the remarkable fit between the properties of living beings and their forms of life was the English clergyman William Paley's (1743–1805) strongest evidence for the power of a Divine Creator. The complex internal structures of living entities, their synchronised and seemingly purposeful activity, their invariant mode of reproduction, and their exquisite adaptedness to the environment provided him with indirect evidence that these creatures had been designed by God (Paley 1802). In his analysis, Paley contrasted the likelihood for the emergence of life's complexities from a pure chance process with the likelihood of life's emergence provided a rational designer, and convincingly claimed that the design hypothesis was vindicated by being the better explanation for life's emergence (Sober 1993).

The argument from design was radically challenged when Charles Darwin published *The Origin of species* in 1859. In it he claimed that the adaptations we observe in nature are the result, not of a benevolent Creator or a chance process, but of natural selection. Like Paley, Darwin asked "How have all those exquisite adaptations of one part of the organisation to another part, and to the conditions of life, and of one distinct organic being to another being, been perfected?" (1859, p. 114). But in opposition to Paley, Darwin provided evidence that the likelihood of life's emerging complexity given natural selection was higher than the likelihood of life's emergence given the design hypothesis, and thus that life could be explained by natural processes without the need to invoke a rational designer.

Darwin's theory had three primary components. First, Darwin observed that all species show considerable natural variation in the forms and behaviours of individual organisms, due primarily to the fact that offspring differ from their parents. Second, he noticed that these variations were heritable, and third, that individuals typically produce more offspring than can be supported by the environment. By combining these conditions Darwin proposed a selective process that would allow evolution; individuals that by their particular variations are better suited to reproduction and survival must leave more offspring than those that, for whatever reason, are less fit to survive and reproduce. This process, which he termed natural selection, could, when operating cumulatively and given enough time, account for the evolution of the diversity and adapted complexity of all living things.

A fundamental assumption of Darwin's theory was the inference that"every single organic being around us may be said to be striving to the utmost to increase in numbers." (1859, p. 119). Nevertheless, since "more individuals are produced than can possibly survive, there must in every case be a struggle for existence, either one individual with another of the same species, or with the individuals of distinct species, or with the physical conditions of life." (1859, p. 117). He clearly saw the necessity in restraining the massive increase in organisms, but acknowledged that"What checks the natural tendency of each species to increase in number is most obscure. Look at the most vigorous species; by as much as it swarms in numbers, by so much will its tendency to increase be still further increased. We know not exactly what the checks are in even one single instance." (1859, p. 119).

At the time of writing Darwin had some knowledge about epidemic diseases, but since the microbiological reframing of disease was still 20 years into the future, his understanding of microbiological agents as causes of disease was rudimentary. But even though his treatment of epidemic diseases as checks for the growth of individuals was unrefined, he clearly stated the principle that there is an active struggle between parasites and host in which both take part.

> When a species, owing to highly favourable circumstances, increases inordinately in numbers in a small tract, epidemics – at least, this seems generally to occur with our game animals – often ensue: and here we have a limiting check independent of the struggle for life. But even some of these so-called epidemics appear to be due to parasitic worms, which have from some cause, possibly in part through facility of diffusion amongst the crowded animals, been disproportionably favoured: and here comes in a sort of struggle between the parasite and its prey. (Darwin 1859, p. 122).

This view entails that organisms are never perfectly adapted to their environments. If they were, no parasite would ever breach the host's defences and, likewise, no potential host would ever escape infection.

The elucidation of the mechanisms responsible for the struggle between parasites and their prey was hampered by a variety of reasons, some related to the advancement of science and others to ingrained beliefs. Amongst the latter was the belief in spontaneous generation, the view that it is possible in the world that now exists for inanimate matter to transmute into living forms by a natural process not requiring or involving divine intervention (Harris 2002). The idea was not dismissed until the end of the 19th century, largely due to accumulated evidence against it. As late as 1838 Theodor Schwann, one of the discoverers of the cell, put forward a theory in which he argued that the cells of the body were spontaneously formed in the extracellular fluid by a process akin to crystallisation. The process started when minute inanimate particles aggregated into larger particles that eventually developed into whole cells. The leading anatomical pathologist in Europe at the time, Karl Rokitansky, promulgated these ideas in an almost unmodified form in his influential *Textbook of pathological anatomy* published between 1842 and 1846.

Rokitansky's ideas were put to doubt by observations of cell division, and the German pathologist Rudolf Virchow publicly took issue with Rokitansky by claiming that cells, rather than being the by-product of some nondescript forces, are the basic units of both normal life and disease. But even though Virchow was crucial in establishing a relationship of cells to the pathology of the tissues, he did not see cells as defenders of the organism. Rather, he believed that the inflammatory cells were causing the disease. In a similar mode Robert Koch, the leading German bacteriologist who had seen bacteria within phagocytic cells, believed that cells localised at the site of inflammation were a vehicle for the spread of bacteria in the body (Golub 1997).

It was the Russian biologist Elie Metchnikoff who first convincingly demonstrated that phagocytic cells are true defenders of the multicellular organism (Tauber 1990). Metchnikoff was initially critical to the competitive principle in Darwin's theory, but his research gradually convinced him of the importance of the Darwinian concepts. So while on an expedition to Messina in 1883 he came to observe phagocytic cells surrounding a foreign body underneath the skin of transparent starfish larvae, he was quick to give it a Darwinian interpretation. He compared the response in the starfish larvae to the accumulation of cells in inflammatory lesions of humans, and hypothesised that the phenomenon was a general one that had been conserved throughout evolution. He imagined, contrary to Virchow, that the inflammatory phagocytes were attempting to engulf and destroy the infecting bacteria, just like the cells that were attempting to engulf the foreign body in the starfish. This observation suggested to Metchnikoff that the critical struggle in disease was between individuals of different species.[1] And more importantly, he remodelled Darwin's size-disparate struggle between parasite and prey into a more size-equal struggle between parasite and phagocyte.

But Metchnikoff also recognized that phagocytes are effective at eliminating, not only infectious agents, but also cells of the body that are either superfluous or harmed. He summarised these thoughts on the integrity-shaping activity of the phagocyte in a remarkable paper entitled *The struggle for existence between parts of the animal organism*. Having observed that phagocytes were present in all organs, including the spleen, liver, blood and brain, he concluded that they maintained "a considerable degree of independence while directing their activity to exterminating every weak member of the colony" (Metchnikoff 1892, p. 213). To

[1] Darwin believed that conflicts between individuals of different species would be most pronounced between closely related species since they tend to occupy the same ecological niche, and that the struggle would be most severe between individuals of the same species. The reason for this was that individuals from differing species impinge less on each other's lives than individuals from the same species. Richard Dawkins (1989, p. 67) expressed this idea succinctly: "Moles and blackbirds may compete for worms, but blackbirds and blackbirds compete with each other for worms *and* for everything else."

explain how phagocytes differentiated between cells to be spared and cells to be attacked, he surmised that healthy cells would emit special substances that would repel phagocytes while sick or weakened cells would fail to produce the same substances. He thus envisioned a "war between cells and phagocytes", the outcome of which would be a "considerable strengthening of the entire organism" (p. 213). Nevertheless, he also acknowledged that the phagocytic cells can do significant harm if they escape control. Therefore, the "analysis of the factors that strengthen or weaken phagocytes should lead to their manipulation to increase the potential benefits and decrease harmful effects." (p. 215). Thus, the speculative seeds of what we now describe as autoimmune disease and immunotherapy had been planted.

TOWARDS A UNIFIED VIEW

The side-chain theory

Metchnikoff was the first to state clearly that the body actively defends itself against invading microbial agents.[2] His view that the phagocyte was the organism's sole defender was, however, challenged when Emil Behring and Shibasaburo Kitasato in 1890 reported that protective substances were present also in the cell-free fraction of the blood. When they injected rabbits with an amount of tetanus toxin too small to cause disease, the animals were able to survive greater and greater amounts of toxin at subsequent challenges. The substance that protected against disease was called antitoxin, and later on antibody. This remarkable observation led to an experiment in which they transferred small amounts of blood from vaccinated mice to disease-free mice, who following injection with a lethal dose of toxin survived. The experiments showed for the first time that there was an inducible substance in the blood that conferred specific protection against the inducing agent.

The findings by Behring and Kitasato, accompanied by previous demonstrations by Edward Jenner and Louis Pasteur of the efficacy of vaccination, provided a practical foundation for specific prevention and treatment of infectious disease. But while the findings allowed the establishment of serological tests that contributed immensely to diagnosis, follow-up, and epidemiological control of infectious disease, there was little theoretical understanding of immunity. Metchnikoff's phagocytosis theory, at the time the unrivalled theory of active immunity, completely failed to accommodate the novel observations, and speculations as to where and how antibodies were formed and of what accounted for their exquisite specificity, were soon initiated.

The side-chain theory, proposed by Paul Ehrlich in 1897, was the first theory contesting Metchnikoff's theory. It combined Metchnikoff's insights, that the

---

[2] The historical evidence surveyed in this and the following sections is selected from various treatises on the history of immunology. These include books and papers by Arthur Silverstein, Alfred Tauber, Pauline Mazumdar, and Edward Golub enlisted in the references. The references to Metchnikoff are taken from Tauber's work except where indicated.

immune response is active and founded on a cellular response, with the humoralist idea of antitoxins. Ehrlich proposed that the antitoxins, which he termed side-chains, were formed by cells. He further surmised that the surface of antitoxin-producing cells consisted of side-chains that could react with antigens and thus lead to production of more side-chains. On subsequent encounters with the antigen, more and more side-chains would be formed, until their number would be so large that they would begin to be shed from the surface of the cell and start to circulate in the blood. The theory was thus built on the principle of natural selection; the cell naturally produced a variety of side-chains, the side-chains possessed their specificity from the start, and the antigen selected the correct specificity of the side-chains.

Ehrlich was strongly committed to the idea of absolute specificity, the belief that there are sharp discontinuities between organisms of various species, and hence that each microbial species is antigenic distinct (Mazumdar 1995). He thus believed that there was a limited number of microbial agents that the side chains would have to recognize, and he convincingly argued that the hundreds of different side-chains on the surface of each cell were sufficient to discriminate between the different infectious agents. However, within a few years it became known that a single intruder could bear thousands of antigenic determinants, which in turn could initiate the production of thousands of different antibodies. In addition, infecting microorganisms were shown to have overlapping properties – while some properties were more or less common to all microbes, other properties were fairly restricted to certain kinds of microbes. These new observations made it clear that neither the belief in absolute specificity nor the belief in a limited antibody repertoire were compatible with the Darwinian concept of natural selection, and so the credibility of the side-chain theory was weakened.

Instructive theories

The wakening of the phagocytosis and side-chain theories marked a revitalization of the passive theories of immunity. But unlike the earlier passive theories, for example Pasteur's depletion theory, the new theories were transformed to accommodate the observations and insights of Metchnikoff, Behring and Ehrlich. The observational data of antibodies and antigens were put into a new theoretical framework; instead of claiming, as did Ehrlich, that the instructions for antibody-formation came from within the cell, the novel theories proposed that the instructions came from the antigen.

Between 1930 and 1960 the accepted explanation for the large repertoire of antibody specificities was that antigen somehow acted as a template to "instruct" the specificity of the globulin forming mechanism. The antigen transmitted the information for its structure to the antibody, thus making the antibody adopt a shape and a chemical affinity that were complementary to that of the antigen.

Although instructive theories were capable of explaining the large number of antibodies and their extreme specificity, they could not explain how antibody production could persist in the absence of antigen, why the secondary antibody response was stronger and qualitatively different from the primary reaction, or why not antibodies were formed against self-components. Along with discoveries in the 1950s that protein structure was predetermined by the DNA sequence, enough theoretical and experimental evidence had been gained to dismiss the instructive theories altogether.

The clonal selection theory

Generation of diversity

By the early 1960s instructive theories had largely been abandoned for a comprehensive but simple theory that fit the immunological data and background biological knowledge better than did its predecessors. The novel theory was sparked by Niels Jerne's natural selection theory of antibody formation (Jerne 1955). In accordance with contemporary molecular theory and the basic elements of the side-chain theory, Jerne proposed that the information for antibody production lay within the organism, and that the antigen producing mechanism somehow generated a wide variety of different antibodies that, upon encounter with antigen, were selected according to their match with the antigen.

A crucial point of Jerne's theory was that selection occurred at the level of the circulating antibody molecules. When an antibody had been selected he surmised it would be transported to antibody-producing cells, which were thought of as factories that were capable of producing all kinds of antibody specificities. This theory, although criticised on mechanistic grounds for not paying enough attention to cellular dynamics, was reworked by David Talmage and MacFarlane Burnet. Like Ehrlich, they suggested that antibody-forming cells, not soluble antibodies as proposed by Jerne, were the structures being selected. But unlike Ehrlich, they postulated that each cell had only one type of receptor on their surface, and consequently that each cell produced antibodies of one specificity only. The increasing concentration of antibodies to a given antigen that occurred following immunisation was thought to be a consequence of the clonal proliferation of the antigen-selected antibody-producing cell; hence the name "clonal selection theory". The unit of immunological specificity thus became the clone of lymphocytes, the millions of lymphocytes with identical antigen receptors.

The theory postulates that the antibody repertoire develops spontaneously in the antibody producing cells, and that different antibodies derive their specificity either because of genetic point mutations (Lederberg 1959) or because of recombination of genetic segments (Dreyer and Bennett 1965) within each cell. The theory was able to explain a variety of hitherto inexplicable biological phenomena, and soon won acceptance as the foundational theory of immunology. It stimulated a broad spectrum of research, and laid the foundation for a remarkable progress of molecular immunology during the 1960s and 1970s (Tauber and Podolsky 1997), with the confirmation of the Dreyer-Bennet hypothesis by Susumu Tonegawa as one of the highlights.

As laid out by Tonegawa (1983), the antigen-binding part of each immunoglobulin heavy chain is made from three different families of variable region gene segments, called V for variable, D for diversity, and J for joining. Each gene family contains from five to seventy or more different DNA segments. Each progenitor B cell (and T cell) randomly recombines one DNA segment from each of these three DNA families to form the rearranged V–D–J segment. This process is mediated by recombinase enzymes which recognize specific recombination signal sequences which flank each segment. Combinatorial diversity is further extended by deletions and additions to the DNA sequence at the segment junctions and by somatic mutation. This ensures that each B and T cell express an individualised combining site.

The functional receptor is constructed by splicing the joined V-D-J segments to one of the several constant-region gene segments. These include the genes that make up the constant parts of the immunoglobulin heavy chains and the constant parts of the T cell receptors. The constant parts do not take part in antigen-binding but form the reaction site of the antigen-binding molecule. This construction allows the receptors to detect variability in the antigenic world while still being able to signal the encounter through conserved cellular pathways. The receptors thus combine the propensity for individual somatic novelty with germ-line selected constancy.[3]

The clonal selection theory has remained a basic premise for immunologists working with basic research as well as in the clinics, and is still regarded as the best conceptualisation of the immune response (Silverstein 2002). Although much remains to be explained, especially concerning the regulation of the immune response, the basic premises of the theory have proven valid. It is also noteworthy that the broad implications of the theory paved the road towards unification of the previously disconnected cellular and humoral aspects of immunity.

The hapten-carrier puzzle

In his original formulation of the clonal selection theory, Burnet proposed that antigen directly activates the adaptive immune system. "When an antigen is introduced it will make contact with a cell of the corresponding clone, presumably a lymphocyte, and by so doing stimulate it to produce in one way or another more globulin molecules of the cell's characteristic type." (Burnet 1959, pp. 54–55). During the 1960s experimental results which were in conflict with Burnet's

[3] Information that is transmitted across the generations in germ cells is termed germ-line information, while information that has been obtained from individual learning is called somatic learning, from the Greek word for body, *soma*.

simplistic view began to accumulate, and hypotheses were generated which suggested that in order to become activated, the B and T lymphocytes had to receive two signals. Signal 1 was delivered through the antigen specific receptor, while signal 2 was delivered by a helper cell. For B cells the helper cell was believed to be a T cell (Bretscher and Cohn 1970), while T cells were though to receive help from antigen presenting cells (Lafferty and Cunningham 1975). In the new conceptualisation that emerged, the delivery of signal 1 alone was considered to lead to death or inactivity of the lymphocyte, whereas lymphocytes that received both signals would be rescued and thereby allowed to differentiate and proliferate.

This model allowed the elucidation of a conundrum in immunological theorising, the question of why it is necessary to couple small antigens called haptens to larger protein molecules in order to elicit immune responses against the hapten. Karl Landsteiner, who made extensive use of this principle in his studies of immunoglobulin specificity, realised that the protein was not involved in the specific reaction and that protein therefore acted as a carrier for the hapten. While the carrier's immunological role was unclear, Landsteiner (1945, p. 161) believed that it "affords the antigenic stimulus in the immunisation".

The solution to the hapten-carrier puzzle had to await insights related to the conveyors of immunity, the lymphocytes. The function of lymphocytes had remained a mystery until the early 1960s, when it was realised that they are essential for immunological functions related to specificity, tolerance and memory (Gowans 1996). The study of lymphocytes was thereafter intensified, and by the end of the 1960s it had become clear that the lymphocyte population was heterogeneous and consisted of at least two sorts, the T and B lymphocytes (Roitt et al. 1969). Both sorts of lymphocytes were thought to originate from lymphoid precursor cells in the bone marrow. These cells either remained in the bone marrow where they differentiated into B cells, or they emigrated to the thymus where they matured into T lymphocytes.

The realisation that T and B lymphocytes co-operate during the immune response (Mitchell and Miller 1968), allowed Mitchison (1971) to experimentally verify that antibody-producing cells were dependent upon help from T cells to produce anti-hapten antibodies. The awareness that T cell activation was restricted by the major histocompatibility complex (MHC) (Zinkernagel and Doherty 1974), and that antigen presenting cells were capable of modulating the T helper cell response (Neveu 1977a), further elucidated the mechanisms behind the hapten-carrier phenomenon. The phenomenon could now be explained as a sequential process in which the uptake of the hapten-carrier conjugate by the antigen-presenting cell was the first step. The second step was the dispatching of information about the carrier from the antigen-presenting cell to the T helper cell. After activation of the T helper cell, the third step, which involved T help to the hapten-specific B cell, ensued (Neveu 1977b).

It is not a prerequisite of the model that the T and B cells recognise the same part of a molecule. Thus, it can be envisioned that the T cell can be activated by the carrier protein and still provide help to the hapten-specific B cell. To see how this may come about, consider first some experiments performed in the late 1920s by Avery and Goebel. They noted that polysaccharide antigens by themselves were poorly immunogenic, but when they were chemically attached to protein carriers the polysaccharide antigens became converted into full-fledged antigens (Landsteiner 1945). These experiments can be explained as follows. First, antigen presenting cells engulf and process the hapten-carrier molecule. The carrier proteins are thereafter exposed within their MHC molecules. Second, T cells that express T cell receptors specific for the carrier will, upon encounter with the carrier-MHC complex, become activated. Third, B cells with antigen-binding receptors specific for the polysaccharide hapten will, like the antigen presenting cell in the first step, bind the hapten-carrier molecule and internalise it. Since B cells are efficient antigen presenting cells they can expose the carrier within their MHC molecules. If the T cell that was activated in the second step comes in contact with this B cell, the already activated T cell recognises the carrier-MHC complex on the surface of the B cell and thus stimulates the polysaccharide-specific B cell to differentiate and secrete antibodies.

Adjuvancy

The pushing back of the self-nonself discriminating capability, from the B cell to the T cell to the antigen-presenting cell, that took place from the end of the 1950s to the 1970s, begged the question of how the antigen-presenting cell could make the crucial distinction between self and non-self. A solution to the question occurred when Charles Janeway (1989) speculated upon the role of adjuvants in immunisation. The term adjuvant, which is derived from the Latin verb *adjuvare*, to help or aid, was introduced in the 1940s when it was discovered that strong immune responses against purified antigens could be obtained if antigen was mixed with killed bacterial debris. Since injection of the purified antigen alone induced little immunity, any substance that, when injected together with an antigen, would increase the specific immune response towards the antigen, was termed adjuvant.

So, in order to produce strong immune responses immunologists first have to purify their molecules and then contaminate them with a solution of bacterial molecules. This trick, which Janeway called the immunologist's "dirty little secret", suggested to him that clonal selection was insufficient to explain immunogenicity. If clonal selection was all that mattered, antigens should be able to stimulate lymphocytes and elicit an appropriate immune response by themselves. Since they do not, antigen is by itself not sufficient for immunogenicity, and there must be something in the adjuvant that provides an additional necessary signal. He thus hypothesised that antigen presenting cells were endowed with certain pattern recognition receptors, highly conserved receptors expressed on macrophages and dendritic cells, that would detect common structures of invading microbes. This model gained considerable support when it was demonstrated that Toll-like receptors, which are expressed on dendritic cells, are capable of sensing various bacteria, viruses and parasites. Further experiments demonstrated that stimulation of the Toll-like receptors by microbial substances activate the dendritic cell, which thereafter delivers costimulatory signals to T cells.

Upon noting that not only non-self but also certain self-antigens, like heat shock proteins and DNA, may bind to pattern recognition receptors, Polly Matzinger (1994) proposed that the initiation of the immune response could be initiated by damaged cells in the tissues as well. In her model, aptly termed the danger model, she proposed that the activating capacity was an inherent property of all tissue cells, and that they upon traumatic experiences would release danger signals that stimulated the antigen presenting cells. Hence, she advocated the idea that the ultimate power to activate the immune system lay in the tissues. Despite much initial scepticism, experimental support for the danger model has been amassed (Matzinger 2002), and observations that crystalline uric acid, which is a degradation product of nucleotides, induces potent activation of dendritic cells forcefully supports the model (Shi et al. 2003).

1.2. The war metaphor

Immunology is an experimentally driven science with a firm basis in foundational principles laid down during its inauguration in the 1890s. For most of the 20th century it was war-like interactions, which are the most pronounced and thus closest to the enquirer, that preoccupied the community of immunologists. The metaphorical[4] conceptualisation emphasised the destructive interactions that take place between hosts and microbes, and ingrained the field with an idiomatic but discomfiting military language that highlighted tracking of enemies, camouflage, battles within, attack rates and victorious outcomes for one or the other combatants. These principles are currently epitomised as *The body at war* (Dwyer 1988) and *The war within us* (Mims 2000).

The emphasis on conflicting interactions and disease was a consequence of the educational bias of the researchers, as most of them were educated in medicine. But it was also owing to a one-sided understanding of natural selection, thus emphasising competition while disregarding cooperation. Yet, Darwin had also provided examples that cooperation and altruism, even between members of distantly related species, could evolve by natural selection. This latter aspect of the theory was, nevertheless, ignored by the early microbiologists. Their attention was focused almost exclusively on the disease-producing microbes, and even though they often

[4] Metaphorical language involves understanding one kind of thing, *e.g.* a landscape, in terms of another, *e.g.* a map. Metaphors are not intended to imply identity of process or function, but should serve to cast light on the phenomenon being investigated. Explanatory metaphors should be handled with care because they may seduce one into thinking that the phenomenon to be explained is indeed the same as that to which it is being likened.

observed uncharacterised microbes that accompanied the pathogens, they largely neglected to question their functional significance (Rosebury 1962).

An unfortunate consequence of the war metaphor's pervasiveness was that enquiry diverted away from the adaptive role that microbial agents play in the functioning of the host organism. In a very real sense the triumph of the war-metaphor sidetracked the evolutionary and ecological role of microbes, and seduced the medical community to believe that health could be gained through winning a war. The few opponents to this metaphorical conceptualisation, including amongst others Metchnikoff, were strongly criticised by their contemporaries.

Metchnikoff's phagocytosis theory, which was sensitive to both the cooperative and the conflicting aspects of Darwin's theory, was truly prescient and in many ways strikingly modern (Tauber 1990). Being an embryologist by training, he came without the medico-cultural bindings of the bacteriologists and serologists. In contrast to their emphasis on host-pathogen interactions, he was more concerned with relating developmental biology to the Darwinian theory of natural selection. By analogy to the Darwinian competition between individual organisms, he came to view the developing organism as a disharmonious entity composed of competing cell lineages that strive for harmony (Metchnikoff 1892). The Darwinian *interindividual* competition was thus turned into an *intraindividual* competition. Since the cells of the body function as an integrated whole only when cooperating, Metchnikoff further argued that there had to be a conflict-mediating mechanism that would bestow harmony in the developing organism. He assigned the function of conflict-mediator to the phagocyte, which he saw as a multitasking entity that was responsible for destroying malignant cells, repairing damaged tissues, removing degenerate cells, and monitoring growth processes (Tauber 2003).

To Metchnikoff, the great puzzle was how potentially unrestricted self-replication at the level of competing cell lineages could be restricted in favour of the interests of the organism as a whole. He came to see this puzzle as being tightly connected to the puzzle of how the host defends itself against microbial insults. In both cases some cells or microbes are envisioned as serving the fitness-related interests of the organism, while other cells and microbes are envisioned as serving their own self interests in conflict with the organism's interests. The two puzzles are being connected through the phagocyte, which is capable of differentiating between friend and foe. Within the developing organism some cells were thought of as redundant and thus removable, while other cells enhanced fitness and should thus be spared. The former included cells with a defined function during embryogenesis as well as cells injured during inflammation, while the latter comprised the germ and somatic cells. Quite analogously, he considered microbes as being either redundant or fitness promoting.

Hence, interacting cells, be they differentiating cells within the developing organism or immune cells and parasites, are not necessarily at war with each other. Defence thus remained but one aspect of immunity; the other aspect, that of defining the integrity of the organism, was equally important. Whether cooperation or war should ensue would depend upon whether or not the conflict-mediating mechanisms functioned properly. This commitment allowed him to hypothesise that microbes are capable of living in harmonious coexistence with host organisms, and that the microbes may be important for the host's wellbeing as well. A practical offshoot of this hypothesis was his promotion of the ingestion of health-bringing microbes as a regimen to outperform pathogenic bacteria in the colon (Bibel 1988).[5] But it also led him to the conclusion that the phagocyte could be responsible for certain pathophysiological processes. This would occur if the organism during the struggle between embryonic components came to appear as a foreigner to itself, as well as if bacterial toxins produced by intestinal microbes activated phagocytes to attack the body's own tissues, thus creating a form of autoimmunity that initiated the ageing process.

Metchnikoff's theory is compatible with the revised theory of natural selection, termed multi-level selection theory, which provides a comprehensive explanation of both co-operative and competing relationships (Michod 1999).The revised theory has brought about an elegant framework for the explaining of co-evolution between microbial agents and their hosts. It sees the question of whether inter-species interactions evolve towards warlike interactions or towards peace and harmonious coexistence as a contingent matter. The outcome depends upon characteristics of the external environment as well as of the relational interactions that take place between the interacting organisms.

The effects of such interactions occurring over evolutionary time are well explicated by relations between the multicellular organism and its intestinal microbes. It is now commonly accepted that the gut has coevolved for millions of years with a vast consortium of microbes, and that the colon has become dependent upon the commensal bacterial impact. This dependence is evidenced by the fact that non-pathogenic bacteria acquired during the early postnatal period are required for the development of tolerance to self as well as towards luminal antigens, and by the observations that resident colonic bacteria induce significant postnatal developmental events in the colon. These include angiogenesis, shaping of epithelial innate immunity, and diversification of the developing mucosal adaptive immune system.

Even though Metchnikoff was amongst the first to provide a scientific theory for cooperative interactions between hosts and microbes, he was not without confederates in promoting the health bringing capacity of microbial agents. For hundreds of years medical doctors had reported cases of malignant tumours that had disappeared spontaneously, often following episodes of fever and infection. In the 1890s William Coley utilised this knowledge to promote and successfully practice "immunotherapy" of patients with malignant disease (Cann et al. 2002). His method consisted of repeated injections of dead bacteria into the

[5] These ideas on the use of probiotics for education of the immune system have recently gained renewed attention with observations showing that helminths, saprophytic mycobacteria and lactobacilli induce immunoregulatory mechanisms in the host, but cause little, if any, harm.

tumour tissue. The dose of bacteria, which was tailored to each patient, was gradually increased to keep the local inflammatory response as well as the systemic feverish response going. Immunotherapy, which gradually fell into disrepute owing to the emphasis on aseptic surgery and increased focus on hygienic procedures, is nevertheless still utilised for the treatment of bladder cancer. When the mycobacterium BCG infiltrates the bladder epithelium, it induces a strong local inflammatory reaction that effectively controls the malignant cells (Lockyer and Gillatt 2001).

Not only did the war-metaphor direct the enquiry away from inter-species cooperation; it also directed the enquiry away from the disease-inducing capacity of the immune response itself. Immunity was initially thought of as a response that served to protect the organism from external aggressors, and the thought that the immune system could harm the organism was barely considered. The belief in the protective aspects of immunity was so entrenched that Paul Ehrlich in the late 1890s proclaimed that self-reactive antibodies that would destruct the integrity of the organism lacked biological plausibility.

Despite the demonstration of self-reactive pathogenic antibodies by Karl Landsteiner in 1904, unfounded beliefs on the nature of immunity effectively dismissed autoimmunity from the research agenda until the 1950s. While pathologists had long accepted the idea that chronic and degenerative diseases were owing to self-destructive inflammatory processes, immunologists were reluctant to accept that such phenomena could be caused by the immune system (Parnes 2003). So entrenched was the principle of non-reactivity to self that when Ernest Witebsky and his students discovered autoantibodies against thyroid tissue in the early 1950s, they initially refused to believe the observations and even withheld the publication for more than three years before they felt confident with the results (Maslloréns 2000).

In a similar manner, the discovery of anaphylaxis by Charles Richet in 1902 and of allergy by Clemens von Pirquet in 1906 presented striking challenges to the conceptualisation of immunity in terms of organismal defence. The concept of allergy, which was based on the idea that external harmless agents would induce in the organism a faulty, distorted and confused reaction, stirred the immunological community. But although it would seem that antibodies were involved in the allergic reactions, the pervasiveness of the war-metaphor excluded alternative explanations, and allergic reactions were thus simply ignored by the immunologists (Jackson 2003).

As evidenced by the lives saved by antibiotics and vaccines, the concern for the wellbeing of individual human organisms as they fight infectious agents has been a successful medical enterprise. Hence, the war-metaphor has served a useful role. But its success has also hampered our understanding of the immune system as it relates to the world. These deficits become especially pressing as we encounter infectious agents that can not be fought off with conventional antibiotics or vaccines, and where the only prospects for a cure are to be found within the resources of the immune system, either at the individual or population level.

1.3. Metaimmunological musings

SPECULATIVE THINKING

Two opinions

Science is a process that discovers new data and tests hypotheses about the organisation and operating principles of the world. But science is also a creative and constructive process in which the world becomes categorised and conceptualised by the scientist. The English expression "science" (Lat. *scientia* – having knowledge) captures only the first of these ideas, while the German word "wissenschaft" (wissen – knowledge, schaft – creation), and the related Norwegian word "vitenskap" (viten – knowledge, skap – creation), capture both connotations. The two aspects are tightly integrated; observations rarely make sense on their own, and theories are seldom accepted if not empirically grounded.

While the descriptive aspect of science rarely stirs heated controversies, the use of observations to generate new hypotheses often does. A spectacular case of this was vehemently demonstrated in disputes between the Nobel laureates Paul Ehrlich and Jules Bordet as the 19th century turned into the 20th (Christ and Tauber 1997). Ehrlich, who was a speculative thinker that sought to integrate factual knowledge with theoretical descriptions, postulated a multitude of reactants and even visualised antibodies in a pictorial form. To him a fact could not exist without belonging to a theoretical matrix in which mechanisms of action were embedded. This view contrasted sharply with Bordet's conceptions of science, according to which there should be a marked segregation between description and explanation, between fact and theory. Bordet's response to Ehrlich's endeavours was thus critical, and he accused Ehrlich of transgressing the line between fact and artefact, of constructing reality.

The dispute between Bordet and Ehrlich was not so much related to observational data as to communication and interpretation of data. Their conflict was thus at what we may term the horizontal, not the vertical level (Figure 1.1).

Scientists stand in a communicative relation with each other; their training to become scientists consists of learning the cumulative wisdom of their discipline, and when they have become established scientists, they stand in a collegial relation with other scientists concerning the interpretation of results. The horizontal aspect thus concerns how two or more conscious subjects communicate with each other. The vertical aspect, on the other hand, relates to the scientist's activity as he investigates nature, be it immunology or some other field.

While scientific research may be understood as a subject-object relation, between an independent scientist and the discovered world, the other aspect of science, that of interpreting and convincing others about the interpretation's correctness, presupposes a subject-subject relation. Such intersubjectivity is essentially mediated through the medium of language, in contrast to subject-object relations which are mediated through observation and experimentation.

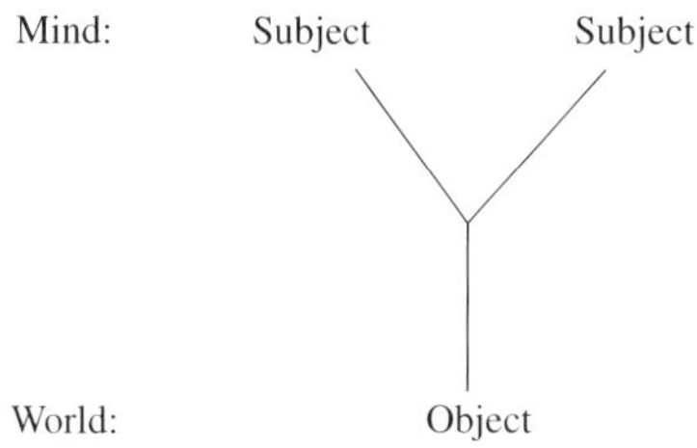

Figure 1.1. Natural scientists stand in a subject-object relation to their investigated object and in a subject-subject relation to fellow scientists and the lay public. In the horizontal dimension this may be communication of results and their meaning, and in the vertical dimension it may be experimentation and observation of effects

But even though the relations depicted in Figure 1.1. give a fairly comprehensive image of the scientific project, and comply with the tendency to think that subject-subject relations belong to the sciences of man while subject-object relations belong to the natural sciences, the figure still gravely misrepresents the complexities involved by ignoring the temporal aspect of science and by depicting false dichotomies. Thus, two modifications of the contents of Figure 1.1 are of central importance for the understanding of immunology. First, insight requires a communicative relation built upon shared presuppositions. Truth is not something reached by an individual scientist alone, it is the "opinion which is fated to be ultimately agreed to by all who investigate" (Peirce 1992, p. 139). Truth is thus a long term product of the community of investigators. And second, there is no sharp distinction between the subject and the object; as will become clear in the next section, the subject is already out with the objects, relating to them understandingly.

Self and other

Metaimmunology

As emphasised by Mazumdar (1995), the dispute between Ehrlich and Bordet was partly owing to their different philosophical views. They belonged to seemingly irreconcilable scientific traditions that interpreted the world and scientific investigations differently. The philosophical basis of their conflict, which goes back to the old Greeks, took its modern form from the philosophical investigations of Rene Descartes (1596–1650). In *Discourse on method* (1637), which laid a theoretical foundation for much of modern science, Descartes described how he methodologically set out to doubt everything that he hitherto had believed. Upon this endeavour he found reasons to doubt anything but the fact that he himself, since he was the one who doubted, had to be an existing individual. Descartes thus uncovered himself as a subjective self distinct from the world. Following this inference, he reasoned that reality is partly *res cogitans* (mind) and partly *res extensa* (objects). This dichotomisation had further consequences for how reality should be investigated; while the mind was amenable to investigation by introspection and self-reflection, the world, including the bodies of animals, should be investigated by the use of scientific methods grounded in mathematics.

By separating the knower from the known, Descartes laid the foundation for scientific dualism and the two philosophical fields epistemology (gr. *episteme* – knowledge; gr. *logos* – learning), which deals with the quality of evidence, arguments and inferences, and ontology (gr. (*to*) *on* – the being) which studies the various modes that beings in the world can take. Upon this model, science is concerned with the extraction of facts, a subject-object relation, whereas philosophical reflection, a subject-subject relation, takes care of the validity of claims to knowledge. So when Ehrlich concerned himself with a multiplicity of distinct biological entities, he was concerned more with the ontological side of knowledge, while Bordet, in his criticism of Ehrlich' methodology, was more concerned with the epistemological side.

Although the fields of epistemology and ontology appear distinct, they are in many ways reciprocally dependent. This is made explicit by the tendency of modern intellectual culture to think that the question of what something is (ontology) can be resolved in terms of how it is known (epistemology) (Taylor 1995), irrespective of whether the knower is the mind or some other sort of biosystem (Plotkin 1982). Furthermore, since knowledge is something that exists, it follows that knowledge has an ontological foundation. Martin Heidegger was amongst the first to deliberately investigate this line of thought. His main project was to investigate the ontology of human knowledge; instead of investigating epistemological questions he performed an ontological investigation concerning what kinds of beings humans are and how this being is bound up with the intelligibility of the world (Heidegger 1927; Dreyfus 1991).

But if we can have ontological knowledge about knowledge, it also follows that we can have ontological knowledge about the knowledge of knowledge. This sort of reasoning, if deliberately pursued, leads to an endless regress that may not be solved within the confines of epistemology or ontology. Questions concerned with boundary conditions of awareness and ultimate reality, including what is appearance and what is reality, is the world one or many, what is individuality, what is being, what is life, and what is self, are investigated by metaphysics (gr. *meta* – after, beyond, and gr. *phusis* – nature), a branch of philosophy that is closely associated with ontology (van Inwagen 1993). Questions at the metalevel are raised by ontological and epistemological queries and investigations. But while entities addressed by ontology are, at lest in principle, amenable for scientific investigations, metaissues can not, at least not presently, be solved by science. However, since we revise our views about what is amenable to empirical investigation as we change our understanding of the world, metaproblems may not always have to be so.

Metaimmunology investigates the conditions and limitations for what could possibly be the contents of immunological science, including questions concerned with the ultimate workings of the immune system. This largely unexplored field owes its importance to the fact that immunologists continuously have to face boundary conditions of knowledge. For example, unlike most other natural scientists, immunologists have to face a double layer of epistemological reflections. This is made explicit in Efroni and Cohen's (2002, p. 23) claim that immunologists would like to "understand how the immune system understands what it should preserve and what it should attack". Thus, the immunologist needs to acknowledge his own epistemological standing as he investigates the epistemological problems faced by the immune system. The central problem for the immunologist, which is to understand how the immune system obtains true "understanding" of the antigenic world, therefore needs to be extended with an investigation into whether true understanding has the same meaning within immune epistemology as it has within mentalistic epistemology. But this again implies that the immune system is a subject, albeit a non-rational one, and paradoxically that the immunologist stands in a subject-subject relation in addition to a subject-object relation to the system (Figure 1.2).

Mediating dualism

Scientists are still working within the confines of classical epistemology, according to which knowledge is understood as a relation between what is "out there" and various inner states that represents the external reality. This model, termed "the epistemological construal" (Taylor 1995), is tightly bound up to three related notions. The first of these is the idea of the subject as a disengaged observer, being free from irrational bindings to the outside world and thus performing its reasoning without interferences. The second notion, which is derived from the first, is the idea of a punctual self; a self distinct from and independent of the outside world. From these follows the third notion, that science should be concerned mainly with explanations in terms of individualistic properties.

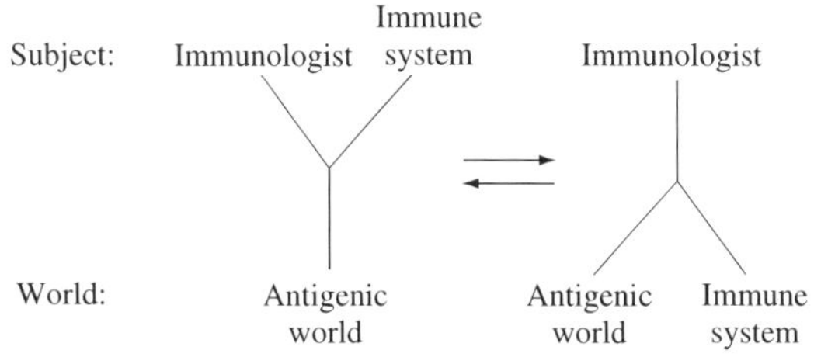

Figure 1.2. Two alternative depictions of relations between the immunologist, the immune system and the antigenic world. The immunologist as well as the immune system stand in a subject-object relation to the antigenic world but in a subject-subject relation to each other (in case of the immune system it stands in a subject-subject relation to the immunologist only artificially, for example in cases of transplantation). The immunologist also stands in a subject-object relation to the immune system

While the epistemological construal has propelled outstanding scientific successes in the natural sciences, several researchers have come to question its fundamental premises. They see the distinction between the knower and the known, and the related questions of how to establish contact between the *res extensa* and the *res cogitans*, or between the *res cogitans* of two individuals, as quasiproblems. This is because, they maintain, Descartes' made an erroneous inference when he disconnected the soul and the body. The critics emphasise that the fundamental assumption of the construal is not knowable on the basis of experience, *a posteriori*, but is made independently of experience,*a priori*. From this follows that the construal is but an untested construal and, accordingly, that the really deep problem posed by the construal is not how the separated instances can establish contact with each other; the problem is the distinction itself.

The German philosopher Immanuel Kant (1724–1804), who became aware of the shortcomings of the construal, argued that Descartes' mistake was to insist that subjects come naïve into this world. Contrary to this, Kant claimed that subjects come equipped with innate ideas that serve to organise the external reality, and that individuals thus already have some foreknowledge about the world's existence.[6] By this, he not only brought to attention the limitations of the epistemological construal, he also provided an ingenious solution to the construal's foundational problem.

In his discussion Kant utilised what has come to be known as transcendental arguments. Unlike arguments that relate to real entities, transcendental arguments do not directly concern entities of the real world. They relate to how these entities constitute themselves for the mind; they are conditions for the possibility of understanding entities in the world. The foreknowledge possessed is seldom questioned nor brought to conscious attention, even though it is the basis upon which all our derived knowledge is built. This is because the foreknowledge, being a precondition for all knowledge, is invisible to the naïve subject.

To understand the problematic involved, consider how to go about to investigate extraterrestrial life forms. A challenge for investigators embarking on these kinds of studies is that they, in contrast to investigators of terrestrial matters, have little if any valid pre-understanding of their subject matter. Astrobiologists may well ground their research on the belief that the physical properties of the elements and the laws of physics are universal, but they still have no scientifically grounded idea of what physical realizations life in other worlds may have. Paradoxically, extrapolation of terrestrial principles of life may even preclude any chance of finding new principles of life. This is because a preoccupation with one case may blind the investigators to other aspects of reality and thus bring about a state in which relevant observations become treated as irrelevant.

---

[6] Kant's claim was that we come equipped with innate knowledge of time, space and causality. These ideas were brought further in the 1990s when evolutionary psychologists claimed that humans come equipped with mental modules that unconsciously guide our adaptive behaviour in specific situations (Barkow et al. 1992).

Any investigation, including investigations of extraterrestrial life, requires a preunderstanding of what the object of enquiry is. Before we set out to investigate unknown life forms, we therefore need to come to grips with conditions for the possibility of detecting such life. The preunderstanding sought cannot rely on anybody taking Descartes' position of an all-knowing deity who sees the world like it really is. The idea of the disinterested spectator who sees everything in a timeless perspective (*sub specie aeternitatis*) is a construction, for how can anybody seize eternity if bound up in earthly space and time? Or stated differently; since nature continuously changes, how can something be seen in a timeless perspective, when it, in fact, is time that makes a difference?

Kant's form of argumentation was taken further by Georg Wilhelm Friedrich Hegel (1770–1831) and Martin Heidegger, who both criticised the Cartesian view for not taking into account the role of temporality during the shaping of the subject. By elaborating on Kant and taking account of history as a transcendental condition, they decisively demonstrated that the epistemological construal and its related notions should be seen as they were; but one aspect of knowledge. This was also the view of Hans Skjervheim (1959, 1996) and Charles Taylor (1995), who independently provided seminal arguments that effectively served to undermine beliefs in disengaged subjects, punctual selves, and individualism. They maintained that subjects are not detached observers; they are already engaged in their world. This has to be so because subjects must already be engaged in coping with the world to interpret it in one way or the other.

Despite forceful criticism, the problematic of the Cartesian distinction resurfaces whenever attempts are made at enforcing the study of man into a naturalistic framework. That man is a product of evolution by natural selection is uncontroversial, but to claim that man can be understood as any other biological entity is not; the heated discussions that followed the publication of Edward Wilson's book *Sociobiology* (1975) is evidence to this. In her book which summarises these controversies, Ullica Segerstråle (2000) convincingly argued that the critics of sociobiology were upset, not so much by the observations of animals that Wilson described, but rather by the implications Wilson's theory had for man's role in the world. The critics accused Wilson of being ideologically naïve, and for not understanding that scientific theories of man's role in society can not be isolated from the society that man inhabits. Wilson's mistake was, accordingly, that he failed to see the intimate relations between the *res extensa* and the *res cogitans*.

While Skjervheim and Taylor convincingly argued against the epistemological construal and against the dualism proposed between the *res cogitans* and the *res extensa*, they did not discuss dualism within the *res extensa*, for example between the living and the dead. The idea that life was breathed into matter by supernatural forces is one aspect of this dualism, albeit currently an uninteresting one. A more interesting dualism within the *res extensa*, between the organism and its environment, is sometimes found espoused in ideas about natural selection. According to that view the organism's role is to produce variability and the environment's role is to select the best fit alternatives; or stated differently, nature proposes and environment disposes. The organism-environment dualism, which has much in common with the epistemological construal, has been criticised for giving a simplified and false picture of natural selection (Lewontin 2000).

That Lewontin's criticism of the organism-environment dualism has something going for it is seen more clearly if one considers the standing of the organism according to multilevel selection theory (Keller 1999). The theory holds that traits of biological entities are selected at multiple levels, including the level of the gene, the cell, the organism and the group to which the organism belongs. Differential selection pressures are imposed at each level during interactions with the environment. The environment includes both the external environment as well as different levels within the organism, and the individual levels thus become both subjects and objects of natural selection. Since there is no way to disentangle when the level exhibiting the trait is subject and when it is object, it appears that principled dualism is as untenable for the biological sciences as it is for the sciences of man.

Participant and spectator

Man's sense of identity is tightly coupled to his developmental history. Looking back, a trajectory of decisions abounds, each of which gives different experiences and differential premises for future decisions. Quite analogously, an individual's immune system is what it is because of its history of decisions. But while we frankly acknowledge the role of decisions when it comes to shaping our mental selves, the same role is not easily ascribed to the immune system. This may be owing to methodological reasons, but may also be a result of historical constraints and prejudices on behalf of microbiologists and immunologists.

In a crude but credible caricature we may, with some confidence, claim that microbiologists see an extreme complexity in the microbial world whereas they envision the immune system as a more or less fixed object that reflexively responds to the microbial stimuli. And with equal buoyancy, we may picture immunologists as scientists that envision the immune system as a complex organ that can be probed by relatively fixed microbes. Hence both fields take their own study object to be active and the other to be passive. That this can not be so is evident for both groups of scientists, but since the abstraction is methodologically less demanding than the alternative, it is still adhered to in areas where it seems fit.

That there are cases where the abstraction does not hold is self-evident, and these cases thus need to be investigated in their full complexity. Such complex investigations must take into account that the individual is a being within an ecosystem where all actors are active. Such a task would require that we expel Cartesian dualism of all variants. As laid out in the previous section, the existentialist philosophers have already made the necessary groundwork. They see the Cartesian problem of how subjects enter a sphere of communicative interaction as a non-problem; subjects are already out there, they are beings-in-the-world. And as beings-in-the-world, they are already actively engaged in others.

The existentialist view has consequences for how we should envision not only microbial life-forms, but also the immune system and its relation to the microbial world. The immune system can not be seen as an ordinary organismal object, on par with hearts, kidneys and muscles. Rather, it is more like a subject, albeit a non-rational one, that responds adaptively to environmental stimuli. But this also implies that investigations into the performance of the immune system must be carefully carried out so as not to violate this special standing of the system. Research carried out by seeing the immune system from the perspective of the immune system itself, not from the perspective of the detached spectator, makes clear that immune system development is formed as if it was a trajectory of decisions made upon close encounters with significant others. The participant perspective allows a deeper understanding of how the immune system is being shaped and shapes the external world, or more generally; of how the immune system becomes a being amongst others in the world.

The participant perspective also has consequences for how we should view pathology. By claiming that some pathology is a matter of being with others in a deviant way, the war-metaphor emerges as but one interpretative perspective. Upon the spectator perspective, the immune system fails because it lacks resources, most often of a genetic kind, while the participant perspective sees the malfunctioning immune system, not as a deficient but rather as a biased system; its decisions are biased in one or the other direction owing to the impact of previous decisions. Thus developmental history becomes of essence, and diseases of the immune system can no longer be investigated from a real-time perspective only. Rather, they are bound up in developmental time and ecological space and should be investigated as such.

These thoughts also have consequences for therapeutic strategies. Allergy caused by grass pollen can, for example, be treated either according to the spectator or the participant perspective. When an allergic individual is treated symptomatically, for example by antihistamines or immunosuppressive medication, this is done according to the spectator perspective; the therapist can remain ignorant of the immune system's perspective. When, on the other hand, the allergic individual is treated with hyposensibilisation, for example with a vaccine that deviates the immune response in a beneficial manner, the therapist takes the participant perspective; he sets himself in the role of the immune system. Both procedures may provide the same result, making the allergic free from symptoms, but they fulfil this deed in different ways (Frew 2003).

Whether we should describe the person that takes the participant perspective as therapist or pedagogue is a matter of convenience. Treatment is anyway installed to "educate" the immune system to perform a novel pattern of reactivity upon encountering noxious stimuli. But the pedagogue view is appropriate because it makes it meaningful to ask whether common childhood infections teach the immune system to be allergic or not (Folkerts et al. 2000), and whether "immunoeducation" by the means of vaccines could be useful in preventing allergy without the need to immunise against all of the allergens that are potentially encountered during an individual's life time (Matricardi and Bonini 2000). Such immunoeducation, although currently far beyond reach, can only be accomplished if we envision the immune system according to a participant perspective.

By envisioning the immunotherapist as a pedagogue, the armamentarium of remedies used by the therapist can be envisioned under a novel perspective. As any pedagogue knows, the targets of their activity can learn by being persuaded, manipulated or convinced. The choice of one or the other of these strategies would have to be guided by the goal of the education. Persuasion, which presupposes a subject-object relation, gives knowledge but not necessarily insight; manipulation, which also presupposes a subject-object relation, is risky in that the manipulators themselves risk to become manipulated. Conviction is, in contrast, based on a subject-subject relation, and proceeds as an open communicative interaction during which insight is reached. Quite analogously, the immune pedagogue can utilise the same teaching techniques. Persuasion, for example in the form of symptomatic treatment, gives short-lived relief. Manipulation, for example by antigen-specific hyposensibilisation, may lead to counterstrategies that in turn lead to manipulation of the manipulator. The final strategy, the immunoeducation advocated by Matricardi and Bonini (2000), would alter the manner of the immune system to respond functionally to any stimuli.

An interesting aspect of the existential perspective is that one cannot choose away engagement; as a being-in-the-world one is already engaged in something. Even individuals that have deliberately chosen to, or have been enforced to take on the spectator role, are in a sense participants. But their participation is often inferior and may even be pathological if they are not being with others in the correct way. For example, the immune system of individuals infected by *Trypanosoma brucei*, the protozoan that causes African sleeping sickness, are in a sense made into impotent spectators. Following infection, the microbe continuously changes its antigenic profile. Because of the vast repertoire of immunologically distinct antigens that develop during illness, immune responses to the surface molecules of the protozoan are not effective at protection. In this manner, *T. brucei* enforces the immune system to take the role as a spectator. The microbe thereby effectively turns the immune system into an object that can be manipulated. It attacks the immune system's freedom to decide its own development and thereby takes on the role as its master.

Abduction

Neither deduction, which is the logical process whereby the inference goes from secure premises to a certain conclusion, nor induction, the inferential procedure that tests the validity of a hypothesis by investigating predictions based on the hypothesis, add anything new in terms of concepts. The two procedures are merely activities that test the validity of already known hypotheses, and nothing new is being revealed in the conclusion that was not already present in the premises. The generation of new concepts and theories requires speculative thinking, termed abduction by Charles Sanders Peirce. "Abduction is the process of forming an explanatory hypothesis. It is the only logical operation that introduces any new idea; for induction does nothing but determine a value and deduction merely evolves the necessary consequences of a pure hypothesis". (Peirce 1998, p. 216).

Peirce held induction and abduction to be opposite pools of reasoning. Both procedures lead to acceptance of a hypothesis because observed facts are such as would necessarily or probably result as a consequence of the hypothesis. But where induction starts from a hypothesis and proceeds to collect facts to support or falsify the hypothesis, abduction starts from facts and proceeds to generate an acceptable hypothesis that would explain the observed facts.

According to Peirce, any hypothesis purporting to explain surprising observations should fulfil three considerations. In the first place the hypothesis must be capable of experimental testing. It is not enough to invent a hypothesis, one needs to be able to extract from the hypothesis a testable prediction and then to discern whether that prediction is true or false. In the second place the hypothesis must be such as to explain the surprising facts encountered. And thirdly, the hypothesis must be parsimonious in the detailed structures it incorporates.

The importance of abduction was acknowledged by Karl Popper (1989), who advocated the idea that scientific knowledge grows by unjustified (and unjustifiable) anticipations and tentative conjectures. To ensure that the conjectures have their roots in the real world, he further advocated that they should be controlled by criticism, by attempted refutations. And although the conjectures can survive the attempts at refutation, and thus be accepted for the time being, they can never be proven true.

But Popper's claim that scientific theories are falsifiable by possible or conceivable observations poses a difficulty – an observation can only falsify a theory if it is itself conclusively certain. Popper was well aware that observations are not necessarily true representations of the world. He held every observation to be fallible, but was reluctant to admit of degrees of fallibility. Even so, he held that observations that are admitted as evidence "are accepted as the result of a decision or agreement; and to that extent they are conventions" (Popper 1935, p. 88). "From a logical point of view, the testing of a theory depends upon basic statements whose acceptance or rejection, in its turn, depends upon our *decisions*. Thus it is *decisions* which settle the fate of theories" (Popper 1935, p. 91).

The ideas of Peirce and Popper were well received by Peter Medawar, one of the founders of modern transplantation immunology. In a lecture before the American Philosophical Society, Medawar (1969, p. 46) claimed that "...science, in its forward motion is not *logically* propelled." Upon citing Peirce's view on the generation of hypotheses, Medawar continued by claiming that "The process by which we come to formulate a hypothesis is not illogical, but non-logical, i.e., outside logic. But once we have formed an opinion, we can expose it to criticism, usually by experimentation; this episode lies within and makes use of logic, for it is an empirical testing of the logical consequences of our beliefs". The formulation of a hypothesis is a creative act. However, "Scientists are usually too proud or too shy to speak about creativity and "creative imagination"; they feel it to be incompatible with their conception of themselves as "men of facts" and rigorous inductive judgements." (Medawar 1969, p. 55).

The hypothesis introduced by Charles A. Janeway (1989) to explain the initiation of an appropriate immune response against infectious agents bears the typical marks of an abductive inference. His new hypothesis, while accepting the capability of the clonal selection theory to explain immunological specificity, moved beyond the theory in being capable of explaining immunogenicity as well as the mysterious properties of adjuvants. Hence, the hypothesis explained surprising observations in a parsimonious manner and, furthermore, it was susceptible to experimental testing.

Charles Sanders Peirce viewed scientific activity as a long-term project that gradually titrated in on the truth. As such, the title of Janeway's essay, "Approaching the asymptote? Evolution and revolution in immunology", is truly Peircian. It matches Peirce's idea that "It is a primary hypothesis underlying all abduction that the human mind is akin to the truth in the sense that in the finite number of guesses it will light upon the correct hypothesis." (Peirce 1998, p. 108).[7] Hence, scientific activity is a self-correcting and long-run activity. Man's abductive powers are, according to Peirce, successful because they are embedded within a web of background knowledge that has been derived through the process of natural selection. The background knowledge guides our decisions and actions, and allows us to respond to some information without having to scrutinise the whole field of information. Owing to this innate knowledge, our decisions are right more often than wrong.

In a remarkable paper entitled *The complete solution to immunology* Niels Jerne (1969) proclaimed a quite different metaphysical view. He was apparently so absorbed by the clonal selection theory and the solution of the mechanisms of antibody specificity that he believed that immunology would be completely solved within a short while. After the complete solution had come, he contemplated: "I wonder what scientists will be doing during the following thousand years. In fact, I think they will be doing nothing, because all scientific knowledge worth knowing will be known. The scientific era will draw to an end," (Jerne 1969, p. 7). When Ronald Germain (2004) thirty-five years later asked whether immunological knowledge had brought us closer to the end in terms of conceptual advances, he surmised that "despite the cataloguing of nearly all genes in mice and humans, we are still far from knowing which ones contribute to immune activity, much less how

---

[7] Ironically, Peirce supported man's capability for abduction by utilising an inductive inference – he used historical evidence to corroborate the hypothesis that man has made great progress in science, thus making it probabilistically true that man's abductive powers are sufficient to reveal nature's true being. One drawback of this argument is, of course, that the sample space of true versus false hypotheses is unknown. As with all inductive arguments, if you do not select the group to be studied randomly, you may risk a bias when the results are extrapolated inductively to the population. This is a well-known problem in statistics. The validity of an inductive generalisation from investigated group to population is critically dependent upon the degree of similarity between the group being investigated and the population. If the investigated group is biased, the inductive generalisations are not valid.

they do so." (p. 1315). There should therefore, contrary to Jerne, be ample room for speculative thinking within immunology for the years to come.

EVALUATION OF HYPOTHESES

The thought collective

Ludwik Fleck, who studied the vigorous debates that took place amongst immunologists during the first decades of the 20th century, came to the conclusion that scientific facts are profoundly influenced by the thought style of the scientific community. Fleck acknowledged that to perform rationally "the true expert must free himself from the shackles of authority and justify his first principles again and again until he establishes a purely rational system." This ideal is, he acknowledged, rarely accomplished, largely because "the specialist is already a specially moulded individual who can no longer escape the bonds of tradition and of the collective; otherwise he would not be an expert" (Fleck 1935, p. 54). Hence, the acceptance of a fact may depend less upon its truth than upon the willingness of the leaders of the field, the thought collective, to acknowledge it.

Fleck's ideas inspired Thomas Kuhn (1962) to investigate scientific transitions on a more general level. Kuhn challenged the view that science is a process whereby indisputable facts are gathered in a cumulative fashion. For long periods, he argued, the scientific community performs "normal science" in which puzzle solving is their main activity. This solving of puzzles is performed within a paradigm; a commonly held belief among scientists in a given discipline. The paradigm need not be correct, only accepted. If a person or a group produce data that can not be interpreted within the standard paradigms, the scientific community experiences an anomaly. Anomalies may bring science into a crisis and a call for new explanatory paradigms. Paradigms are not changed lightly, and when they do change, a revolution is being initiated. Such an incident took place when the Copernican model of the solar system supplanted the Ptolemaic view. Based upon examples like this, Kuhn further claimed that it is not always possible to decide whether the post-revolutionary theories are more correct than the older. He has therefore been criticised for being a relativist as concerning the truth of theories.

In his review of the history of immunology, David Talmage (1988) provided evidence that progress in immunology has taken place in innumerable small steps, without any revolutionary changes. Anomalies that lead to crises have thus far not been experienced, and contemporary immunology is still performed within the same explanatory paradigm that has been with the field since the 1880s. Not even the clonal selection theory marked a revolutionary change; the germs of the theory were already implicit in previous theories, and Burnet did not amass novel observations to support his theory. Its acceptance relied less upon experimental evidence for or against it than upon the totalities of shared beliefs within the scientific community (Talmage 1986).

Thus, both Fleck and Talmage emphasised the impact of the thought collective. The differences between their views and that of Kuhn probably stems from the fact that immunology, unlike physics, is weak on theory. Immunology is an experimental more than a theoretical science, and since small differences in experimental setups may be of overwhelming importance, immunologists know that their experimental truths may be local and not easily transferred from laboratory to the field. Talmage's observation, that experimental evidence has played such a restricted role in changing the attitudes of immunologists, is thus likely related to the thought collective's awareness of the non-universal and context-dependent status of immunological results.

Since immunological results are weakly embedded within a theoretical matrix, the thought collective has gained a relatively strong influence on the sorts of speculations that are allowed. For example, the presence of suppressor cells, originally proposed by Richard Gershon in the 1970s (Gershon 1975), was initially met with scepticism. But as the suppressor phenomenon was identified by functional assays in a wide variety of laboratories around the world, scepticism changed to acceptance. Still, the considerable work devoted to establish the phenotype of the putative T cell and to understand the mode by which the suppressor cells functioned, were feebly rewarded. Coherent explanatory models were difficult to assemble, and the idea that suppressor cells existed as ontologically distinct entities thus gradually fell into disrepute (Keating and Cambrosio 1997).

The manner by which suppressor cells could be transformed from non-existence into experimental existence, and thereafter their relegation to a status of being non-existent artefacts, serves to testify to the importance of experimental results in the erection of immunological science as well as to the lack of theoretical guidelines. But this is also why the thought community of the 1990s so quickly accommodated as a fact novel observations of cells with a similar regulatory potential as the dismissed suppressor cells (Sakaguchi et al. 1995). To demonstrate their scepticism to the previously described suppressor cells, but also to acknowledge the fact that the relationship between the entities described as suppressor cells in the 1970s and in the 1990s remains unclarified (Athanassakis and Vassiliadis 2002), immunologists of the 1990s renamed the cells regulatory T cells (Treg).

While the continuing story of the suppressor/regulatory cells clearly demonstrates the willingness of immunologists to let themselves be guided by observational data, it also reminds us that the pervasiveness of the thought collective may be counterproductive to science if immunologists become subjugated to group thinking and fashionable interpretations, and let the "sexiest" ideas gain impact by their sheer sexiness. If so, questions may never become settled; they just go in and out of fashion. It also reminds us that immunologists create fictional characters, and that these sometimes have an ephemeral being. The thought collective should be aware of these dangers as it attempts to regulate immunological activity through refereeing scientific papers and through organising specialist and general scientific meetings.

The frame of reference

According to Peirce, any hypothesis purporting to explain surprising observations must be capable of experimental testing. However, if scientists were to follow Charles A. Janeway's (2001, p. 7467) advice "to never accept what looks like a plausible explanation until you have explored all possible alternative explanations", they would probably spend the rest of their careers testing futile hypotheses. The pragmatic implications of the advice, which are similar to Sherlock Holmes' maxim in the *Beryl Coronet* "that when you have excluded the impossible, whatever remains, however improbable, must be the truth" (Conan Doyle 1976), would be disastrous for the scientific enterprise.

Free-floating abductions signify madness more than method. Dr. Watson was probably aware of this when he in *The Reigate Square* reassured the police inspector that "I have usually found that there was method in his madness." The inspector did apparently not agree as he muttered "Some folk might say that there was madness in his method" (Conan Doyle 1976). Holmes emphasised the need for pursuing several possible lines of explanation that take account of the facts, and in *The Hound of the Baskervilles* Dr. Watson revealed Holmes' method as one in which "he weighed every particle of evidence, constructed alternative theories, balanced one against the other, and made up his mind as to which points were essential and which immaterial." When being accused of guess work, Holmes replied that what may appear as guessing is "rather, into the region where we balance probabilities and choose the most likely. It is the scientific use of the imagination, but we have always some material basis on which to start our speculation."

Method in madness is thus instantiated when abductive inferences are restricted to the best explanations, in which the "best explanation" is taken to be best among the competing alternatives that fit the facts. The best hypotheses are those that are compatible with the data to be explained as well as the referential frame in which the observations have been made. Hypotheses that fit the context are the only hypotheses that should be counted among the "possible alternative explanations". There is a caveat here, though; the problem of unconceived alternatives. Owing to shortcomings of the scientific imagination, there is always the risk that the truth lies in the vast space of theories to which we are blind, rather than in the small areas that we are able to survey.

The importance of contextual testing of scientific hypotheses was strongly emphasised by Quine (1951). He argued that hypotheses taken on their own do not have observational consequences. The hypotheses must be conjoined with auxiliary assumptions if they are to be tested. It is not the hypothesis H in isolation that predicts whether the observational fact O will be true, it is the conjunction of H and auxiliary assumptions A that has this property. If it turns out that a prediction fails to come out as true, if (H&A) entails O, and not-O turns out to be true, then we may choose to abandon either H or A or we may abandon the logical principles involved in the inference (Figure 1.3).

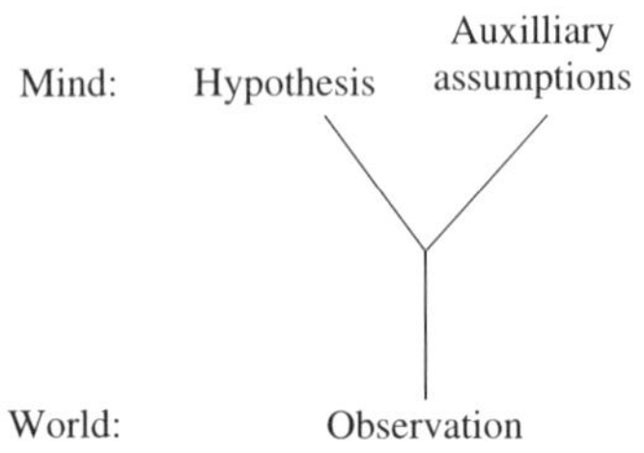

Figure 1.3. Evaluation of hypotheses is not performed as a two-way relation between a hypothesis and an observation, but as a three-way relation between a hypothesis, an observation and auxiliary assumptions

In his evaluation of Quine's arguments, Sober (1999) argued that Quine overlooked the important fact that auxiliary assumptions are often independently tested and merely function as a frame of reference when new hypotheses are tested. It is for example central to experimental methodology that controls and other background knowledge are not independently tested in the experiment. Rather, scientists try to find auxiliary assumptions that they have good reasons to believe are true regardless of what the experimental observations may bring to bear on the hypotheses. Thus, if not-O turns out to be true, we should abandon H but not A.

Background assumptions have made their way into probabilistic reasoning through the use of Bayes' theorem. Bayes' theorem is a mathematical consequence of the definition of conditional probability. Bayes' formula is defined as follows:

$$\Pr(H|O) = \frac{\Pr(O|H)\Pr(H)}{\Pr(O)}$$

The conditional probability Pr(H|O) is the posterior probability of the hypothesis H given the observation O. The conditional probability Pr(O|H) is the probability of O given H. This conditional probability is also called the likelihood of H. Pr(H) is the probability of H before O has been gathered, and Pr(O) is the probability of O without knowledge about H.

The relation between these probabilities can be demonstrated by an example from medical diagnostics (Ulvestad et al. 2001). When a physician considers the probability that a patient has rheumatoid arthritis (H), tests for rheumatoid factor (O) are often requested to back up or challenge this diagnosis. The physician knows from experience that the prior probability for rheumatoid arthritis, Pr(H), is 1% in his patient population.[8] Hence, the context in which he performs the test is the patient

[8] There is almost always a certain arbitrariness about which reference class is chosen as a base for the prior probabilities. By varying the contrasting alternatives, genuinely different testing problems are formulated. The larger the reference class the more reliable the statistics, but the less relevant they may be for the predictive problem at hand. When the reference class is made very specific the probabilities become more relevant but less general. This problem is solved in diagnostic medicine by referring the patient's pre-test probability for disease to a differential-diagnostically relevant group.

population. If he applies the test to another population, say an out-patient clinic at a hospital instead of general practice, the context changes and so does Pr(H). He also knows from textbooks that the probability of a positive test for rheumatoid factor, Pr (O) in the population is 4.1% and that probability of rheumatoid factor in patients with rheumatoid arthritis, Pr(O|H), is 82.4%. By entering these values into the formula of Bayes' theorem he then predicts the posterior probability of rheumatoid arthritis to be 16.9%. Thus, the test result increased the disease probability from 1% to 16.9%.

Several auxiliary assumptions, which are not independently tested, are encountered in this example. One of these is the assumption that tests for rheumatoid factor provide unmediated access to the presence or absence of rheumatoid factor in the patient. But as any clinical immunologist is aware of, the level of measured substances depends on characteristics of the assay being used. These characteristics, which include the true and false positive rate of the test as well as characteristics of the population upon which the test is applied, are seldom investigated during testing of sera. Their characteristics are accepted as a given; as the background upon which the testing is performed.

The application of Bayes' theorem on cases where prior probabilities can be given objective probabilities, as in the rheumatoid factor example, is uncontroversial. However, when posterior probabilities of scientific hypotheses are investigated using Bayes' theorem, it is often not possible to give objective prior probabilities. It has therefore been argued that the probability of a hypothesis being true can not be evaluated by Bayes' theorem and that Bayesian inferences should be abandoned.

As an alternative, the plausibility of scientific hypotheses can be evaluated probabilistically without the use of prior probabilities. This can be done through comparisons of their likelihoods (Royall 1997). If we compare two hypotheses, their posterior probability is determined by two factors; their likelihoods and their prior probabilities. This can be seen from a rearrangement of Bayes' theorem. In the light of a given observation, H1 is more plausible than H2 when $\mathrm{Pr(O|H1)\,Pr(H1)/\,Pr(O) > Pr(O|H2)\,Pr(H2)/\,Pr(O)}$. This rearranges to

$$\frac{\mathrm{Pr(O|H1)}}{\mathrm{Pr(O|H2)}} > \frac{\mathrm{Pr}(H2)}{\mathrm{Pr}(H1)}$$

If prior probabilities are fixed, it is the likelihood ratio that determines the plausibility of the hypotheses in light of the observations. According to likelihoodism, O strongly favours H1 over H2 if and only if H1 assigns to O a probability that is much higher than the probability that H2 assigns to O (Sober 1993). This occurs when $\mathrm{Pr(O|H1) > Pr(O|H2)}$. The likelihood ratio gives a quantified estimate of the strength of this contrastive testing.[9]

[9] An interesting feature of the likelihood ratio is that it can be calculated from multiple independent diagnostic tests of the hypotheses H1 and H2. The results from the tests, which can be quantitative or binary, can be combined into a multivariate likelihood ratio vector. It can be shown that the

Unlike classical frequentist statistics, likelihoods can be applied also to singular observations. As a result, even unique historical events can serve to statistically discriminate between two rival hypotheses. This aspect of likelihoods is possible because the rival hypotheses are evaluated within the context of an explicit model in which the background assumptions that need to be met are identified.

A likelihood rationale

Adaptationism, the idea that natural selection is the major causal factor for the evolution of a trait, has been criticised for being so flexible that it can be maintained no matter how many of its specific models are invented and refuted (Gould and Lewontin 1979). One way of avoiding this accusation is to take into account the possibility that the character would have evolved without there being selection for it. Instead of just looking at evidence for adaptation caused by natural selection, one should also consider evidence in favour of alternative hypotheses for the same adaptation. In a likelihood framework, support for the adaptive hypothesis is best captured by the ratio of the likelihood that the trait would have evolved via natural selection to the likelihood that it would have evolved without such selection (Baum and Donoghue 2001). Hence, by investigating the predictions issued by the hypotheses substantial insights into occurrences that happened in the past can be revealed.

A consequence of likelihood reasoning is that if the contrasting hypotheses under test assign equal probabilities to O, the predictions they issue are identical and O can therefore not be used to discriminate between their likelihoods. An example of this would be:

H1: The immune system (O) was designed by a benevolent creator.

H2: The immune system (O) is the product of natural selection.

Since the two hypotheses confer the same probability on O, the likelihoods of the hypotheses can not be used to differentiate between them. But one can still claim that H1 is more plausible than H2. If so, this must according to Bayes' theorem be owing to differential prior probabilities of the two hypotheses, that $Pr(H1) > Pr(H2)$. And as stated previously, prior probabilities are set on the basis of other background information and independently tested auxiliary assumptions.

---

likelihood ratio for the tests x_1, x_2, to x_n, where X is a combined vector of the tests x_1, x_2, to x_n is: $LR(X) = \exp(\alpha_0 + \alpha_1 x_1 + \alpha_2 x_2 + \alpha_n x_n)$. Values for the coefficients α can be derived using logistic regression analysis in which the best-fitting values are found by finding the maximum likelihood values of the parameters – the values of the parameters that maximise the probability of the data. Maximum likelihood estimation involves testing several hypotheses. The hypothesis that best explains the result is chosen as the maximum likelihood estimate. If only two hypotheses, H1 and H2, can explain the observations (O), then H1 has maximum likelihood if $Pr(O|H1) > Pr(O|H2)$. But likelihood ratios can also be generalised to the multivariate case, where the information from several tests is combined (Ulvestad 2003).

Some scientists do, nevertheless, believe that the credence of the two hypotheses can be tested by a likelihood comparison. Michael Behe, a leading proponent of the intelligent design movement, has argued that the immune system is an irreducible complex system that can not function without all the interacting parts being present. Since natural selection is a process that proceeds cumulatively with small gradual changes in the evolving organisms of a species, Behe's main argument for endorsing the design hypothesis is the fact that the presence of say 10% of an immune system is invisible to natural selection if those 10% are non-functional. "Since natural selection can only choose systems that are already working, then if a biological system cannot be produced gradually it would have to arise as an integrated unit, in one fell swoop, for natural selection to have anything to act on." (Behe 1996, p. 39). And of course, if the immune system arose "in one fell swoop", it is by default intelligently designed by a benevolent creator. Accordingly, H1 confers a higher probability to O than does H2, and consequently, the likelihood of H1 is higher than the likelihood of H2.

While it is true that the immune system is extremely complex, it is questionable whether the immune system is irreducible complex. Behe's definition of an irreducibly complex system is "a single system composed of several well-matched, interacting parts that contribute to the basic function, wherein the removal of any one of the parts causes the system to effectively cease functioning" (Behe 1996, p. 39). Although Behe argues that the definition gives an appropriate description of the immune system, the degree to which observational data in the literature support his claim is questionable.

Several of the interacting units of the immune system have an independent function, sometimes not even related to defence, in invertebrate animals. Hence, the nuts and bolts of the immune system have an evolutionary history that may be quite unrelated to their current usage, and there is therefore no need to invoke "one fell swoop". Behe also fails to take into account several mechanisms other than natural selection. Both sexual selection and evolutionary drift may be powerful agents for evolution. Furthermore, gene duplications may allow the differential derived copies of an ancestral gene to become integrated into other regulatory networks without affecting the functioning of the original gene complex.

It is quite clear that Behe has omitted important observations in order to increase the likelihood of H1, and as it turns out, his epistemology is similarly flawed. Like all other argumentation, reasoning by likelihood comparisons can be manipulated, or as claimed by Peirce (1998, p. 108), "experience shows that likelihoods are treacherous guides". A major omission in Behe's work, like Paley's before him, is that he does not discuss the likelihood of H1. His main project is to dismiss H2 and thus by default to argue that H1 is best supported by the data. This method, which is analogous to significance testing within classical statistics, has some serious shortcomings as to what one should believe on the basis of observations (Royall 1997). Put simply, classical significance testing accepts the null hypothesis and rejects alternative hypotheses if the null hypothesis fits the observations better. The truth or falsity of the null hypothesis is not questioned, even though we know that most null hypotheses are obviously false. Since the range of null hypotheses is so wide as our minds are fertile, new null hypotheses can always be postulated, and our confidence in such hypotheses should, accordingly, not be very high without evidential support.

The strategy of likelihood testing is different from significance testing in that there is no favoured null hypothesis. The two hypotheses are tested against each other on the basis of observations and background assumptions. Sine the design hypothesis entails almost any observation deductively, it can be argued that no single observation can be used to differentiate the likelihoods of the two hypotheses. Hence, if Behe would like to claim that H1 is more plausible than H2, this must be owing to differential prior probabilities of the two hypotheses, and hence that $Pr(H1) > Pr(H2)$. But this is a case for unconditional belief. The inequality does not include observational data and the differentiation of the hypotheses is therefore no longer within scientific discourse.

THE MISHAP OF THEORY

Theory in immunological practice

We often take for granted that there is a connection between practical and theoretical science, and that theory is required for practice. This view probably stems from the idea that modern science's main preoccupation lies in finding the truth, in the sense of what is true irrespective of time, place, culture and individuals. According to this view applied science is merely deductive application of the true knowledge derived from pure science. However, this interpretation is too simplistic and does not fit the historical evidence (Nordin 1999). The two appear to interact in a more subtle manner than this. That practice may be prior to theory is, for example, well exemplified by the history of modern immunology, which started out as an offshoot of vaccinology.

While some form of immunisation had been performed long before the advent of modern science, the scientific study of vaccines, vaccinology, started out at the end of the 1790s when Edward Jenner learned that some ingredient in the scars of patients with cowpox protected susceptible individuals against smallpox. He deliberately injected ingredients from cowpox into naive patients and after some months challenged them with material from patients with smallpox. He thus noticed that the cowpox procedure was effective in protecting against smallpox. Nearly one hundred years later Louis Pasteur proposed to call the procedure vaccination, after the Latin word for cow, *vacca*, as homage to Jenner.

Neither Jenner nor his famous successor Louis Pasteur had any theoretical knowledge of the immune reactions invoked by the vaccines. Despite the lack of theoretical knowledge, vaccinology turned out to be a tremendous medical success, with the eradication of smallpox in 1997 as its greatest success to date. As of now, three million deaths are prevented annually by vaccines, one million alone through vaccination against measles, and millions of people and animals are annually protected against debilitating diseases like rabies, tetanus and diphtheria. It is likely that even more success-stories can be added to this inventory as eradication of poliovirus and measles may be accomplished in the 21st century.

The science of immunology became firmly established some 100 years after the establishment of vaccinology. As it became known that the immune system is involved in some way or other in almost every imaginable disease, high hopes were launched that theoretical insights into immune system functioning would lead to tremendous success in all areas of medicine and pathology. These potentialities have not been fulfilled, and the merits of immunology for the practice of medicine have so far been relatively modest. While immunology has contributed a great deal to practical matters like diagnostics and transplantation surgery, and in some cases has been crucial for a theoretical understanding of disease mechanisms, the field has not lived up to its therapeutic or predictive potential.

In retrospect, these disappointing results should not have come unexpected. Immunologists are devoted to the study of the regulatory intricacies and complex genetic interrelationships that unfold as the immune response develops, and are therefore prey to the whole cascade of context-dependent solutions that coevolution between host and pathogen has shaped. Vaccinology takes advantage of the fact that the immune system is devoted to the eradication of infectious agents, and thus aims to get the proper output from a proper input. That the predictive success of vaccinology has been attained despite little explanatory insights into regulatory and contextual mechanisms, makes evident that explanation and prediction are different and disconnected concepts.

Explanation and prediction

The functioning of the immune system, which is complex and highly individualised, has evolved through a variety of selective pressures. The fact that the adaptive immune system differs between every organism, even between identical twins, suggests that theoretical immunology may be predictive only on a case-by-case basis. Universal predictive models will therefore be hard to establish.

That immunological theory has a high explanatory but low predictive relevance has some serious consequences for the status of immunological science. Ideally, a science should be able to explain past observations and predict future results. In the physical sciences this is done using quantitative mathematical models that make testable predictions that may or may not strengthen the interaction of hypotheses with evidence. The low predictive accuracy of immunology may suggest that the theoretical models utilised by immunologists are too narrow or, alternatively, that the complexity of the immune system is too complex to handle.

The differences between explanation and prediction can be illustrated by an example. Let us imagine an anachronistic immunologist being asked by a pharmaceutical company to design a vaccine against smallpox. Being an immunologist, he will be obsessed with the specificity of the immune reaction, and will therefore concentrate his vaccine-efforts on the infectious agent that causes smallpox. Furthermore, since previous exposure to smallpox protects against re-exposure, he knows that induction of a memory response is crucial. Since there are no animal models available, and since smallpox is a disease restricted to humans, he decides to investigate the human immune response to the infecting agent. During the course of these investigations he observes that patients respond to a few important molecules on the virus, and that cytotoxic cells are responsible for clearing the infection, with antibodies taking a secondary role. Based on this information, his advice to the company will be that they should concentrate their vaccine efforts on creating a cytotoxic cellular response against the infecting agent.

In this case, the immunologist's obsession with specificity precluded his chances of utilising other infectious agents as a vaccine, and he did therefore not come up with Jenner's vaccine. Maybe the principle of cross-reactivity should have been a part of the immunologist's theoretical tool kit. But although Jenner was successful in utilising the cross-reactive potential of two viral species, cross-reactivity does not always induce protective responses. The vaccine against the virus that causes poliomyelitis needs to contain three different strains of the virus to be effective, and the vaccine against pneumonia caused by pneumococci must contain the most prevalent types since immunity to one serotype does not confer protection against the other 90 serotypes. The principle of cross-reactivity may even be dangerous to utilise in some cases, since immunity against one virus may lead to increased disease susceptibility if a closely related virus of the same species later infects the individual. There is for example currently an urgent need for a vaccine against dengue virus, but the risk that a vaccine against one serotype could enhance the risk for hemorrhagic fever following infection with a different type, has been a serious impediment for its development.

The anachronistic immunologist's idea of designing the vaccine based on a recreation of the normal immune response to the infecting microbe is also theoretically sound, but may not be practically useful. By focusing on the naturally unfolding immune response, the immunologist did not come to consider the idea that induction of antibodies could have had a better preventive effect than induction of cytotoxic cells. A vaccine is all about preventing a subsequent infection with an infectious agent, and it may be that antibodies are better at inactivating and eliminating the infectious particles as they enter the body than are cytotoxic cells, even though cytotoxic cells are better at eliminating the infection once it has occurred. It is true that natural infection sometimes elicits very poor functional antibody responses. Still, several observations indicate that antibodies can be effective in preventing or limiting infection by many pathogens (Burton and Parren 2000). This includes antibodies against vaccinia virus, which appear to be protective by preventing viral dissemination (Bray 2004).

A further problem with the naturally unfolding immune response is that it may not even hint at the major targets that need to be attacked. For example, the natural immune responses against the bacteria *Clostridium tetani* and *Corynebacterium diphteria* do not reliably confer immunity to recurrence of diseases. Even so, Emil Behring and his co-workers managed to produce effective vaccines against disease caused by the same microbes. They did, however, not investigate the natural immune response towards the microbes. They knew that the bacteria produced toxins, and utilised this knowledge by vaccinating individuals with inactivated toxins from the bacteria. Vaccination thus created an immune response that involved different mechanisms than those elicited during the natural infection. This observation led Casdevall and Pirofski (2003) to put forth the view that some vaccines protect against disease by eliciting "unnatural" immunity. The same phenomenon has also been noted in attempts at vaccinating against cancer, in which modified self-antigens perform better than the natural antigens (Yu et al. 2004).

While theory is not necessarily applicable for practical purposes, there are nevertheless several examples in which theoretical immunology has contributed substantially to vaccinology. An important example is provided by the carbohydrate-protein conjugate vaccines that also elicit "unnatural" immunity. The idea of enhancing the immune response through conjugation came from immunochemical research performed in the 1920s (Ada and Isaacs 2003). Infants under the age of two years have an immature immune system and mount poor immune responses to polysaccharide antigens. They are therefore susceptible to severe infections caused by infectious agents with an outer capsule consisting of polysaccharide. Vaccinations with purified polysaccharides do not induce protective immunity in small children, but if the polysaccharide is coupled to a protein molecule, the child induces a stronger and protective immune response against the polysaccharide. By utilising this procedure, an effective vaccine against *Haemophilus influenza* type b has been developed.

It has recently been demonstrated that vaccination can have subtle effects other than protection against a specific agent, and the risk of diverting the immune response through vaccination needs to be considered as well. Epidemiological studies have indicated that infections with measles may protect against allergic disease, whereas vaccination does not have this effect (Shaheen et al. 1996). In measles, both vaccination and infection have profound and long-lasting effects on the immune system. But apart from the fact that both stimuli protect against subsequent measles infection, the stimuli differently switch the T helper profile; vaccines towards a Th2-mediated antibody response and infection towards a Th1-mediated inflammatory pathway. Hence, vaccines do not only protect against infection, but actually replace the infectious agent as a stimulus that may educate the immune system. The vaccinated immune system may thus experiences an "input deprivation syndrome" (Rook and Stanford 1998). Hence, vaccines should be designed not only to protect against specific disease, but also to maintain a proper cytokine network.

If our anachronistic immunologist had been asked to prepare a vaccine against a novel infectious agent, he would be at risk of missing the target if he followed the strategy of previous vaccine efforts. The successful vaccines that have been introduced so far have been mostly against microbes with stable antigens towards which the responding organisms make a protective immune response. Some microbes, for example *Trypanosoma brucei* do, however, exhibit a highly variable antigenic structure during infection. Given such constantly shifting targets, it is perhaps unsurprising to learn that conventional approaches to vaccine design, involving whole, killed organisms or purified surface proteins, have yielded little success with this microbe.

But there are also cases where we have a thorough theoretical insight but no efficient vaccine. For example, the bacterium that causes tuberculosis, a major disabling infectious malady, has been known since 1882, and the whole genome was sequenced in 1998. Despite intensive efforts there is still no efficient vaccine available (Flynn and Chan 2001). The researchers that try to make a vaccine against HIV have experienced similar problems. Despite the facts that the entire genome of HIV has been known since 1985 and the immune response to HIV has been most extensively investigated, the prospects for a vaccine are still dim (Wei et al. 2003).

Predictive accuracy

Simplicity versus complexity

Scientists that affirm the existence of objective truth and who claim that theories created by science should aim at being true representations of the way the world is are called realists. Another group, the instrumentalists, claim that science is best looked upon as a useful tool for coping with and controlling the environment. Owing to the complexities of the world, instrumentalists hold that our scientific theories can not aspire to report the way the world is. Instead scientists should strive to make empirically adequate theories that function as useful instruments for organising experience and predicting the consequences of events. The two metaphysical views are not of equal magnitude in the various sciences, and it has been argued that sciences may be classified in proportion to their commitments to realism or instrumentalism (Rosenberg 1994). On this scale, biological science is an instrumental science to a much greater degree than physics and to a lesser degree than the behavioural sciences.

Differential commitments to instrumentalism and realism are vehemently espoused in the immunological theorising of Melvin Cohn, Irun Cohen and their associates (Efroni and Cohen 2002; Langman and Cohn 2002; Cohn 2003; Efroni and Cohen 2003). While they seem to agree that the immune system is an extremely complex organ and that truth matters, they differ in their commitment to instrumentalism. Cohn (2003, p. 138) is an instrumentalist when he argues that "mathematics and programming are tools for biologists" that help them get their concepts of reality right, whereas Cohen takes a realist attitude when he thinks of hypotheses and models as being true descriptors of reality.

While both Cohn and Cohen utilise theories and hypotheses to explain and predict observations, Cohn seeks a minimal model (Langman and Cohn 2002) where Cohen seeks a complex model (Efroni and Cohen 2002). Cohn argues for a minimal hypothesis with universality of explanation as the goal; his approach to understand immunity is to meet immune system complexity with a simplistic hypothesis. Cohen, in contrast, argues that the understanding of complex systems can be achieved only through the erection of complex models: "To understand complex systems, we need to build models of its networks. To build such models, we can no longer treat models as we treat hypotheses" (Efroni and Cohen 2002, p. 29).

Parsimony and fit to reality

As made clear in the discussion on likelihood, choice of hypothesis should be guided by observations more than by the experimentalist's metaphysical views. Whether one selects a minimal or a complex hypothesis is thus not a matter of convenience. But fit to observations can not be the only factor to consider when selecting hypotheses. As any experimentalist knows, data are often beset with inaccuracies. Hence, if a scientist constructs a hypothesis that perfectly fits the data, he will be at risk of adapting it to noisy data. There is thus a risk that the hypothesis overfits the empirical observations, and thus that it fails at generalising to new observations. Therefore, the scientist that selects a hypothesis has to find a way of trading off fit with generality.

The way fit to data relates to generality can be explained by invoking Ockham's razor. This is the maxim that theories should be parsimonious, and that their parismoniousness is relevant in deciding their truth-value as well as their capability to make accurate predictions. Although complex hypotheses tend to have a higher likelihood than simpler hypotheses, it is also often a fact that simpler hypotheses are more predictively accurate. In order to estimate the trade-off between explanatory fit and prediction, Forster and Sober (1994) employed Akaike's theorem, which utilises the goal of predictive accuracy to estimate the relative importance of simplicity and complexity in hypothesis selection.

Within this framework of thinking, there is a distinction between a model and a hypothesis; a model is a family of hypotheses. Consider the two equations I and II, which are both models:

$$\mathrm{I}: y = \alpha_0 + \alpha_1 x_1 + \alpha_2 x_2$$

$$\mathrm{II}: y = \alpha_0 + \alpha_1 x_1 + \alpha_2 x_2 + \alpha_3 x_3$$

A model has some number of adjustable parameters. For example equation I has 3 adjustable parameters: α_0, α_1, and α_2, while equation II has four. But when the parameters are fixed on the basis of observational data, the parameters are no longer adjustable – they are adjusted. The two equations are then no longer two families of hypotheses, they represent two hypotheses.

Predictive accuracy is obtained through a two-step process. The scientist first uses observations to find the most likely hypothesis of each model, L(H), and then uses L(H) to predict new observations (Sober 2002). Akaike's theorem says that the predictive accuracy of a family of hypotheses can be estimated by considering two factors – how well the likeliest member of the family (H) of models fits the evidence at hand (O) and how many adjustable parameters (k) the family contains.[10]

$$\text{Predictive accuracy} = \text{Log}\,\Pr[O|L(H)] - k.$$

[10] For technical reasons the model's fit to the evidence is measured by the logarithm of the likelihood.

By utilising Akaike's theorem the two adjusted hypotheses can be compared both with regards to their closeness to truth as well as their complexity. The likelihood that the likeliest hypothesis confers on the data – its closeness to truth, is penalised by the number of adjustable parameters included in the hypothesis – its complexity. This means that hypothesis II is penalised for having one more parameter ($\alpha_3 x_3$) than hypothesis I. Thus, if I and II fit the data about equally well, one should expect the simpler hypothesis I to be more predictively accurate. For the more complex hypothesis to have a higher predictive accuracy it must fit the data a lot better than the simpler model.

There is a caveat here, though. Since closeness to truth is estimated indirectly through predictive accuracy, Akaike's theorem entails that a false theory may sometimes be more predictively accurate than a true theory. Hence, when the adequacy of Cohn's minimal and Cohen's complex models are being evaluated within this framework it becomes clear that their closeness to truth can not be settled until the values for the likelihood and complexity parameters are entered into Akaike's theorem and their predictive accuracies calculated and compared.

Instrumental realism

While predictive accuracy, closeness to truth, complexity and simplicity are important parameters for the evaluation of the validity of scientific hypotheses, their relevance for immune system functioning is quite another matter. The predictive accuracy framework has been elaborated for scientific epistemology, and may or may not be relevant for immune epistemology.

Efroni and Cohen (2002, p. 23) contrast scientific and immune epistemology by arguing that scientific understanding depends on clarity of reasoning from first principles, whereas immune system understanding "is bereft of clarity, reasoning, and first principles; the immune system operates solely by observations and experiments". While this is at bottom an empirical claim, Efroni and Cohen seem to use it as an unfounded metaphysical premise in their view of immune epistemology.

Scientists use theories to organise, explain and predict observations, and theories are validated by their success at the same parameters. If it were the case that the immune system operated solely through observation and experiments, the immune system would be like a scientist without a theory, to whom observations would be meaningless. Like the scientist without theory, an unconstrained immune system would respond in a confused and irreproducible manner to stimuli. However, as evidenced by the specificity and reproducibility of immune responses across individuals within the same species, antigens are definitively meaningful to the immune system, thus indicating that Efroni and Cohen's view of immune system epistemology is misguided.

While the immune system is capable of giving an extremely diversified response to infectious agents, the unfolding immune response is generally focused on a few significant antigens. There are several reasons for this. First, although it is clear from the Dreyer-Bennet hypothesis and Tonegawa'a verification of it that the immune system has at disposal a nearly unlimited mechanism to recombine genetic segments, several checks serve to restrict the realised repertoire of receptors. In the parlance of the predictive accuracy framework, the genes of the immune system receptors are adjustable parameters, while the genes that encode the rearranged receptor are adjusted parameters. Adjustment of the parameters is a matter of chance, but selection of the adjusted parameter is done according to how the world is. Thus, the checks serve to restrain and fix the adjustable parameters and thereby to focus the response towards the infectious agent in a realistic manner. Second, the focusing and adjustment of the response parameters are adjusted in response to signals derived from antigen presenting cells that make up the context of the response. This imposes method for madness.

Analogous to the fitness of scientific models, immune system fitness can be measured according to the system's capability to predict an incident with a high degree of accuracy. The fitter organism is the one that is capable of predicting the infectious agent's next move up the coevolutionary ladder. Whether or not it recognises every aspect of the infectious agent is non-consequential, what matters is whether it is capable of focusing on significant aspects of the agent or not.

Goodness of fit is important for immunity, but simplicity in obtaining the specificity is also relevant. These two are to a certain degree traded off against each other. If the generative mechanism is very complex, the fit needs to be very much higher to compensate for the costs of increased complexity, while if the generative mechanism is simple, the fit can be lower while still being functional. This may occur if a high number of degenerate receptors with overlapping specificities are generated. If the immune system makes highly specific receptors with little degeneracy, there is a risk that the high fit will not allow the response to generalise towards a new agent with slightly modified antigens. Hence, degeneracy ensures that the animal possesses the ability to make immune responses to protect against essentially any foreign antigen (Edelman and Gally 2001). Degeneracy is therefore not a problem but a virtue for the functioning immune system.

Constraint can be viewed as a limitation on the response, or alternatively it may be viewed as a pragmatic requirement for real-time decision making. To be useful, the immune system needs to act on information in real time. To be fast, it must be frugal in what it takes into account – the search procedure must be bounded. The bounded rules may be seen as evolutionary generated rules; those that increased the fitness of its bearers became dominant while those that did not became redundant. Simple rules that do not lead to appropriate responses are not adapted to their environment, and thus lack ecological validity.

That false belief may sometimes have a higher predictive accuracy than true belief has a corollary in immune system epistemology, for example when vaccination with cowpox results in an immune response that generalises and protects against smallpox as well. Accordingly, the immune system, like the scientist, operates according to the principle of instrumental realism. It is instrumental in generating protective responses based on evolutionary learned dispositions that realistically represent the infectious world. That the principle of instrumental realism is applicable to immune system functioning, not only to scientific reasoning, further testifies to the way that epistemology and ontology reciprocally depend upon each other when systems attempt to explore the intricacies of a diversified and continuously changing world.

CHAPTER 2

Immunobiology

In 1903 Arthur Grünbaum proclaimed that Metchnikoff's and Ehrlich' theories were "only rays of light refracted at different angles from the same source" and that the two theories would "ultimately be made to fuse into one brilliant cellulo-humoral theory of immunity" (Grünbaum 1903, p. 776). Still, nearly one hundred years passed before results emanating from comparative studies of immunity, combined with a conceptual reorientation, made clear that the adaptive immune system is built upon contrivances of phylogenetically elder non-adaptive immune systems. Metchnikoff's hypothesis, which was proposed from investigations of invertebrates, and Ehrlich' hypothesis, which was suggested from investigations of vertebrate immune systems, could thus be unified by considering the phylogenetic emergence of the defences.

Another reorientation that took place during the latter half of the twentieth century, and which challenged the ingrained conceptions of how hosts and pathogens interact, occurred when behavioural ecologists began to view hosts and their pathogens from a wider biological angle. By emphasising interspecies conflicts, evolution of virulence and behaviour in groups, they revealed that the immune status of an organism has consequences for its chances of finding mates and thus its ability to reproduce. Along with emerging knowledge that the immune system is influenced by and influences endocrine, nervous and cognitive functioning, the fields of evolutionary ecology and immunology were, like humoral and cellular immunology, ready for integration into a unified model.

While this second unification has yet not been accomplished, its establishment should be encouraged and pursued. This is because an integrated understanding of immune system functioning would enable researchers to provide explications of why and how organisms grow to give flexible responses to changing microbial environments and of how organisms differentiate between constituents that are to be spared and constituents that are to be eliminated. These tasks would require immunologists to provide explanations of the defence system's evolutionary history, its current function within a relevant ecosystem, the development of immunity over the organism's life-span, and the mechanisms used to regulate activation and suppression of the defence related components.

The challenges encountered would further require immunologists to explain how groups and individuals relate to each other and how groups and individuals influence the evolution and ecology of intra- and interspecies relations. These complicated issues are amongst the most debated topics in contemporary biology. And since the debate concerns the relation between the world and the scientific mind's conceptualisation of it, it is a rich source of philosophical disputes as well. Elucidation of these explanatory tasks requires meticulous collection of historical, observational and experimental data, but also necessitates that scientists muster and master a collection of concepts and metaphors that can serve as descriptors of the revealed reality.

The task of explaining intricacy and causality in the biological world must thus proceed simultaneously with an explication of how the world's complexity is represented within the scientific mind. It is a major assertion of this book that immunology's conceptual core lies buried somewhere in-between these criss-crossing, interactive and intertwined spatio-temporal networks. This chapter provides some glimpses into central issues that have puzzled biologists, immunologists and philosophers throughout the ages. The examples and discussions, which serve as preambles for topics brought up in the following chapters, all relate to the theoretical and biological status of defending organisms. These unique entities, the most striking features of the living world, interact with each other both as individuals and as groups. Continuously cooperating and competing, at multiple and simultaneously instantiated levels, they and their defence mechanisms gradually change with time. The ensuing variability, which enhances the fitness of the defending organisms, is simultaneously a veil of complexity that scientists must break through to enable explanations and prediction of the occurrences.

2.1. The received view

PATTERNS OF INTERACTION

The innate and adaptive subsystems

Contemporary immunologists dichotomise the immune system into its innate and adaptive subsystems. The innate system, which consists of molecules and cells that function as constitutive or inducible defenders of the organism, is conserved throughout the animal kingdom. Orthologs of some innate defence molecules have even been demonstrated in plants. The adaptive system, on the other hand, is generally believed to be unique to the jawed vertebrates as it comprises functions associated with lymphocytes and their diversifying receptors.

While the innate immune system is the result of the evolutionary germ-line learning of the species, the adaptive immune system is a result both of germ-line learning and individual learning during organismal development. An interesting consequence of this is that organisms of species with adaptive immune systems should be phenotypically more diverse than organisms of species without adaptive immune systems. Whether this makes one or the other species better suited for life is, nevertheless, a contingent matter. Simple solutions to environmentally posed problems may sometimes be more functional than complex solutions, for example in terms of error rates and maintenance problems.

While the entities that make up the innate system come in predetermined configurations, the entities of the adaptive system come in non-functional pieces that need to be assembled anew in each individual organism. Still, entities of the innate system are not predetermined in their functioning. Single cells within the innate system are in principle capable of responding adaptively to noxious stimuli. This may occur if they have multiple receptors on the cell surface, and if the triggering of each receptor results in a different response. Another example of indeterminate actions of innate molecules would be the signalling network of cytokines. When they were first discovered, it was believed that each cytokine would convey a unique signal for a defined cellular response. This idea has now been replaced by the idea that most cytokines have multiple and diverse biological functions. The production and expression of cytokines and their receptors is now known to be embedded in a cytokine network under complex biological control, in which the outcome of single cytokine signalling is modified by the simultaneous presence of other cytokines and their receptor interactions. This complexity, combined with pleiotropic and redundant effects of several cytokines, has made it almost impossible to predict the outcome of inactivation experiments of genes encoding cytokines or their receptors (Callard et al. 1999).

The possibility that many organisms other than the jawed vertebrates might generate diversity through somatic diversification has recently questioned the validity of the innate/adaptive dichotomy along a phylogenetic gradient. The observation that molluscs secrete somatically diversified fibrinogen-related proteins that bind to parasites (Zhang et al. 2004), the identification of extremely diverse leucine-rich repeats on lamprey lymphocyte-like cells that are shed during infection and bind to as yet unidentified microbial products (Pancer et al. 2004), and the somatic diversification of immunoglobulin-superfamily molecules in insects (Watson et al. 2005), all suggest that a form of adaptive immunity exists in other species than the jawed vertebrates. The new data suggest that instead of grading the adaptive/innate distinction along a phylogenetic gradient, the distinction should be made on the basis of organismal development; those organisms that diversity their defence mechanisms during the individual's own life time possess adaptive immune systems, while those that do not, possess innate immune systems.

Stratified security

During infectious disease, decisions made by the immune system are met with counter-responses from the infecting microbe. The outcome of the interaction between the two species depends upon the balance between rates of birth and death of leukocytes and infectious agents, as well as on the rates by which inducible inflammatory mediators and microbial toxic products change. If the host is capable of inducing defence mechanisms fast enough, the infectious agent may be cleared before the onset of symptoms; if the infectious agent is the quicker one, disease may ensue and transmission of the infection becomes more likely.

The rates of birth and death of interacting entities are guided by molecular interactions between molecules and cells of the interacting organisms. These include the host's innate and adaptive immune systems as well as virulence factors of the infecting agent. From the host organism's perspective, the defence mechanisms can be grouped into three sequentially activated defence layers – the constitutive (skin, mucous barriers), the rapidly induced (acute phase reactants) and the slowly induced (lymphocytes). The defences may be further divided as to whether they function locally or systemically, and according to the mechanism by which they defend against variable pathogens.

The most ancient defence mechanisms play major roles, not only as first barriers of defence, but also because they rapidly and efficiently provide adaptive immunity with specific information about infectious as well as neoplastic danger. The innate immune system may also be important in adverse inflammatory, allergic and autoimmune diseases. This is because misdirected or uncontrolled innate immune mechanisms may lead to abnormally enhanced inflammatory reactions and enhanced tissue damage, mainly because the innate immune system directs the magnitude and type (Th1 or Th2, IgG vs IgE) of the adaptive response.

Design principles

Being unsatisfied with the application of cause-effect models to explain the dynamic interactions of organic systems, von Bertalanffy (1962) began, in the 1950s, to envision the organism as an open system in which a relatively constant internal structure was being maintained through a continuous exchange of component materials between the organism and the environment. He further acknowledged that complex systems, which may be identified throughout the various sciences, tend to share a number of properties; they consist of components that interact, they are dynamic, they interact with their environments and adapt their internal structures as a consequence of such interaction, and the nonlinear interactions taking place among the components can lead to emergent behaviours not attributable to the properties of the individual interacting entities.

For system-investigators, the interesting properties of networks are those that emerge from the organisation of the individual components. The components, by themselves, are not their primary focus. Still, much useful information has been collected by reductionistic approaches, and the erection of general systems theory was thus not intended to dismiss other scientific approaches to knowledge. But von Bertalanffy did set forth his argument to contend that such knowledge does not disclose the complexities of the systems as they function *in vivo*.

The strategy used by immunologists when studying regulatory aspects of the immune system, which involved investigating connections and linear causal relations between isolated components *in vitro* under strictly controlled conditions, would fall prey to the critique raised by the system theorists. By taking account of this, immunologists of the 1960s came to realise that the reductionistic strategy provided a too simplistic view of immunity. They were thus prompted to promote an explanatory strategy that envisioned the immune system as a network consisting of a large collection of assorted decentralised components with extensive non-linear and criss-crossing connections. This view would entail that causality is distributed; there are no central control units, and all behaviour is thus defined and controlled by the microenvironment within which the lymphocytes are embedded. That various defence-related components are bound up in a systemic network, now termed the immune system, is thus a rather recent notion (Moulin 1989).

Probably the most influential theoretical construction that grew out of these studies was the view that the immune system is regulated through interacting variable regions of the antigen-binding receptors – the idiotype network theory[1] (Jerne 1974). Much experimental research was initiated to test Jerne's theory, but despite confirmative results in select experimental systems (Reitan and Hannestad 2002), the theory is no longer believed to account for immunity in the manner hypothesised by Jerne. Much of this scepticism is owing to some grave theoretical deficiencies of the theory. These shortcomings were unintentionally brought to the fore by Jerne (1993) in his 1984 Nobel lecture, in which he gave a brief view of his theory by comparing the immune system with linguistic systems. In the lecture Jerne claimed that the reactivity of the immune system, like linguistic sentences, is complete and open-ended. In accordance with the post-modernist view on language, words are defined in part by their relationship to other words which again are defined by still other words. This goes on until one gets back to a definition containing the original word and the circle is completed. But of course, if this is so, words need not have any connection to reality – language may thus be performed in an experiential vacuum. And similarly, the critique goes, a complete idiotypic network would be "absurd" because it would make the immune system causally impotent and thus inconsequential to biological function (Langman and Cohn 1986).

In spite of the explanatory failure of the idiotypic network theory, the systems view has by now permeated almost all thinking about immunological phenomena. It has, for example, become common knowledge that, as a result of the extensive feedback and cross-regulatory activities that takes place during the immune response, small differences in the nature of signal and context can lead to markedly different responses. The networks thus created can prevent the reversion of established phenotypes, even in the presence of significant perturbations and fluctuations. An example of this would be a patient who has developed an allergic phenotype, in which case the internal workings of the immune system stabilizes at a level from which it is very difficult to destabilize it and thus rewind it to the non-allergic phenotype. Thus, the self-organising behaviour of the developing immune system shapes traces of behaviour, termed memory, which alters the response characteristics of the system.

[1] Antibodies vary at both their constant and variable parts. When antibodies against antigen-binding receptors detect variability in their antigen-binding regions, the structural bearers of the variability are termed idiotypes, whereas antigenic determinants that impose variability in the constant region are termed isotypes (or allotypes).

These ideas fit nicely into recent molecular data which have verified that defence mechanisms first elaborated in unicellular organisms have been used repeatedly in multicellular organisms (Forsdyke et al. 2002). It has also become clear that vertebrate defences are mosaics of mechanisms first evolved in invertebrates (Salzet 2001) and that more recently evolved defences owe their capacity for activation and regulation to the phylogenetically elder defence mechanisms.

To provide a framework for thinking how the immune system works, Orosz (2001) proposed that four basic design principles were needed. These were the principles of phylogenic layering, parallel processing, dynamic engagement and variable connectivity. *Phylogenic layering* refers to the principle that phylogenetically elder systems are activated before the younger systems. While this principle alludes to how the immune response is guided towards its completion, there are multiple response options along the pathway, and there is therefore no single prototypic immune response to any given set of foreign antigens.

The principle of *parallel processing* seeks to explain the broad-based and wasteful multiple response strategy employed by the immune response when initiated by an antigenic stimulus. The principle accommodates the response constraints that any immune response is dependent upon, including the nature and amount of antigen, the profile of immune resources available, tissue-specific restrictions and any prior experience with the same or a similar antigen.

In close association with the previous principle is the idea of *dynamic engagement*, which subsumes the notion that lymphocytes continuously enter the inflammatory site, function for a short period, and then depart or die. The outcome of infections in vertebrate organisms depends upon dynamic interactions that take place between host and parasite. And once the inciting stimulus has been eliminated, the immune reaction needs to be shut down to avoid immunopathological damage. The principle of dynamic engagement ensures the instantiation of this transient, albeit wasteful, activation of lymphocytes.

Since networks are composed of a large number of interacting elements, which are sometimes stochastically distributed in the organism, the interconnections between the elements need to be loose and easily established or broken. This ensures that the network can adjust its function by varying the degree of connectivity among its elements. The principle of *variable connectivity* expresses the thought that the immune system connects and disconnects its interactive elements. This allows the establishment of coherent immune responses at any inflammatory site. Importantly,

> the fact that each immune response is a novel exercise in network construction provides a significant advantage to the immune system (although it is a clear disadvantage to human investigators), since it allows the immune system to accommodate "surprises" offered by environmental pathogens that are constantly evolving strategies to subvert the immune system. (Orosz, 2001, p. 143).

EQUIPPING THE ADAPTIVE TOOLBOX

Lymphopoiesis

The T and B lymphocyte lineages, which together constitute the adaptive toolbox of the vertebrate immune system, are derived from a common progenitor cell – the haematopoietic stem cell. These multipotential cells are, at least in mouse and man, originally formed in the yolk sac, from where they sequentially seed the developing liver, spleen and bone marrow via the blood circulation (Palis and Yoder 2001). The haematopoietic stem cell is capable of both self-renewal and differentiation, and forms all types of blood cells, including erythrocytes, granulocytes, monocytes, platelets and lymphocytes. In adult animals, the primitive haematopoietic stem cells reside next to the inner surface of bone, from where they migrate towards the blood vessels at the centre of the bone marrow cavity as they mature and differentiate (Zhang et al. 2003).[2]

The haematopoietic system, which produces about one trillion blood cells each day in adult humans, develops by a stochastic process in which lineage determination is caused by alternate activation of lineage-specific transcription factors (Abkowitz et al. 1996; Cantor and Orkin 2001). During maturation haematopoietic stem cells lose the ability to self-renew and gradually become committed to specific lineages. One committed type is the common lymphoid progenitor cell, which can give rise to all classes of lymphocytes, including T, B and NK cells (Rothenberg 2000). The B cell precursors develop in the bone marrow while T cell precursors migrate to the thymus before they differentiate into mature cells of the CD4[3] or CD8 lineage. The differentiation into CD4 or CD8 T cells appears to be directed by the recognition by the T cell receptor of MHC class II or class I molecules, respectively (Itano and Robey 2000).

The antigen-binding regions of both immunoglobulins and T cell receptors are assembled by the recombination of two or three gene segments (V(D)J recombination) selected from a large pool of segments to join. The proteins produced by the recombination activator genes (RAG) RAG1 and RAG2, which are expressed only in B and T lymphocytes, initiate the recombination. These proteins co-operate to make double-strand DNA breaks at specific recognition sequences. Broken ends are then processed and joined with the help of other molecules. This assembly is not

[2] Until recently it was believed that cells move forwards along their respective differentiation paths, but never backwards. The possibility has now emerged that mature cells can be "reprogrammed" or "transdifferentiated" for a different function. For example, fibroblasts can be reprogrammed to express T cell functions if grown in cell-extracts from T cells (Hakelien et al., 2002).

[3] CD stands for cluster of differentiation and designates specific cell surface molecules. Thus, CD4 designates a molecule that is expressed mainly on T helper cells but also weakly on monocytes, while CD8 is expressed mainly on cytotoxic cells. The CD nomenclature originated from workshops in which a wide variety of monoclonal antibodies were tested out. Those antibodies that detected the same molecule were assigned to the same cluster. It was only later that the CD denomination was transferred to name the surface molecule.

adapted to the needs of the organism and is essentially a random process that creates a high variability. It has been estimated that if recombination occurred unrestricted, the combinatorial and mutational possibilities would allow a potential repertoire of 10^{10} different B cell receptors (Milstein and Neuberger 1996) and 10^{12} different T cell receptors (Livak and Petrie 2002). The available repertoire is, nevertheless, considerably smaller than the potential repertoire because of biased combinatorial joining efficiencies and potent post-recombination selective mechanisms.

Selection within

Even though all lymphocytes in a given organism are being created from the same genetic material, knowledge of the animal's genome is not sufficient to predict the unique biological properties of the immune system. This is because the lymphocytes that develop from the common progenitor cell become selected according to their phenotypic characteristics. Those lymphocytes that react too strongly or too weakly with the selective filter during the shaping of the primary receptor repertoire are deleted. Those that interact too strong receive an apoptosis inducing signal, while those that interact too weakly do not get the proper survival feedback signal. The individual cells that provide the system with the plasticity needed to rapidly adapt to changing environmental situations are those that interact neither too strongly nor too weakly with the selective filter.

Several lines of data indicate that newly produced lymphocytes compete with each other as well as with resident lymphocytes for fitness-enhancing organismal resources (Freitas and Rocha 2000). These resources include ligands for the antigen receptor, costimulatory and adhesion molecules, and soluble growth and differentiation factors. Analogous to what happens during natural selection, selfish as well as co-operative cells may be formed during the process of developmental selection. Several mechanisms serve to limit the expansion of selfish cells, including apoptosis and the generation of regulatory T cells. In addition, there is a niche-dependent control of the lymphocytes in that T and B cells and their various differentiation stages occupy different locations in the lymphoid organs. There is also an element of chance since newly produced cells appear to migrate randomly between the blood and the lymphoid organs (Miller et al. 2003).

The self-organising properties of the immune system have been taken all the way to the levels of the variable parts of the antigen receptors of B and T cells, and it appears that peptide loading into the MHC is a selective process as well. While the antigen processing machinery is the same within every human being, most non-related individuals have different combinations of MHC molecules. Hence, the same antigen processing machinery must create peptides that fit within every MHC groove. This task is accomplished by generating peptides at random, and thereafter by selecting those with highest binding affinity for presentation (Serwold et al. 2002). The peptides are thus not produced to fit within the MHC groove; they are selected according to their fit.

This means that all three types of adaptive antigen binding receptors, including the T cell receptor, immunoglobulins and MHC molecules, function according to a trial and error selective process in every organism. The truly remarkable observation, that the immune system of different individuals manage to perform the same functions with the same degree of precision despite multiple risks of failure, attests to the robustness of self-organising systems.

Selection by antigen

Despite an enormous daily production of lymphoid cells, the concentration of lymphocytes in the peripheral circulation is relatively stable. This suggests the presence of dynamic mechanisms for close regulation of cell proliferation and death in the self-organising immune system. The mechanisms, which appear to work simultaneously and at multiple levels, are to a certain degree genetically influenced, but are also affected by the microenvironment within the lymphoid tissue as well as by environmental stimuli (Amadori et al. 1995; Tanchot et al. 1997; Hall et al. 2000).

As demonstrated by the nearly normal embryological development of organisms born with a deficient adaptive immune system, the immune system is a relatively autonomous module within the vertebrate organism. The developing immune system is, however, not autonomous in the sense that it develops independent of external conditions. Quite to the contrary, the internal organisation of the developing system is adaptively shaped by external conditions. The antigens encountered during development thus have a decisive influence on the outcome of the developmental process. But although the external conditions restrict some actions among the range of possible actions, the conditions are not determining. Hence, the developing immune system exhibits a relatively independent and open-ended reactive potential.

The influence from the antigenic world starts shortly after conception. Neonatal vertebrates have a limited capacity to synthesise immunoglobulins, and maternally produced antibodies therefore provide the primary form of antibody-mediated immune defence for offspring. Maternal antibody transmission occurs through the placenta, milk or yolk in mammals, while fish, reptiles and birds transmit antibodies through the deposition of antibodies in eggs (Grindstaff et al. 2003). Maternal antibodies appear to influence the development of the diversity of the B cell repertoire as well as the number and response intensity of the B cells in mice (Malanchere et al. 1997; Lemke and Lange 1999), and may even affect immune function across multiple generations (Lundin et al. 1999). In addition, one or more as yet undefined maternal lymphocyte factors transferred through breast-milk function to enhance the antibody production in new-born mice (Shimamura et al. 2003).

The diversity and quantity of specific antibodies transmitted to offspring reflect characteristics of disease-inducing pathogens in the local environment, thus ensuring that the offspring receives updated information about, and protection against, the infectious ecosystem that it is about to enter. This transgenerational mechanism therefore ensures that offspring become adapted to their local environment prior to birth, and hence that fit mothers deliver fit babies. This further suggests that maternal antibody transmission has evolutionary potential, and that it may lead to changes in the repertoire of antibody genes in subsequent generations (Lemke et al. 2004).

But the antigenic world also makes an impact on the adult immune system. For example, a reduction in environmental antigens drastically reduces concentrations of IgG and IgA, and it thus appears that the production of IgG and IgA, but not IgM, depends on stimulation of the immune system by external antigens (Hashimoto et al. 1978; Benner et al. 1981).

There are at least two stages where external antigen can induce selection of lymphocytes and stimulation of immunoglobulin production. During the primary immune response there is selective proliferation and differentiation of antigen reactive cells. Following elimination of the antigenic stimulus, these cells are subject to selective death and survival in the form of memory cells. This round of selection will thus lead to an alteration of the available repertoire. The development of a secondary immune response to the same or a similar antigen probably follows a pattern analogous to the primary immune response (Milstein and Neuberger 1996).

Such observations demonstrate that the response patterns of the developing immune system changes upon antigenic experience and that the realised immune system becomes established through dynamic and collective interactions between host cells and antigens, thus evoking the emergence of novel properties in the system. As a consequence of randomly generated antigen-binding receptors, identical twins start out with non-identical immune systems, and their immune systems' antigenic experience leads to further diversity in patterns of reactivity. The feasibility of identical twins for studies of immune mediated disease is therefore questionable (Salvetti et al. 2000; Singh et al. 2002).

EXEMPTION EXPLAINED

Memory

Immunological memory, which is the capacity of the immune system to respond differently, in terms of specificity, strength, and temporality at successive exposures to the same antigen, allows the organism to adjust its behaviour on the basis of ontogenetically derived information. The phenomenon of memory is closely associated to the process of learning, which designates the organism's flexible adjustment of behaviour as it experiences the world, and with the process of decision making, which involves both discrimination between signs as well as the capability to decide upon which sign is optimal according to a given criterion. Learning and decision making both depend upon evolutionary selected rules that allow the organism to adapt to environmental stimuli, and hence, to increase its propensity to survive and reproduce.

Most changes induced on organisms by the environment, like effects of injury and disease, are usually non-adaptive. Hence, a Lamarckian mechanism that was capable of incorporating information about the adult phenotype into the genetic message of the organism, and which thereafter could be transmitted to the next generation, would lead to deterioration and not adaptation of the offspring. This is, nevertheless, not the case with offspring of vertebrates. They come equipped with a defence system that is already adapted to the environment, but not through a Lamarckian mechanism. The phenomenon that allows adaptedness is immunological memory generated in the mother, and which is transmitted as maternal antibodies to the offspring.

Immunological memory has generally been held to be a unique characteristic of the adaptive immune system, but this view is probably too simplistic. Both viruses and bacteria employ sophisticated learning mechanisms that enable them to adjust physiological processes to changes in their environment (Casadesus and D'Ari 2002), and accumulating evidence supports the hypothesis that invertebrate defence mechanisms also exhibit some form of specific memory (Kurtz and Franz 2003; Moret and Siva-Jothy 2003). Hence, past experience with an infectious agent can provide individual invertebrates, and maybe their descendants, enhanced immunity. Although the mechanisms and consequences of memory in invertebrates are still not well characterised, the functional similarity between vertebrate and invertebrate immune responses may point to some common constraints and selective pressures that guide their evolution.

The memory response in vertebrates is a property of previously activated antigen-specific B and T cells that have undergone a transformation from their naive state. Upon first time stimulation by antigen, the naive lymphocytes become activated to proliferate and differentiate into either effector or memory lymphocytes. While the former combat the infectious agent, the latter are long-lived quiescent antigen-specific cells with potential for rapid activation upon new exposure to antigen, often at a stage before disease develops. This property is partly owing to the different requirements for costimulation by naive and memory T cells. Whereas activation of naive $CD4^+$ T cells requires potent costimulatory signals delivered by specialised antigen presenting cells, memory $CD4^+$ T cells can respond to several types of antigen presenting cells and are less dependent on accessory cell costimulation.

As demonstrated for CD8 cells, memory cells are probably not generated from a separate lineage (Wherry et al. 2003). They are more likely generated through a stochastic process in which a few effector lymphocytes become randomly selected for survival and differentiation into memory cells (Kaech et al. 2002). The signals that permit continuation of memory cells in the quiescent stage are, nevertheless, unclear, and especially the role of antigen for the memory response is controversial. One line of thought proposes that intermittent re-stimulation of memory lymphocytes is necessary to maintain memory responses (Zinkernagel 2002), while the contrasting opinion holds that memory cells represent a specialised state of long-lived quiescent cells that do not require further stimulation by antigen (Maruyama et al. 2000). In between these views come demonstrations that memory may be maintained by polyclonal activation (Bernasconi et al. 2002) or by cross-reactive stimulation (Antia et al. 1998).

The ability of the immune system to respond to the vast spectrum of foreign antigens located in the tissues is dependent upon an efficient distributing mechanism for the antigen receptor repertoire. Since immunoglobulins produced by B cells circulate freely in plasma and intercellular fluids, the functioning of the humoral immune system is not dependent on the capacity of B cells to circulate. In contrast, T cells mediate their functions locally, and have therefore developed an elaborate and extensive circulatory capacity. Circulation to the tissues is highly regulated, so as to provide selective lymphocyte subsets to particular microenvironments. It appears that naive and memory T cells follow distinct circulatory pathways. Naive lymphocytes preferentially migrate from blood into secondary lymphoid tissues through high endothelial venules. This emigration involves adhesion receptors that bind to ligands on the endothelial cells. Memory T cells, on the other hand, emigrate into the inflammatory centre owing to the increased the expression of adhesion molecules on endothelium in inflamed tissues. The rationale for different migration pathways for naive and memory T cells appears to be coupled to their functional potential, and especially to their differential requirements for activation.

While much effort has been devoted to investigations into the functional characteristics of memory cells, less effort has been devoted to discussions concerning what sorts of relation there is between the exemption phenomenon and the cells that instantiate it in the organism. To understand this, it may be useful to interpret memory as an organismal characteristic that becomes manifested as non-susceptibility to disease, or health (H), upon reexposure to infectious agents (I). The primary immune response can thus be characterised by the generalisation *If I then D then H*, in which D denotes disease (Figure 2.1). Memory, which is an organismal property experienced during a secondary infection, can then be characterised by the higher level generalisation *If I then H*, since D does not ensue. The if-then conjunctions involved in memory are not law-like relations, and there are for example several cases of chronic infections in which the conjunction *If I then H* fails to be instantiated. But even though chronic infectious diseases do not lead to the memory phenomenon, cells with characteristics of memory cells can still be identified in the tissues of the chronically infected patients. The lack of organismal memory despite evidence for immunological memory is visualised as a broken arrow between m_2 and H in Figure 2.1. Although agent a_2 induces memory cells m_2, these cells are not protective, thus explaining the lack of organismal memory at the higher level.

A range of observations has verified that the generation of memory-cells depends upon a constellation of characteristics related to the infectious agent and the host. These include type of antigenic stimulus, antigenic dose and route of entry. For example, memory responses to non-cytopathic versus cytopathic viruses differ, and too high or too low antigenic doses may lead to tolerance rather than memory. It has also been demonstrated that antigen delivered subcutaneously gives different memory responses

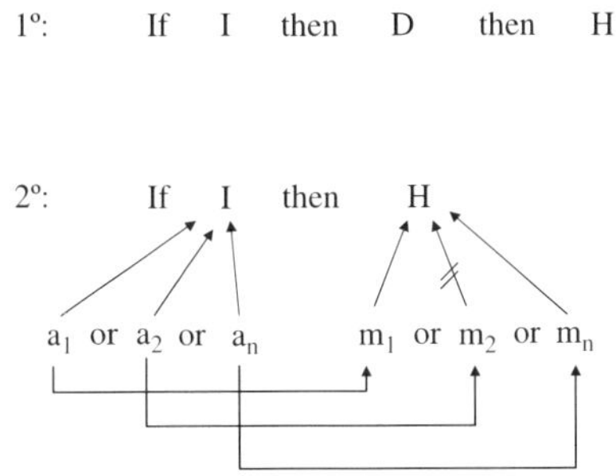

Figure 2.1. Schematic representation of the difference between a primary (1°) and a secondary (2°) infection where memory is involved. Relations between the lower and higher level properties in the secondary infection are indicated. At the higher level, (I) represents infection, (D) disease and (H) health or non-susceptibility to disease. The lower level property (a) represents properties of various microbial agents that lead to infection whereas (m) designates various constellations of antigen-experienced memory lymphocytes

from antigens that have a mucosal entry (Ahmed and Gray 1996; Zinkernagel 2002). The general pattern of memory at the organismal level is thus dependent upon a variety of constellations at the lower level. The I and H properties at the higher level can therefore be characterised as being multiply realisable by the lower level constellations.

By looking upon memory as a higher level phenomenon that may be multiply realised, several inconsistencies in the literature may be explained. As pointed out by Zinkernagel (2002), some of the inconsistencies may be owing to methodological issues. Measurement of neutralising immune responses, for example in the form of neutralising antibodies (m_1 in Figure 2.1), are tightly correlated with higher level memory, while measurement of specific antibodies of unknown functional significance (m_2 in Figure 2.1) may not be. In the literature, however, both types of antibodies are subsumed under the same functional umbrella, that of a memory immune response. Vaccine designers have for years been aware of the fact that memory immune responses need not be protective. Thus, the way to study the efficacy of a vaccine is to use Jenner's method – first vaccinate a group of individuals and then investigate whether they contract the disease or not upon challenge with the infectious agent.

Vaccinologists have also known for years that the protective effect of a vaccine does not hinge on whether all individuals in a population are being vaccinated or not. If the number of vaccinated individuals is high enough,[4] infectious agents may be unable to breach the population barrier against infection, thus ensuring that unvaccinated individuals also become protected. This effect, which is a sort of

[4] "High enough", which is a rather imprecise estimate, is a way of stating that the number of individuals that need to be vaccinated depends, amongst other, upon transmission characteristics of the agent, on whether the agent is able to survive in more than one host, and on the protective efficacy of the designed vaccine.

group-level memory, may be so effective that it leads to the eradication of infectious agents, as was the case with smallpox in the 1970s.

The intricate mechanisms involved during activation of the immune system need to be balanced against each other, and sometimes the benefits associated with one mechanism needs to be traded off against the benefits of another. The balancing of the immune response allows several lower level explanations to be given for the lack of higher-level memory against an infectious agent. For example, in some cases antigenic exposure may lead to tolerance, a process by which the organism reacts weaker and not stronger on reexposure to the infectious agent. There are several mechanisms that may lead to tolerance against an infectious agent, including the establishment of antigen-specific T suppressor cells (Belkaid et al. 2002; McGuirk and Mills 2002). Paradoxically, this form of tolerance is a kind of memory at the lower level, because the immune system remembers to respond, but in a non-responsive manner. In a sense therefore, memory involves forgetting and ignoring.

Although the generation of immunity towards an infectious agent is usually beneficial to the organism, the memory response sometimes introduces a bias in the organism's immune response. This may occur if the organism encounters a new infectious agent that is similar but not the same as the previous one. In this case the host immune system may not recognise the new pathogen as novel and may thus respond with a memory response towards the old infectious agent. In this case the immune system is biased, it does not see the new situation as different, and responds with an improper memory response. Such biased responses, which are best known from infections with various strains of influenza viruses, are called "original antigenic sin".

The memory response involved in the original antigenic sin also exemplifies how an infectious agent may evade the immune response by causing a disintegration of the memory responses of T and B cells. The phenomenon may occur if a novel infectious agent has the capacity to re-stimulate memory T lymphocytes generated against a previously encountered infectious agent. The memory response, of both T and B cell type, will then be directed at the original agent, thus leaving the novel agent secluded. This mechanism may, for example, be the reason why infections with the same infecting agent sometimes run markedly different courses in different individuals. Such variability in pathogenesis has often been linked to genetic makeup or the physiological state of the infected organism. But if the mechanism that leads to the original antigenic sin phenomenon is more general than previously thought, it is likely that previous exposure to infectious agents may alter the course of the host's immune response against novel agents. Maybe this is the reason why many viral infections, like measles, mumps and infectious mononucleosis, are far more symptomatic and pathogenic in teenagers and young adults than they are in young children (Welsh and Selin 2002). This may also be a mechanism whereby chronic diseases of unknown etiology, like multiple sclerosis and systemic lupus erythematosus, can be generated.

The memory response is like a double-edged sword, in that it comes with benefits as well as costs. Because cross-reactions with antigens presented in the wrong context could be detrimental, the adaptive immune system has evolved mechanisms that direct the immune response towards a functional end result. The immune system faces the dilemma of being sufficiently cross-reactive so as to maximise the likelihood of responding to a wide variety of infectious agents, while at the same time being sufficiently specific to avoid unwanted cross-reactions. Primary and secondary immune responses with well-adjusted specificities may help the immune system to avoid mistakes. Thus, specificity is of essence for the functionality of memory responses.

Specificity

In 1891 Paul Ehrlich published a remarkable report on differential immune reactivity against the plant toxins ricin and abrin, in which he demonstrated several important characteristics pertaining to the immune response. Firstly, he demonstrated that the immune response was developmentally shaped, as immune reactivity was not detectable before the sixth day after immunisation. Furthermore, he established that repeated injections of toxins result in higher concentrations of antibodies. And finally, he demonstrated absolute specificity[5] – animals immunised with one toxin do not produce antibodies against the other toxin. This observation was further interesting because the two toxins, that can be readily differentiated by their differential reactivity with antisera, can not be differentiated by their toxicity (Silverstein 1999).

Since Ehrlich' report, a major goal of immunological research has been to explain the mechanisms that allow the immune system to discriminate between closely related antigens in a precise manner. During the 1960s and 1970s, elucidation of the molecular mechanisms responsible for specificity took centre stage, and some authors even claimed that solving the riddle of specificity would bring about the final solution to immunology (Jerne 1969). Given conflicting interpretations of the nature of specificity, and the heated debates that various notions of specificity had created between immunologists since the 1890s (Mazumdar 1995), it is understandable that Jerne put so much emphasis upon solving the specificity riddle. Although much has been learned about specificity since Jerne's paper, the riddle is still tangled. This is much because the word specificity harbours conceptual no less than empirical issues, as evidenced by the many conceptions of specificity that exist in the literature, including the goodness-of-fit view, the relational view, the comparative view, and the multiply realisable view.

Structural studies of immunoglobulins, performed during the 1960s and 1970s, provided evidence for the *goodness-of-fit* view implicitly alluded to by Ehrlich. It was then demonstrated that specificity was mediated by differential insertions

---

[5] Specificity, which is derived from the word species, describes what pertains to and characterises a biological species. Species means a sight, outward appearance, shape, form, visible or sensible presentation, and is derived from the Latin verb *specere* – to look.

of amino acids within a region of the antibody that became known as the hypervariable region. High-resolution X-ray crystallography studies of the three-dimensional structure of antibodies and their complexes with antigens, demonstrated that antibody molecules interact with antigen via a surface that is complementary to that of the antigen. The different amino acids in the hypervariable region possessed different physicochemical properties, and thus allowed the creation of different conformations within the antigen-binding part of the antibody. These observations allowed specificity to be defined molecularly as goodness-of-fit, as how well the antigenic determinant fit into a pocket created by the antibody's heavy and light chain hypervariable regions.

Molecular specificity was originally discussed in the context of antibodies, but it has later been demonstrated that the receptor for antigen on T cells exhibits antigen specificity with a similar structural basis. But in contrast to antibodies, which can recognise a broad spectrum of different native antigens including proteins, peptides, carbohydrates, small molecules, and DNA, T cell receptors recognise antigens only when these are bound to MHC molecules. Amino acid sequences as well as X-ray crystal structures of T cell receptors and their interactions with MHC-peptide complexes have demonstrated that T cell receptors are fundamentally immunoglobulin-like, and that the specificity of the reaction between T cell receptors and antigen is mediated via the receptors' hypervariable regions.

Since tight binding between antigen-receptors and antigen is a prerequisite for immunological response, one could imagine that binding affinity would be a good measure of specificity. However, it has turned out that there is no necessary link between specificity and affinity. One may find that antibodies with the best fit to antigen are the least specific, and that antibodies of low affinity to antigen may exhibit a higher specificity than antibodies with a higher affinity for antigen (Van Regenmortel 1998). So while biochemical investigations have provided knowledge about the molecular basis for binding, the same studies have apparently not solved the specificity riddle. Since specificity is not something that can be bestowed to an antigen-binding receptor irrespective of the nature of the antigen, goodness-of-fit cannot be the only relevant consideration when trying to understand specificity. Accordingly, the notions of specificity derived from Ehrlich's experimental results, in which he demonstrated absolute specificity to ricin and abrin, combined with his metaphysical views, may have imposed a misguided notion of what it is that constitutes immunological specificity.

Ehrlich' metaphysical views were formed in the late nineteenth century in the midst of a heated struggle between two philosophical schools, the pluralists and the unitarians (Mazumdar 1995). The pluralists, amongst whom Ehrlich became a central person, were convinced that nature was discontinuous and that properties of bacterial species' were constant. The apparent continuities that were observed among bacteria were explained as methodological distractions, not owing to a natural property. The unitarians, in contrast, amongst whom Landsteiner became a leading scientist, claimed that nature was a continuum of forms. They were influenced by Charles Darwin's refutation of the essentialist doctrine of species and his claim that no clear line could be drawn between the various species.

While Paul Ehrlich envisioned the presence of innumerable different antibodies, each reacting only with one defined antigen, Karl Landsteiner viewed the serological cross-reactivity that exists between microbes from different species as a documentation of the fact that there is a series of gradual transitions with no sharp boundaries or discontinuities between antigens. Landsteiner also demonstrated that the immune system could react with almost any antigen one could imagine, some of which would never be expected to exist anywhere in nature. He believed this to be an effect of antibody multispecificity, and came to view the specificity of antibodies as a consequence of more or less good fit, not an all or none recognition phenomenon (Mazumdar 1995).

While Landsteiner's view was compatible with the goodness-of-fit view, it also comprised the *relational view* of specificity. This view is typically expounded in studies of interspecies interactions, where one of the most fundamental properties of any parasite species is defined by its host specificity (Combes 2001). The specificity of a parasite is measured not only as a function of how many host species it can exploit, but also of how closely related these host species are to each other (Poulin and Mouillot 2003). An infectious organism that infects only one host, like the smallpox virus that infects *Homo sapiens* and no other species, is highly specific – it is a relation between one host and one parasite. An infectious agent that infects more distantly related species, like the influenza A virus that infects humans, pigs and birds, has a lower specificity – it instantiates a one-to-many relation. In this relational conceptualisation the infecting agent thus becomes more specific to the degree it reacts with fewer species, much like an antibody which is of higher specificity to the degree it reacts with fewer antigens.

According to the unitarian metaphysics, there is no cut-off separating antigen-binding receptors that are specific from receptors that are not. And just like nature provides no sharp boundaries, so it is with the concepts that describe nature. It is therefore not to be unexpected that the relational view of specificity is, in many ways, tightly coupled to the *comparative view*. Since antibody specificity "acquires a meaning only with respect to the antibody's capacity to react differently with two or more antigens and thereby to discriminate between them." (van Regenmortel 1998, p. 43), specificity is also a comparative concept – one antigen-binding receptor is more specific than another receptor is.

That specificity is as much a relational and comparative concept as it is a goodness-of-fit concept, suggests that specificity should be understood locally, not globally; there can be no absolute and general account of specificity, the reason being that the degree to which specificity accounts for the properties of receptors, immune systems or micro-organisms is dependent upon the placement of a cut-off criterion. And as made evident in the discussion of signal detection theory in chapter 3, the placement of a cut-off criterion represents a value-laden decision criterion and not a scientifically neutral placement that flows from the properties of the interacting entities. In science, the placement of the decision criterion is determined by the researcher on the basis of some preconceived ideas of usefulness, while the placement of the decision criterion by the immune system and the organism is evolutionary and developmentally shaped on the basis of biological fitness. The cut-off criteria of the researcher and the biological system need not coincide.

To further complicate matters, some researchers have come to view antigen-binding receptors as being both specific and degenerate (Cohen 2000; Eisen 2001). A receptor is said to be degenerate to the degree that it reacts with many antigens, and it is specific to the degree that it can discriminate between closely related antigens. Since some antigen receptors can bind to several different antigens while at the same time discriminate exquisitely between closely related antigens, an antigen-binding receptor can be both highly specific and degenerate. Furthermore, there appears to be a hierarchy of degeneracy amongst the receptors. The T cell receptor on CD4 T cells is more degenerate than the T cell receptor on CD8 T cells, while antibodies appear to be least degenerate, probably as a consequence of the phenomenon of affinity maturation that occurs in the latter (Wilson et al. 2004). The individualisation of the immune system is further enhanced by the fact that T cell receptors and immunoglobulin that react with the same antigen, do not have the same set of degenerate antigens (Hoffmann and Kappler 1972).

Degeneracy is the capacity of elements that are structurally different to perform the same function or to yield the same output. Accordingly, degeneracy may be viewed as a property of both antibodies – several different antibodies may bind to the same antigen and eliminate it; and of antigens – several different antigens can elicit production of the same antibody. The functional significance of degeneracy may be that it equips animals with the capability to make antibodies that protect against almost any infectious agent (Edelman and Gally 2001). Thus, in a sense the immune system takes advantage of the evolutionary interrelationships between species and of the multitude of highly conserved proteins. Since T cell receptors are highly degenerate in their binding properties, a positively selected T cell with modest affinity for self stands a good chance of recognising similar peptides from infectious agents.

The fourth conceptualisation of specificity, the *multiple realisable view*, is used when discussing the different levels at which specificity may be encountered. Degeneracy of antigens and low specificity of antibodies can, for example, be instantiated at the molecular, cellular and organismal levels. Degeneracy and low specificity at the molecular level may even be a prerequisite for specificity at the higher level. Talmage (1959) noted that the mixture of antibodies with overlapping specificities and differing affinities that occur during an immune reaction might be more specific than the reaction between a single antibody and its antigen. Whereas an antigen might fit only partially into the combining site of a particular antibody present in relatively low concentration, the antigen could be well recognised by the totality of all combining sites in a heterogeneous antiserum. If so, the immune response can not be reduced to the specificity of its receptors. Immunological specificity is thus, like immunological memory, a multiply realisable property – the two levels are sometimes co-extensional, but sometimes they are not.

Tolerance

That antibodies are difficult to generate against self-molecules has been a long time observation. As the 19th century turned to the 20th, Paul Ehrlich claimed that self-reactive antibodies, if they existed, were dysteleologic, and Karl Landsteiner (1945, p. 18) noted that "the formation of antibodies, which would react with the proteins of the immunised animal, is suppressed, or if such antibodies are formed, they must be neutralised in the body, the procedure then being equivalent to an absorption experiment *in vivo*."

The first glimpses into mechanisms of tolerance came from studies performed outside immunological science. In the early 1940s, Ray Owen made the seminal observation that red blood cells of dizygotic cattle twins that had shared blood supplies *in utero* became red cell chimeras for life (Owen 1945).[6] This observation suggested to him that "the critical interchange is of embryonal cells ancestral to the erythrocytes of the adult animal. These cells are apparently capable of becoming established in the hemapoietic tissues of their co-twin hosts and continuing to provide a source of blood cells distinct from those of the host, presumably throughout life." (Owen 1945, p. 401).

These observations were recognised by Macfarlane Burnet and Frank Fenner in their 1949 book *The production of antibodies* in which they proclaimed that tolerance was generated during foetal life against self but not against foreign antigens (Brent 1997). Billingham, Brent and Medawar (Billingham et al. 1953) further elaborated the phenomenon of tolerance when they realised that tolerance to self, like immune reactivity against non-self, is an active process. Their experimental results verified that actively acquired tolerance is antigen specific, and they also provided insight into the mechanisms of tolerance when they demonstrated that tolerance develops when foreign antigens encounter the immature but not the mature immune system. When antigen encounters the mature immune system, immune reactivity, not tolerance, ensues.

By the end of the 1950s the concept of tolerance had become central to immunology, and theories of antibody production had to take tolerance into account because it could explain non-reactivity to self and hence the regular absence of autoimmune disease. In *The clonal selection theory* Burnet (1959) explained

[6] In a later work Ray Owen et al. (1946) used the broader term of mosaicism when describing red cell chimerism in dizygotic twins of cattle. The terms are currently used such that an organism that consists of cells derived from more than one organism, usually of different genotypes, is called a chimera, while an organism that consists of cells of more than one genotype derived from a single zygote, is called a mosaic. Hence, Owen's mosaics would now be termed chimeras. The term mosaic is, however, also used more broadly to describe organisms of mixed genotype, whatever the initial cause. Under this broader definition a chimera would be a subset of mosaics. Mosaicism also has an evolutionary meaning when the genome of an individual is viewed as a mosaic of discrete segments, each with its own unique history and relatedness to different contemporary and ancestral individuals (Paabo 2003), or when the defence systems of vertebrate organisms are envisioned to be mosaics of invertebrate defence systems (Salzet 2001).

tolerance as an actively learned process in which non-reactivity to self resulted from the elimination of self-reacting antibody-producing cells during the embryonic period. If this process failed, the result would be breakdown of tolerance and the release of forbidden clones that could cause autoimmune disease. While Burnet emphasised clonal deletion as a means to the acquisition of tolerance, Lederberg (1959) introduced the possibility that self-reacting antibody producing cells could merely be rendered tolerogenic rather than destroyed.

The chief mechanism of T cell tolerance is the deletion of self-reactive precursor T cells in the thymus. The negative selection ensures that T cells with receptors unable to recognise self, or that recognises self too strong, are deleted. Without this purifying process, randomly mutated receptors would generate many cells with receptors that reacted strongly with self. B cell tolerance is also achieved by clonal deletion of self-reactive maturing cells that occurs in the bone marrow. Mature B cells, on the other hand, are made tolerant by the phenomenon of receptor editing or owing to lack of help from T cells. The editing of antibody receptors, which leads to changes of the receptor specificity, occurs when the receptor recognises self-antigens (Casellas et al. 2001).

By the mid-1960s, Burnet's clonal deletion paradigm of tolerance had been firmly established. There was, nevertheless, still a theoretical problem to be solved. If the generation of diversity in antigen binding receptors results from a random process, there had to be a mechanism that allowed the self-nonself discrimination to occur, not only during embryonic development, but throughout the lifetime of the organism. The two-signal models of Bretscher and Cohn (1970) and Lafferty and Cunninham (1975) addressed these theoretical challenges when they proposed that signal 1 in the absence of signal 2 rendered the B and T lymphocytes tolerogenic. For T cells the maturational stage of dendritic cells is also of essence, in that tolerance appears to be induced by partially matured dendritic cells while fully mature dendritic cells induce immune reactivity (Lutz and Schuler 2002).

The mechanisms that induce tolerance are often envisioned as belonging either to the central or to the peripheral tolerance models. Burnet's clonal deletion theory is concerned with central tolerance, while the two-signal theories are concerned with peripheral tolerance. A further distinction can be made between recessive and dominant mechanisms of tolerance. Recessive tolerance is a passive form of tolerance, in which self-reactive lymphocytes are deleted, while dominant tolerance is defined as tolerance that can be transferred from a tolerant donor to an immunocompetent host. Central tolerance is usually recessive, while peripheral tolerance can be either recessive, like in the two-signal models, or dominant. One example of dominant tolerance is the one induced by regulatory T cells. The cells prevent autoimmune disease and promote graft acceptance after transplantation, but they also suppress beneficial immune responses, such as those involved in rejections of tumours and infectious agents (McHugh and Shevach 2002).

Investigations into the mechanisms responsible for tolerance have made a large impact on clinical immunology owing to the rich therapeutic prospects of such knowledge. Researchers of autoimmune disease are, for example, trying to restore peripheral tolerance by interfering with T cell receptors and molecules associated with co-stimulation, as well as through immunising with self-molecules. Similarly, allergologists are trying to shift the immune response away from debilitating inflammatory reactions, while transplantation immunologists are trying to induce tolerance by creating stable bone-marrow chimerism in which donor and host cells tolerate each other (Starzl and Zinkernagel 1998; Rotrosen et al. 2002). Although it seems certain that the maturity of the immune system as well as its state of activation are important parameters for tolerance induction (Anderson and Matzinger 2001), the fundamental problem of understanding how to balance between signals that induce tolerance and signals that induce immunity remains to be solved.

2.2. The integrated view

REVISING THE VIEW

Scientific activity is guided by the quest for explanations. To attain general knowledge scientists usually study a variety of objects under different experimental and observational conditions. When it comes to immunology, it is striking that generality has been sought by studying the defence mechanisms of one species only, that of *Homo sapiens*. Although the mouse and to a certain extent the rat immune systems have been studied as extensively as that of man, they have not been studied in their own right; as devices that orient rodents in their dealings with the natural environment. Rather, the rodents have been challenged with a limited repertoire of human-relevant pathogenic microbes, and are thus treated as mimics of the analogous system in man.

The mouse has been studied because of its experimental tractability, its short generation time and its adaptability to a life in captivity, and the studies have been designed and performed to generate knowledge about the defence systems of man. Such extrapolation is not without its problems; different species populate different environments with different selective pressures, and it is quite possible that the selection that occurs within the different species focuses on different signalling displays with different life-history trade-offs as well. Hence, by ignoring the ecological diversity that has selected the defence systems, we have most likely introduced certain biases in our understanding of the immune system as it functions in health and disease.

The anthropocentric bias in studies of defence mechanisms may be looked upon as a natural consequence of man's evolutionary history. Like any other animal species, man has evolved a natural concern for his own being-in-the-world. When living as hunter and gatherer, man was absorbed in activities that would enhance his prospects for survival and reproduction. This natural attitude was still prevalent when man adopted sedentary habits and later on urbanised life styles, and it was therefore natural that the first scientific studies of interactions between macro- and microorganisms were performed to investigate infectious diseases that afflicted humans or their domesticated animals and plants.

Nevertheless, while the achievements of the scientific mind certainly have a basis in man's natural history, it is equally clear that knowledge serves as an instrument that transcends mere self-preservation. Hence, as knowledge of vertebrate defence mechanisms began to accumulate, man's curiosity and willingness to explore unknown territory made him advance to study economically less interesting species, such as flies and earthworms. These studies soon provided some surprising facts as to the functioning of vertebrate defence mechanisms as well. It turned out that vertebrate immunity was derived from phylogenetically elder mechanisms of defence, and that a component of the vertebrate defence system still utilised molecular mechanisms analogous to those employed by the invertebrate defence mechanisms. These revelations necessitated a re-conceptualisation of the vertebrate defence mechanisms.

Yet, the anthropocentric bias still prevails in contemporary immunology. For example, although infectious agents have increasingly been recognised as agents of ecological and evolutionary importance, there is surprisingly little interaction between immunologists and evolutionary ecologists. While immunologists are concerned primarily with studying the fine details of the immune system as it functions in health and disease, evolutionary ecologists treat the immune system as if it was a black box that provides unmediated access and quantitative answers to fundamental evolutionary and ecological questions. Also differing from immunologists, who investigate the functions of the human, mouse and rat immune systems, evolutionary ecologists study immunity in mammals, birds, fish and insects. And whereas immunologists publish their experimental investigations in specialised medically orientated journals, evolutionary ecologists publish their observational studies in journals that address evolution and ecology at a more general level. The opposing research strategies and the fact that the two groups seldom congregate at the same congresses have led to a sometimes confusing use of terminology and to the erection of incomprehensive explanatory models. Hence, unification of the disparate fields is much needed.

The word immunobiology, which would be a proper name for the integrated fields, has traditionally been used by immunologists to designate the integration of cellular and humoral immunology. For example, Silverstein used the term immunobiology to denote that immunologists in the 1960s no longer viewed the phenomenon of immunity as a chemical problem; their task became to explain how the cells of the immune system matured to discriminate self from non-self through dynamic interactions and Darwinian competition. He thus contrasted the immunobiological period with the biomedical era that lasted from about 1880 to 1910 and the immunochemical era that lasted from about 1910 to 1960 (Silverstein 1991).

This restricted use of the term immunobiology is still prevalent. In their edited book *The biology of immunologic disease* Frank Dixon and David Fisher (1983) envisioned disease largely from an intraorganismal perspective. And in a similar manner, Janeway and Travers (1994) in their textbook *Immunobiology, The immune system in health and disease*, surveyed the bits and pieces of the immune system as it develops to fight infectious disease. While both books pay much attention to medical aspects of immunity, evolutionary and ecological aspects of immunity are scarcely mentioned. Immunobiology as understood by contemporary immunologists is thus a misleading term.

To live up to its credentials, immunobiology should comprise the aspects that concern biologists. These include investigating similarities and differences between individuals within groups, and the investigation of similarities and differences between groups of individuals. By redefining the task of immunological research as being the investigation of how organisms come to give flexible responses to a changing environment, the full sphere of immune system functioning should hopefully be explicated. Since some of the most striking examples of adaptive phenotypes resulting from context-dependent development are seen in the defence responses to parasites (Tollrian and Harvell 1999), immunology should profit immensely from integrated studies of defence mechanisms. Such investigations, which attempt to elaborate the costs associated with use and maintenance of defence mechanisms, have recently emerged under the term ecological immunology (Sheldon and Verhulst 1996; Norris and Evans 2000).

The same considerations have also been seized by other groups of experimental and theoretical biologists, and integrated investigations of evolution, ecology and development, which sometimes go under the sobriquets evo-devo, eco-devo, or evo-eco, have been advocated in order to combine genomic information with morphological change in organisms over time (Raff 2000; Gilbert 2001). In accordance with these views organismal development is seen as a plastic process that depends as much upon decoding of the genome as upon responses to environmental perturbations. And since it is the developmental process that produces the phenotypic features of each generation in evolving lineages, evolution cannot be understood without understanding the interactions that take place between the developing organism and its environment.

IMMUNOCOMPETENCE

An organism's capability to prevent or control infectious disease, its immunocompetence, is a complex trait that subsumes a variety of aspects. While the resources that make up an individual's immune system are the most important aspects of immunocompetence, organisms may be immunocompetent despite harbouring deficient immune resources. And conversely, immunocompetence may be reduced despite the presence of well-functioning immune systems. The complexities involved in defining immunocompetence are well exemplified by infections with schistosomes, which are important parasites of humans and other animals. These parasites require components of the host immune system to complete their development (Davies et al. 2001). They thus succeed in immunocompetent hosts but fail to thrive in immunodeficient hosts. This, by implication, identifies the immunocompetent as immunodeficient, thus underscoring the difficulties of finding a global definition of immunocompetence. Accordingly, immunocompetence is a relational property that transcends the boundaries of the organism. And being a multidimensional concept, it is neither easy to measure nor to define.

Hosts as well as parasites have evolved to require the presence of some aspect of the environment for their normal development. For humans, the developmental aspects of immunocompetence were dramatically spelled out in nineteenth century Europe when the major causes of morbidity changed from infectious to life-style diseases. The decreases in death rates owing to diphtheria, smallpox, tuberculosis, pneumonia and measles were not effects of improvements in measures designed to prevent diseases from spreading or to variations in the infectious agents themselves. Rather, the simultaneous development of decreased virulence and increased immunocompetence was a consequence of a common cause; economic growth and associated changes in human life-style (Golub 1997). As people were offered better working conditions, became better nourished and could afford better clothing, they became better able to recover from the further stress conferred by infectious disease. This change was also a major impetus for the decreased childhood mortality and increased fecundity that gradually lead to an escalation of the population growth rate.

It thus appears that costly life history decisions that extract resources from a limited resource pool need to be traded off against each other; an increment of resources allocated to one trait necessitates a decrement of resources to another trait. The inhabitants of nineteenth century Europe should, accordingly, have experienced a trade-off between investment in immunity and physical suffering; if too much of the available resources were devoted to reduction of stressful events, there should be fewer resources available to maintain immunity. As availability of resources increased, the trade offs would still take place, but more resources could then be allocated to maintenance of immune defences.

The trade-off hypotheses have been corroborated by several investigations, and it appears that both the innate and the adaptive immune systems are subject to modification by organismal perturbations. Malnourished patients and individuals exposed to prolonged and heavy training are for example more susceptible to infectious disease than well nourished and fit individuals (Marcos et al. 2003; Gleeson et al. 2004), and individuals exposed to severe psychological stress develop increased levels of hormones that suppress the immune response (Ader et al. 1995). There are even data indicating that close social relations to kin and friends are associated with increased resistance to upper respiratory illness (Cohen et al. 1997), thus suggesting that psychological factors may be important resources for adaptive immune system functioning.

The importance of providing integrated explanations is well explicated by attempts at explaining why males and females exhibit differences in the quality and strength of their immune responses. This phenomenon has been proximally explained by invoking the differential roles that sex steroids take in males and females. Since estrogens often enhance immunity while testosterone functions as a natural immune suppressor, the hormonal differences are also to some degree responsible for the increased rate of autoimmune diseases in females as compared to males (Gaillard and Spinedi 1998), and for the higher susceptibility of post-puberty males to die from infectious disease than age-matched female conspecifics (Owens 2002).

While it is true that differences in disease susceptibility, disease severity and response to therapies might be elucidated by investigating how immune effector proteins vary in response to stimuli by sex hormones and microbial infections (Yu and Whitacre 2004), the results can not explain why the sexes differ in immune responses in the first place. Such understanding requires clarification of origin and evolutionary mechanisms. To see why, consider a population in which selection acts simultaneously on two traits, immunity and sexual ornamentation. There are four combinations of traits with associated fitness to consider (Table 2.1).

High immunity and showy ornamentation is the best combination of traits because it optimises viability and fecundity, while having a low immunity and ordinary ornamentation is the worst. The two other combinations are arranged according to which of the two components of fitness, reproduction or viability, are considered as more important. In the given example viability is ranged as more important and the fitness of high immunity and ordinary ornamentation is thus given the fitness value 3. Showy ornamentation, which enhances fecundity, combined with low immunity, which reduces viability, is given the fitness value 2.

If the two designated traits assort independently, natural selection may reach optimal fitness by selecting for each component separately. If on the other hand the traits are antagonistically linked, so that high values for one trait are correlated with low values for the other, selection may not reach the optimal fitness value. In this case, selection for one trait must be traded off against selection for the other trait. Such is the case with the hormone testosterone, which simultaneously suppresses the immune response and enhances male secondary sex characters (Figure 2.2 A). In their attempt to account for the differences in immunity between the sexes, Folstad and Karter (1992) launched the immunocompetence handicap hypothesis, according to which females that select males according to the quality of their ornaments, simultaneously select for lowered male immunocompetence. Hence, they explained the differential immune responsiveness between males and females as being the result of sexual selection, the process whereby one of the sexes selects members of the other sex for mates.

In their model "low-quality" males are not capable of producing both showy phenotypes and good immune responses, and when they divert energy to phenotypic display they simultaneously become more susceptible to infectious agents. Thus,

Table 2.1. Immunity, ornamentation and their combined values of fitness

		Immunity	
		High	Low
Ornamentation	Showy	4	2
	Ordinary	3	1

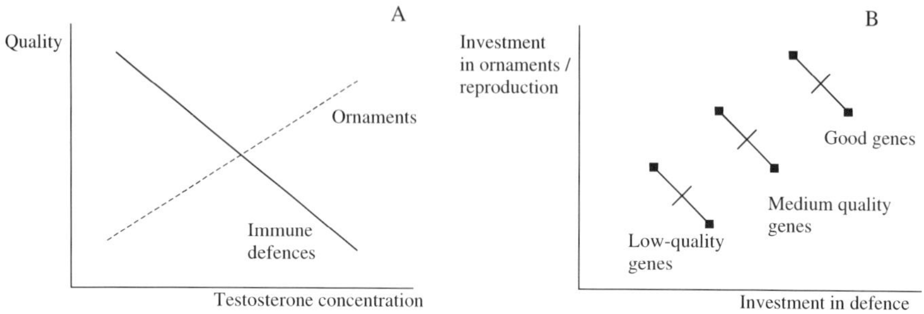

Figure 2.2. A) In the male population there is a positive correlation between testosterone and ornament quality and a negative correlation between testosterone and immunocompetence. B) When male organisms with the same "gene-quality" are grouped together, the trade-off between investment in ornaments or reproduction and immune defences becomes discernible. The figure visualises the observation that organisms with a set of "low-quality" genes possess fewer resources for investment in ornaments/reproduction and immune defences than organisms with "medium quality" or "good" genes

the level of infection is an honest signal of male quality to females (Hamilton and Zuk 1982). By avoiding infected males, choosy females can avoid the risk of contracting infectious disease. They also obtain more parental investment and thus increase the chances of having progeny with genes that mediate resistance against infection.

When females assess potential mates, they should select males with potential "good genes" to increase the viability of their offspring. Nevertheless, "good genes" are seldom directly assessable. Genes are evaluated according to their phenotypic effects, and those genes that impart the highest fitness on the organisms in which they operate are the ones that are being selected. As put forward in the previous example, some genes make their effects visible to selection precisely through their involvement in a pleiotropic signalling network.

So the fact that females select the showiest males would indicate that sexual selection drives the defence mechanisms towards lowered immunocompetence, but that the effects of this, which is lowered fitness as measured in the currency of viability, is traded off against increased reproductive fitness. Hence, by invoking a model that takes account of both facets of fitness and which includes sexual selection as a mechanistic explanation, Folstad and Karter provided a plausible explanation for differences in immunity between males and females.

Surprisingly, while there is a negative correlation between investment in defence and ornamentation at the level of individual organisms, there is a positive correlation between the same variables of investment when analysed at the population level (Hamilton and Zuk 1982; Moller et al. 1999). This contraintuitive result is owing to the fact that males with good immune-related genes can allocate more resources to ornamentation than can low-quality males (Figure 2.2B). Differential quality of genetic resources is thus a variable that can by itself account for the two positively correlated population level traits. Males that are capable of producing high levels of testosterone and thus to develop showy secondary sexual characteristics for display to choosy females, without succumbing to infectious disease, thus simultaneously signal the quality of their defence-related genes

When males are stratified into groups with similar genetic qualities, it becomes apparent that sexual selection drives the male population towards increased immunocompetence, not lowered as would have been predicted from the individualistic trade-off model. Females choosing ornamented males therefore simultaneously select mates with high-quality genes for immunity. In their explanatory hypothesis Folstad and Karter (1992) were thus able to reach a novel explanation of sex disparities by reconsidering the relation between proximate and ultimate causes. Their hypothesis made clear that the key explanatory variable is the quality of the genes that the organism comes equipped with, not testosterone.

The above considerations make clear that the correlation between testosterone and immunocompetence does not tell the one true causal story. In an evolutionary context, the suppressing effects of testosterone become less important than predicted by the proximate effects. Since evolution selects both for increased ornamentation and increased immunocompetence, despite proximate evidence to the contrary, the effects of testosterone on immune defences have become sidetracked in the selective process. Despite a correlation between testosterone and immunosuppression, nature has found ways of reducing the impact of testosterone on immunity.

The example is also an instantiation of a more general principle, according to which biological entities at all levels are capable of adjusting their behaviour in accordance with signs displayed by other biological entities. Accordingly, the immune system may be envisioned as a responder to signals emitted by the microbial universe. By envisioning the immune system as a signal detecting system that responds adaptively to signs by selecting from a range of available responses, the flexible responses to a changing environment can be explained at a more general level.

THE ONE AND THE MANY

Immunology and bacteriology emerged as entwined and complementary sciences in the latter half of the 19th century, shortly after Darwin's expounding of the evolutionary theory. But contrary to most other biological fields, the entwined sciences emerged from a tradition that shared the major concern of medical practice; diagnostics and therapeutics of the diseased human patient. By emphasising the malfunctioning organism, the explanatory models became tied to present occurrences and proximal causes. In addition, disregard for the relation between the singular and the universal, for example between the individual and the group to which he belongs, obscured an important explanatory aspect of life.

The relation between the one and the many, between the universal and the particular that falls in under the universal, has never been considered an easy one. The issue, which has been deemed "the most central of all philosophical problems" (James 1907, p. 50), captivated the Ionian natural philosophers no less than Plato and Aristotle. It took centre stage in middle age philosophy with the questions of whether universals are real or not, and of what kind of existence they have if real. The outcomes of these debates still influence our thinking about how the universal relates to the particular. And although time has buried some of the controversies, several new venues of disagreement have evolved as science has progressed.

The issue continues to stir heated debates amongst biologists and philosophers. This is especially pertinent when it comes to understanding the status of species, the lowest-level classificatory unit for biological organisms. The essentialist tradition, which dates back to Plato (427–347 BC), viewed the particular organism as an illusory appendage to the real entity – the species. Based upon the assumptions that nature is uniform, that there are clear-cut and universally applicable criteria for classification, and that species are spatio-temporally unrestricted classes, taxonomists with an essentialist leaning believed that they could identify properties that were both necessary and sufficient for species membership. This notion, that organisms can be sorted into discrete, non-overlapping kinds on the basis of outward morphological appearances, has been contradicted by a huge amount of data, and essentialism thus no longer plays any significant role in taxonomy (Mahner and Bunge 1997).

Another tradition, which dates back to Aristotle (384–322 BC), opposes essentialism by claiming that universals are inherent in the particulars and that they have no independent existence outside these. Aristotle believed that each organism of a certain species was endowed with a formative element, the *eidos*, which lead to similarity in structure, function, development and reproduction. This view was maintained throughout the Middle Ages, but was strongly contested by William of Ockham (1285–1349) who refused to admit that there is anything in the real world that corresponds to the universal. Instead he maintained that universals are mental concepts that become known through experiencing particulars (Ockham 1990). By insisting that particulars and not universals are the real entities of the world, he took a decisive empirical stance towards evidence.

According to Ockham's conceptualism, the species concept serves as a useful epistemic placeholder that organises our categorisation of the world. This idea is compatible with observations that nature consists of a number of individual organisms. There is no unique way to classify these products, there are many plausible and defensible ways of doing so, and the best way will depend on both the purposes of the classification an the peculiarities of the organism in question. These ideas, which gained a wide influence on scientific thought, served as a foundation for Darwin's nominalistic view that the term species is "one arbitrarily given for the sake of convenience to a set of individuals closely resembling each other" (Darwin, 1859, p. 108).

It is noticeable that Darwin took the reality of individual organisms for granted, without even questioning why this was so. This position, which follows directly from Ockham's philosophy, is not without its difficulties. A variety of scientific observations support the claim that organisms, which appear well demarcated from their surroundings, are actually inhomogeneous entities. They sometimes consist of cells derived from other organisms, in which case the organism is a chimera; or their cells may have been altered during development as happens with cancerous cells or cells of the adaptive immune system, in which case the organism is a mosaic (Pineda-Krch and Lehtila 2004). It is thus clear that our intuitive notions of organismal being are imprecise.

Our intuitions also tell us that there is a clear distinction between species and organisms. But the same intuitions are of little help when it comes to explicating why this is so. It is evident that both organisms and species are situated in time and space; they have a beginning and an end and are located somewhere on planet Earth. And since both species and organisms are real in the spatiotemporal sense, the differential status of species and organisms can not be conferred on the basis of these characteristics. There does, however, appear to be a clear distinction between species and organisms with regards to their capacity of being selected by natural mechanisms. Natural selection works stronger at the level that harbours the higher heritable variation in fitness. And since organisms are more variable and thus stronger selected than species, natural selection should favour traits that are good for the organism more than traits that are good for the supra-organismal group. These traits include, amongst others, cohesiveness and integration. Thus, the best argument for why we hold organisms as being more unified than species is probably provided by the theory of natural selection.

For practical as well as for theoretical purposes it is sometimes useful to introduce a classificatory level in-between the species and the organism. This can be done by classifying the organisms of a species into groups that exhibit the same variety of a given trait. Such traits, which may include anything from behavioural patterns to disease susceptibility, show a wide variation within a given species. Natural groups are of special concern to infectious disease specialists because animals living in groups will generally experience higher local densities of conspecifics than solitary animals, and they may therefore experience higher rates of transmission of infectious agents. Group living is also important because the social systems of group-living animals may facilitate the evolution of increased virulence of infectious agents.

Since no two organisms are identical, allocation of individuals into groups will to a certain degree be conventional. There are many ways to classify groups, and not all classifications necessarily depict real entities. In addition, since there are a wide variety of species concepts (Mishler and Donoghue 1982), the exact demarcation between groups and species may sometimes be difficult to draw. It is therefore of interest to assess whether and eventually the degree to which categorisation of individuals into groups merely represents a convenient and mind-dependent way of aggregating individuals, or whether groups of certain types are indeed real entities. One answer to this question hinges on whether or not it can be demonstrated that groups are causally efficacious.

We intuitively regard something as a cause of if it makes a difference in some regard (Mackie 1980). The idea that organisms may be causally efficacious is widely accepted. But the idea that characteristics of the group or the species may be of a causal nature is far less obvious. Still, if the temporality of causes is taken into consideration, it becomes easier to see how group level causality may come about. While a cause occurs at a specific place and at a specific moment, the consequences of the cause stretch wide into the future. And as a corollary, the effects we experience at one specific spatiotemporal moment were generated at some previous moment. Accordingly, the past causally impinges on the present, which in turn impinges on the future.

But the past cannot impinge directly on the present; it does so in a mediated manner. Thus, diachronic causality, in which the cause is temporally separated from the effect, is different from synchronic causality. Diachronic causality mediates its effects indirectly through a context that impinges on synchronic causality. Since groups provide a context for organismal behaviour, and since it has been repeatedly demonstrated that belonging to a group confers a differential probability of individuals within a group to survive and reproduce compared to individuals in other groups, the requirements of what it is to be a cause are fulfilled also for group level properties.

The idea that the cause of something has both a spatial and a temporal aspect, and that determinants of organismal traits may be found at both the individual and the group level, allow the stipulation of a two-by-two table to differentiate between types of causes (Table 2.2). The causes may be divided according to whether they are efficient in the existing environment or were effective during some previous moment in the history of life. Causes that were effective during the evolutionary history are known as ultimate causes (Mayr 1961), irrespective of whether they made their effects on organismal traits through individual or group level factors. Causes that are effective in shaping organismal traits within the existing generation are conveniently divided into proximal and distal causes (Diez Roux 2004).

When Ernst Mayr (1961) evaluated the concept of cause in biology, he distinguished between evolutionary and developmental biology, and insisted that the two fields differed in methods, explanatory projects and concepts. He further claimed that developmental biologists analyse proximate causes while evolutionists analyse ultimate causes. Immunologists are, like developmental biologists, preoccupied with laying out proximate causes. They investigate and provide sophisticated and elaborate mechanistic explanations of specific immune responses, but take less interest in explaining why traits take the appearances they do. Nevertheless, since proximate causes derive their causal powers from ultimate causes, the latter causes are in a certain sense more fundamental. Ignorance of ultimate evolutionary causes

Table 2.2. The causes of organismal traits, as stratified on ecological and evolutionary circumstances

		Determinants of organismal traits	
		Individual level causal factor	Group level/ecological causal factor
Temporal circumstances	Extant generation	Proximal	Distal
	Past generations	Ultimate	Ultimate

may thus lead to insufficient and sometimes even false explanations, and it is therefore likely that any biological hypothesis would profit from a coalescence of proximate and ultimate causal explanations.

The difference between proximal and distal causes can be illustrated by an example from infectious immunology. Some properties, like the property of being resistant to infectious disease, may result from causal mechanisms taking place at either the individual or the group level. When susceptible individuals encounter an infectious agent like measles or rubella, they first develop disease and thereafter immunity against later attacks. The infectious agent is the proximate cause of disease, and the immune response is the proximate cause of resistance to re-infection. When all these resistant individuals are aggregated, they delineate a resistant group. This resistance is, nonetheless, not owing to a group-level causal mechanism because it is merely an aggregation of individual level resistances.

Herd immunity, on the other hand, is a true group level property. Herd immunity delineates the property that infectious agents like rubella or measles are unable to invade a population despite the fact that several individuals within the group are susceptible to infection. In this case, the susceptible individuals are being protected by being members of a group of individuals where a high proportion of the individuals are immune. If these susceptible but protected individuals move to a new group with a lower frequency of immune individuals they may lose their herd protection. Hence, the group makes a difference and so is causally efficacious. Since the causal effects are mediated by the context in which the protected individuals are located, the cause is designated as being distal as opposed to proximate.

But although groups may be causally efficacious, there are several fallacies that need to be deliberated when the causal effect of the context is evaluated. A most pressing problem posed by varying contexts has come to be known as Simpson's paradox (Simpson 1951). In his original formulation of the paradox Simpson provided an example in which individuals were stratified into two groups, males and females. The individuals within each group received either treatment or no treatment, and the survival or death within the two groups were thereafter recorded. The numbers amassed by Simpson demonstrated a positive correlation between treatment and survival in both groups, but no correlation in the combined population. Surprisingly, what was good for males and females turned out to be equivocal for people!

Hence, interpretation of data collected at multiple levels should involve a careful decoupling of the relations between parts and wholes, and care should be taken when giving multi-level data a causal interpretation. This is because inferential fallacies may occur when one attempts to explain traits at one level with data that have been collected at another level. Several fallacies are so common that they have received their own labels. For example, the ecological fallacy may be instantiated when associations between variables at the level of the group are interpreted as if they were true at the individual level, while the converse fallacy, termed the atomistic fallacy, is instantiated when data on variables collected at the individual level are extrapolated to variables at the group level.

The reality of groups has been a controversial issue in evolutionary biology no less than in ecology. The controversy has lingered on whether or not groups of individuals can be selected, or whether it is the individuals within the group that are selected. For example, individual selection would occur if resistant individuals are selected for, while group-selection would occur if groups that exhibit herd immunity survive and reproduce at a different rate than groups that do not exhibit the trait. If herd immunity confers a fitness benefit on the members of the group, there is a real chance that the characteristics of the different groups as well as the characteristics of the organisms that make up the groups will undergo evolutionary change.

The idea that groups may be selected during evolution has been intensely contested, and during the latter half of the 1960s and up through the 1990s invocation of group selection was regarded as heresy (Sober and Wilson 1998). Elliot Sober (1984) has claimed that some evolutionary biologists deny the possibility of group selection because they fail to see the variegated relations between parts and wholes, and that they therefore commit a fallacious inference of the same type that is laid out in Simpson's paradox. This argument has also been substantiated by proponents of multi-level selection theory, which argues that selection may take place at several levels in the biological hierarchy.

2.3. The lens metaphor

In accordance with the bearings of the received and revised views, the immune system should be conceived of as a relatively independent entity that consists of components bound up in a structured whole that interacts with the environment to perform some function that cannot be performed by the components themselves. The workings of the system, which bestows fitness-related benefits to the defending organism, can be divided into three distinct operational segments, designated as the signal, the decision and the response phases.

The immune response is triggered by antigenic *signs* derived from infectious agents and harmed self-tissues. These signals have differential informational weights as to elucidating the true being of the trigger, and thus for allowing the immune system to make an ecologically valid inference about the experienced event. Signals with high degrees of covariation with the event to be inferred have a large degree of ecological validity. Detection of ecologically valid signals would be selected during evolution because of their survival value, and should be those that become recognised by pattern recognition receptors or that become immunodominant.

In situations of infection the signals consist of conserved chemical patterns, including lipopolysaccharides from gram negative bacteria, RNA from a variety of viruses, and zymosan from fungi, while in situations of harm the danger signals come from dying cells, and include amongst others heat shock proteins and uric acid. When such antigens trigger the innate sensors, they mediate a response that in turn delivers signals to the maturing lymphocytes. In a sense the innate sensors "teach" the adaptive sensors to interact safely with the triggering substance.

Recognition of microbial patterns or danger by Toll-like receptors, either alone or in association with other Toll- or non-Toll-like receptors, triggers the whole cascade of effector mechanisms of the innate and adaptive immune systems (Kawai and Akira 2005). Acting through a restricted portal of four intracellular adapter molecules, a limited number of Toll-like receptors are able to induce the transcriptional activation of hundreds of genes that code for systemic and localised defence factors. These include genes for proinflammatory cytokines, molecules that drive dendritic cell migration to the lymph nodes, and MHC and co-stimulatory molecules.

Upon receiving signals from the environment, the immune system functions as a decision-making unit that provokes a more or less well-adapted response to the signal. The *decision*, which must be frugal in the amount of information that is taken account of and fast in order to keep up with infecting agents, is not uniquely determined by the antigenic stimulus. The adaptive immune response is influenced by properties of both the organism and the antigen. Organismal properties that may regulate the output include genetic makeup, previous encounters with similar infectious agents, and route of infection, while important properties of the antigen include its concentration, the timing and duration of antigen encounter. The temproal association of antigen with expression of host costimulatory molecules is also important to emphasise, because the immune system experiences one antigen at a time only in artificial situations, mostly during experimental investigations and immunisation procedures. Sequential infections may divert the expression of host molecules in diverse directions, and the outcome if latter infections may thus depend upon characteristics of the infectious agents that preceded it.

The dispersed signals that emanate from the innate immune response trigger the components of the adaptive immune system within the secondary lymphoid organs, including the spleen and lymph nodes, where antigen-specific naive T and B cells interact with antigen and antigen presenting cells (Parker 1993). At this point, the extremely complex characteristics of the infectious agents are being met by a similarly complex response performed by the adaptive immune system.

A major event at this early stage is the polarisation of the Th response into Th1 and Th2 cells (Mosmann and Coffman 1989). This diversification depends upon the microenvironment that is being generated at the contact surface between the T cell and the dendritic cell, within which signalling molecules of both cells become congregated. While immature dendritic cells energetically sample their external environments, the degraded molecules are not vigorously expressed on their cell surface MHC molecules. However, upon activation the cells stop internalising antigen and instead concentrate on generating complexes between peptides and MHC class II molecules for T cell activation. Recognition of the MHC-antigen complex activates the T cell, which in turn further activates the dendritic cell to provide more stimulating signals (Boes et al. 2002).

The dendritic cells appear to stimulate T helper cells by transmitting information about the type of infection encountered through the T cell's Notch pathway, an evolutionary conserved signalling pathway that regulates the diversification of Th1 and Th2 cells (Amsen et al. 2004; Tanigaki et al. 2004). It appears that dendritic cells express one of two important ligands, called Jagged and Delta, and that Delta becomes expressed on dendritic cells that have been exposed to Th1 promoting stimuli while Jagged is induced under Th2 promoting conditions. Uncommitted CD4 T cells express Notch receptors that bind to one of the two different Notch-ligands expressed on dendritic cells. When the receptors bind their ligands, the intracellular portion of the receptors is cleaved and can then induce gene expression and T cell differentiation.

But also the body site at which infection or damage occurs is important for the ensuing polarisation of the immune reaction. Different tissues mount different immune responses to the same antigenic stimulus. This is partly owing to the complex pattern of tissue-specific distribution of dendritic cells, but also because epithelial and stromal cells at the site of infection generate Toll-like receptor mediated signals that contribute to the activation of T cells (Sato and Iwasaki 2004). Since costimulatory signals are present in relevant concentrations only in organised lymphoid tissues, a necessary preparatory to understand immune activation is, accordingly, to explain how and in what form antigen reaches the spleen and lymph nodes, and for how long it stays there.

Within the lymphoid organs, the microbial antigens exposed on dendritic cell MHC-molecules trigger the lymphocytes through their antigen receptors, after which lymphocytes proliferate and acquire various effector functions. After a period of several days they are released into the circulation. This usually leads to rapid clearance of the antigen and subsequent death of the clonally expanded lymphocytes. Some antigen-specific cells of both the T and B class do, nonetheless, survive the primary immune response. It is mainly these memory cells, and not naive antigen-inexperienced cells, that patrol the body and maintain immunological surveillance. These lymphocytes remain disseminated throughout the secondary lymphoid organs as long-lived memory cells.

Since the immune system cannot decide away involvement, it is constantly ready to act. And since it cannot step back and overview the alternative possibilities and trajectories, it has to make decisions in real time. If these decisions are made from predispositions built from adaptive representations of reality, there is a real chance that changes in ecological parameters may turn adapted responses into maladaptive responses. Mismatches between the ecological validity of a sign and utilisation of the sign by the immune system may lead to inaccurate judgements and to a maladapted immune response. Such mismatches have been given a causal role in explanation of the increasing prevalence of allergies and autoimmune disease in Western societies.

The *response* may proceed down one of several distinct trajectories that have been shaped by evolution over millions of years and by development over the organism's lifetime. The routes selected are the ones that successfully have allowed the responding organisms to receive fitness benefits. Generally, the adaptive immune response is characterised by being driven in two opposing directions. On the one hand there is a drive towards immunodominance, and on the other hand there is an apparent broadening of the response, especially in immunopathology (Sercarz 1998). Hence, while the preimmune B and T cell-repertoire of a given individual is able to recognize an almost limitless array of molecular specificities, exposure to a specific type of infectious agent results in a pool of highly efficient and specific lymphocytes. This is in part owing to the limited number of antigenic epitopes presented by dendritic cells (Gapin et al. 1998).

The phenomenon of immunodominance, which is a feature found in responding T and B cells, is the consequence of a series of selective sieves that serve to pick out and propagate T and B cells with optimum affinity for the corresponding antigen (Rao 1999). The phenomenon provides evidence that the immune system has been selected to respond with important antigens, and reasonably not with the whole universe of conceivable antigens. One selective reason for this may be that dominant immune responses are easier to regulate than diverse responses. This is important because the risk of self-destruction caused by the potent mechanisms of the innate and adaptive immune systems is overwhelming.

The whole cascade of events, from the triggering of Toll-like receptors by the nearly limitless complexity of microbial agents, via the activation of a restricted set of intracellular signalling pathways that funnel towards a few crucial molecules, from which the signalling pathways again disperse, can be thought of as a set of sequentially placed lenses; the first lens, the innate immune system, focuses and transforms the incoming signals, after which the second lens, the adaptive immune system, diverts the transformed signals (Figure 2.3). In this manner, the "rays of

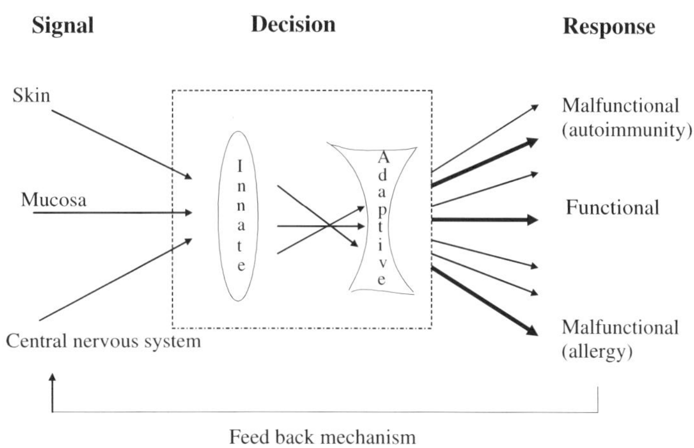

Figure 2.3. The lens metaphor. The combined use of innate and adaptive immune components provides the organism with rapid response kinetics and ability to mount a powerful and appropriate response to a diversity of infectious agents. The immune system receives signals from the microbial world and the peripheral tissues by receptors of the innate immune system. Upon stimulation the receptors trigger secondary events, including release of inflammatory mediators. These secondary signals are made prominent within the lymphoid tissues, and thus constitute the context in which decisions are being made. The focusing role of the innate immune system as well as the contextual parameters sets the lead for the activation of the adaptive immune system. The adaptive immune system generates an immunodominant response towards specific antigens, which may be fit or unfit. Potent feed back mechanisms serve to tailor the immune response to the incoming stimulus

light" that impinge on the lenses are refracted at different angles as a consequence of their origin. The two lenses are tied together by an anatomical support system, which is made up of the secondary lymphoid tissues, thus allowing for the influence of contextual parameters to impinge on the transformation of the incoming signals. The lens metaphor thus takes care of the insight that the immune system connects the experience of the species to the experience of the individual, and thus allegorically visualises the unified theories of Metchnikoff and Ehrlich as well as the integrated views on immunity.

The lens metaphor should not be thought of as depicting an autonomous module isolated from the world. Rather, the metaphor situates the immune system within the organism, which, in turn, is placed within an ecological setting. Hence, instead of viewing immunity as beginning with stimulation of the innate system and ending with the response, immunity should be imagined as extending into the environment in which the organism lives as well as backwards to environments encountered by its ancestors. To envision the immune system as the beginning and end of immunity misses an important fact – that the immune system is a being-in-the-world that copes with significant others in an adaptively plastic manner.

CHAPTER 3

Adaptive plasticity

Organisms daily confront an environment permeated with challenging microbial agents. The outcomes of these encounters have consequences for the organisms' chances of surviving and reproducing. Since the right response in one environment may not be the right choice in other environments, organisms have developed a wide range of possible responses to the challenges encountered. At times the only viable behaviour will be to migrate out of the hostile environment, but most often organisms regain health by mounting active defences.

An organism's environmental conditions vary considerably through time and space, and any given animal will therefore make frequent "mistakes" that may be fatal if does not adapt to those circumstances. The capability to respond with one type of behaviour in some situations and other behaviours in other situations, contingent upon how the world is, is termed phenotypic plasticity. If the plastic trait contributes positively to the fitness of the organisms that embody it, the trait is termed adaptive.

The phenomenon of phenotypic plasticity entails that some phenotypic differences are not uniquely determined by organismal genetics. Evolutionary explanations of plastic traits should therefore include genetic, developmental and environmental causes, and if the trait is to be counted as adaptive, its fitness advantages as compared to alternative patterns of phenotypic expression should be demonstrated as well.

Adaptive plasticity has, in many respects, made *Homo sapiens* into an all time evolutionary winner. The success is not owing to any outstanding anatomical characteristics, but is the result of a sophisticated capability to utilise symbols and thus to reflect upon his own position in the world. As he evolved, man began to question the reasons for his own being as well as for the being of other creatures, and eventually managed to study the questions scientifically. It may for some purposes be useful to look upon science as a derivative of the more general adaptive biological project of keeping the organism alert in its dealings with the world, or as maintained by Karl Popper (1979, p. 261): "Our knowledge consists, at every moment, of those hypotheses which have shown their (comparative) fitness by surviving so far in their struggle for existence; a competitive struggle which eliminates those hypotheses which are unfit."

During man's evolutionary history, reciprocal influences between the mind and the world changed the human mind in several respects (Sterelny 2003). The degree to which these changes have become genetically or culturally engraved is currently much debated; while researchers with a philosophical bending emphasise the socio-cultural aspect (Dupré 2001), neuroscientists (Damasio 1994) and evolutionary psychologists (Barkow et al. 1992) emphasise the neural and biological aspects.

Notwithstanding these disputes, it is obvious that having a superior mind made man less dependent upon his genetic endowment.

Whether or not having a superior mind also liberated man from his epistemological constraints is, nevertheless, not that evident. It has, for example, been claimed that knowledge is tightly coupled to man's biological interests; interest may in fact be taken as a transcendental precondition for knowledge (Habermas 1968). And the novel constructions that he himself built, both the structural and cultural, may sometimes have introduced new types of constraints that may severely have biased his capability to relate objectively to the world. Ludwig Fleck (1935) even claimed that scientific facts have a genesis and development that depend upon the sociological characteristics of the scientific community.

The mind probably evolved to let the organism cope adaptively with challenges encountered in the natural world. However, as science and technology open up new and hidden worlds, we come to experience that our grasp of reality can not be tutored by the same predispositions that helped our ancestors to cope with objects, animals and other people. Inherited intuitions may even hinder understanding if we have to rely on analogies and metaphors that press the old mental faculties into service. For example, modern physics present us with a picture of reality that we have every reason to believe is true. Nevertheless, the picture makes little sense to the intuitions we have about space, time and matter that evolved in the brains of our ancestors. In a similar manner, our intuitions about what it is that constitutes life, consciousness, or the immune and mental selves, may be biased by our prescientific world-views. Hence, although it is clear that the acquisition of mental plasticity adapted man to deal with practical matters in a superior way, it is not self-evident that man's emancipation from nature made him more adapted to the world in a sense that science would applaud.

While it is evident that theory and cultural guidelines do not impinge on the behaviour of biological systems other than the human mind, it is equally clear that evolutionary selected background assumptions play an important role in steering animal behavioural responses. Such predispositions consist of previous "knowledge", and are derived partly from the evolutionary process and partly from experiences made during the organism's own lifetime. Since behavioural predispositions are generated through evolutionary and developmental learning and thus reflect the accumulated experiential knowledge of the lineage to which the organism belongs, predispositions (the universal) serve an analogous role in guiding organismal behaviour (the particular) as the role theoretical knowledge (the universal) plays in guiding human action (the particular).

These ideas on adaptive plasticity and the constraints involved are of relevance for understanding the immune system as it relates to the world. Not only is the immune system an organ studied by science, it is also an organ that serves to orient the organism in its dealings with the environment. A major claim of this chapter is that the immune system, no less than rational man, needs to be studied as a being-in-the-world, and that the types of explanations that have been generated about man's dealings with the world have direct relevance to a deeper understanding of the world of microbes and their interactions with immune systems. Elucidation of the immune system's fundamental relations to the external world is thus seen as an important and necessary preliminary step towards achieving an integrated understanding of the functioning and malfunctioning immune system.

Fortunately, this step does not require groundbreaking philosophical reworking of biological observations. As it turns out, the representation of external reality by the immune system is closely related to an analogous problem encountered in the neurosciences and the philosophy of science; the conundrums associated with how external reality can be represented by internal mental structures in a true manner. These issues and their proposed solutions, which have occupied the minds of Western philosophers for more than 2500 years, have profound bearings on the elucidation of the parallel questions in immunology.

3.1. Being-in-the-world

THE KNOWER AND THE KNOWN

One of the great divides in the theory of knowledge has been between empiricists, typically regarded as thinking that almost all knowledge is acquired from experience, and rationalists, who suppose we come equipped with much knowledge that is innate. Immanuel Kant (1781) attempted to bridge the divide by claiming that although experience supplies us with knowledge, the knowledge is based upon pre-existing mental structures that can not be gained from experience. Our experience of the world depends on pre-experiential knowledge that includes concepts of space, time and causality. These formative elements are therefore necessary constituents of all knowledge.

Kant's attempt to bridge the divide between the knower and the known was strongly criticised by Georg Wilhelm Friedrich Hegel (1807), who claimed that Kant's solution presupposes an incorrect distinction between the observer and the observed. More specifically, Kant's solution took for granted that the world stands on one side and knowledge on the other side, thus precluding our chances of ever reaching the truth about the world. Hegel tried to solve this dilemma by postulating that knowledge of the world is generated through a dialectical process in which scientists follow "the pathway of the natural consciousness which is pressing forward to true knowledge" (Hegel 1807, p. 135). This pathway is historical, and the pre-experiential or transcendental knowledge that Kant claimed to be historically fixed, was in Hegel's opinion capable of being modified by history. History thereby became a necessary part of the constitution of knowledge. Whereas Kant's theory of knowledge was based on a dichotomy between the subject and the world, Hegel looked upon history as a modifying epistemological agent in which various transcendental forms are tested and criticised until unification between the world and our thoughts occurs. Hegel's epistemology was, accordingly, evolutionary and progressive, and his proposed dialectical process between mind and nature provided a mechanistic explanation for the attainable pattern of ultimate knowledge.

Both Kant's and Hegel's theories were proposed as attempts to explain the limits and prospects of human knowledge. Their ideas did not attempt to address the establishment of the cognitive faculties in man; these faculties were taken for granted. Although Kant concluded that we could not possibly gain knowledge about the world without at least some guiding prior knowledge about what to expect, he was not in a position to give a naturalistic explanation of such prior knowledge. This explanatory deficit rapidly changed with the publication of Charles Darwin's (1859) theory of natural selection. Unlike Hegel's theory for the evolution of mental capacities, Darwin's theory for the evolution of life provided a mechanistic process, natural selection, that created an historical pattern, the tree of life. Darwin's idea of natural selection provided the mechanism that allowed life's evolution to proceed without the help of a rational designer and thereby without the rational foresight and deliberation that Hegel's theory invoked.

Darwin's theory spurred Charles Sanders Peirce to hypothesise that man has certain inborn capabilities that makes conjectural guessing productive (Peirce 1955). Individuals with these capabilities were better adapted to deal with various environmental challenges than individuals that were without, and so the traits evolved by natural selection. Konrad Lorenz (1982) similarly argued that the concepts of time, space and causality are products of the human lineage's biological evolution. Pre-experiential knowledge is therefore not a result of the experience of individual organisms, but is rather a hard-won product of the long and arduous evolution of the species. By insisting on the adaptive nature of knowledge, in which knowledge evolves from an interaction with the environment, it must in certain respects reflect the environment in which it evolved. Thus, by reintroducing the epistemological dilemma into a new theoretical setting, Peirce and Lorenz were able to re-conceptualise knowledge and give a naturalistic explanation of the genesis of Kant's prejudices as well as Hegel's dilemma of the correspondence between our thoughts and the world.

Konrad Lorenz' Darwinian explanation of pre-experiential knowledge was an off-spin of his more general investigations into the establishment of species-specific animal behaviour. Beginning about 1931 Lorenz and his collaborators published a series of papers on instinct and innate behaviour in which the learned-innate distinction was seen as dependent upon whether the development of behaviour was determined by environment or by genes (Johnston 2001). Although Lorenz' theory had a widespread influence, partly because of its diagrammatic simplicity and its large number of observations of different species under natural conditions, he was criticised for having failed to understand the relationship between genes and behaviour. Early critics like Daniel Lehrman (1953), a behavioural endocrinologist, claimed that Lorenz not only misconceived the organism-environment interactions, he even used the concept of maturation as a way of ignoring developmental complexity. Lehrman insisted that the use of explanatory categories such as innate and genetically fixed obscure the necessity of investigating developmental processes. These processes are important because they shape novelty, because they are responsible for the

> development of new structures and activity patterns from the resolution of the interaction of *existing* structures and patterns, within the organism and its internal environment, and between the organism and its outer environment. At any stage of development, the new features emerge from the interactions within the *current* stage and between the *current* stage and the environment. The interactions out of which the organism develops is *not* one, as is so often said, between heredity and environment. It is between *organism* and environment! And the organism is different at each different stage of its development. (Lehrman 1953, p. 345).

Lorenz (1965) accepted this critical response and agreed that it is a failure to conceive of the "innate" and of the "learned" as of disjunctive contradictory concepts. In his new theoretical framework he came to regard adaptiveness that was determined by information acquired during an individual's development as learned, and adaptiveness determined by information acquired during the evolutionary history of the species as innate. In addition, he saw the concepts as being integrated at a deeper level since the survival function of learning also presupposes phylogenetic programming.

Even though innateness is a term in common use because it represents a highly intuitive way of thinking about organismal adaptations, several investigators reject the validity of dichotomising between internal and external explanations of behavioural patterns (Oyama et al. 2001). The partitioning of traits into genetic and environmental components has been criticised on both methodological and theoretical grounds (Lewontin 1974; Sober 1993), and many theoreticians have opted for an interactionist view in which genes are seen as context-sensitive difference makers (Sterelny and Griffiths 1999; Jordan et al. 2000; Michelson and Kopan 2002). According to this view, genes produce their effects in a complex network of causes consisting of other genes, the other constituents of the cell, and the external inputs to the developing organism. These ideas have more recently been adopted by proponents of systems biology (Kitano 2002; Hood 2003).

The validity of the previous arguments entail that biological entities should not be studied as if they were Cartesian machines, made up of distinguishable bits and pieces, each with a determined causal relationship to the movement of other bits and pieces. The genotype seldom specifies a unique outcome of development, at most it specifies a norm of reaction, a pattern of developmental outcomes that differ in different environments (Schlichting and Pigliucci 1998). Thus, in their plea for a revitalisation of the concepts of part and whole, Levins and Lewontin argued that:

> What constitutes the parts is defined by the whole that is being considered. Moreover, parts acquire properties by virtue of being parts of a particular whole, properties they do not have in isolation or as parts of another whole. It is not that the whole is more tan the sum of its parts, but that the *parts* acquire new properties. But as the parts acquire properties by being together, they

> impart to the whole new properties, which are reflected in changes in parts, and so on. Parts and wholes evolve in consequence of their relationship, and the relationship itself evolves. (Levins and Lewontin 1985, p. 3).

WORLDLINESS

Increased adaptability is the most pronounced benefit of flexible response characteristics; phenotypic plasticity allows the production of a better phenotype-environment match across more environments than would be possible by producing a single phenotype to all environments. But phenotypic plasticity also has its costs and limitations; the acquisition of environmental information as well as the production and maintenance of the plastic phenotype may come costly, and the usefulness of plasticity is limited if the available information gives poor reliability.

Plasticity is often a nuisance for the investigator, largely because it increases the complexity of the biological world by several orders of magnitude. As a strategy to circumvent this problem, scientists attempt to simplify and objectify the particulars they investigate. But by this procedure, by defining life in theoretical and biochemical terms, they adversely miss out some of the most important aspects of organisms – their embodied drives. And by leaving out the process whereby biological entities understand themselves and their world, biologists commit the same fallacy as psychologists do when they leave out consciousness from behavioural studies. Thus, there are reasons to claim that many scientists in their "craving for generality" take a "contemptuous attitude towards the particular case" (Wittgenstein 1960, p. 18).

In his 1927 magnum opus, *Being and time*, Heidegger forcefully criticised biology for being too concerned with causal explanations, and of being caught up in the Cartesian metaphysics. In his attempt to capture man's contextual relation to the world, and thus to restate the importance of the particular case, Heidegger therefore emphasised the role of the organism as being active in its dealings with the surroundings. The organism responds adaptively to environmental signs and is thus far from a reflexive robot with fixed behavioural characteristics.

Heidegger was explicitly aware of the fact that the present, and so the future of an organism, cannot be understood apart from its history. Organisms do not come "naive" into their world; they are in a sense thrown into the world with a set of coping mechanisms already in place. Thus, unlike Kant, who held that time and space were of equal importance for man's being, Heidegger held time to be of the uppermost importance. Like Lehrman, Heidegger envisioned organismal life as a process of becoming; time and existence are inextricably linked, existence is the "stretching" from birth to death, and organisms live into the future by the means of characteristics derived from their past.

To capture the sort of being that differentiates material objects from animals, Heidegger explicitly distinguished between two types of being; *being-in* (with a hyphen) and *being in*. Whereas the common understanding of *being in* is of something being physically enclosed, *being-in* is not a physical property but relates to the organisms concerned dealings with its surroundings. Thus, the living organism understood as being-in-the-world is regarded as an encapsulated package dwelling inside the environment, like an infectious particle dwells inside its host. *Being-in* is therefore loaded with involvement.

Heidegger's ambition was to show that neither practical activity nor contemplative knowing can be understood as a Cartesian relation between a self-sufficient mind and an independent world. Hence, the goal of philosophy since Plato, of working out a presuppositionless biological science, was seen as misguided. All relations between beings and objects presuppose a basic form of being-with-things, it belongs to the nature of organisms to exist in such a way that they are always already with other things. As such, living entities are always already outside themselves along with the environment, formed by shared practices and absorbed in active coping with the surroundings.

These insights into the dealings of an organism with its environment are extraordinary sophisticated and are highly relevant both for our understanding of biology as well as for our understanding of the scientist's role in pursuing biological questions. Still, there are few references to Heidegger in biological works. This may be owing to Heidegger's intricate style of writing, but also to the fact that he in *Being and time* only rarely introduced examples of or references to experimental work.

Although Heidegger wrote philosophical works for philosophers, he was well aquinted with biological research. And in his lecture notes from 1929/30 Heidegger (1995) clearly stated how the thoughts of Jakob von Uexküll (1864–1944), an Estonian zoologist whose works also inspired Konrad Lorenz, had influenced his thinking about organismal dealings with the world. Being unsatisfied with contemporary Darwinian metaphysics, in which the organism was viewed as passively adapting to a given environment, Uexküll set out to build a new biology in which the individual animal's perspective was retained (Kull 2001). Uexküll started from Kant's idea that animals do not know the world as it is; they know only what they reconstruct from sensory inputs and inborn capacities to interpret signals. This subjective world he termed Umwelt.

The term Umwelt, which is analogous to Heidegger's concept of the world, is basically meant to capture the relational property between the organism and the environment. It designates the contextual elements that are of importance for the successful orientation of the organism, and is thus a meaning-carrying structure that contains a "sign or symbol that members of the same species can understand, but that those of another species cannot comprehend" (Uexküll 1982, p. 77). Uexküll further claimed that thc subjcct on the one hand produces the important symbols in its Umwelt while it at the same time is a product of the same symbols. Hence, the sensory signs it recognises restrict the animal's relation to the world it inhabits, and different Umwelts occur as a consequence of the differential evolutionary history of the various species.

Since organisms and their Umwelten are reciprocally shaped, perceiving is as much about acting on the environment as about receiving signals from it. These ideas have recently received a neurobiological grounding in Antonio Damasio's

(1994) work. In his investigations into human decision making, he observed that responses to environmental signs are heavily influenced by emotional components. And although it is standardly believed that rationality is something obtained when reasoning is performed without emotional "disturbances", Damasio argued that this is a false conception. Instead, emotions should be viewed as essential ingredients in any decision-making process; although any sign may evoke a multitude of response potentialities, only responses with high value become effectuated. This is because the emotions imbue environmental signs with meaning through evolutionary and developmental learning.

In accordance with these views, it can be claimed that neither scientists nor immune systems stand in an objective relation to their world, in the sense that they are detached observers that investigate an observer-independent reality. Rather, they are interactors in the game of life; the immune system in defending life, and the immunologist in explaining the defending of life. The immune system utilises a variety of strategies to defend, as the scientist utilises a variety of strategies to explain. And like evolution has put constraints on the various defence responses that can be realised, the scientific community has put constraints on the kind of questions that can be investigated in a meaningful way. Furthermore, various immune systems interpret antigenic stimuli differently, as do scientists that interpret observations differently.

THE SELF

The tripartite self

The idea that immunologists investigate the discrimination between self and nonself has been an implicit part of immunology since Paul Ehrlich introduced the term *horror autotoxicus* in the late 1890s. The term delineates the organismal catastrophe that would ensue if the immune system reacted with self-constituents. Still, formal expression of the self-nonself dichotomy was not made until introduced by Macfarlane Burnet in the 1940s (Tauber 1994).

Even though the self-nonself dichotomy has served a useful role (Cohn 1998; Nossal 2001), debates about the self continue to stir controversies between theoretically minded immunologists. The disagreements are partly owing to a mixing together of three notions of the self, henceforth referred to as the organismal, the immunological and the immune selves (Table 3.1). Understanding the distinctions between the three notions will not remove all controversies, but some progress towards a more consistent use will hopefully be made.

The organismal self is typically invoked when immunologists discuss conflicts between hosts and their parasites. During these interactions the defending organism is regarded as self and the infecting agent as nonself. Since such interactions are reciprocal, the terms can be reversed. The infectious agent can thus be conceptualised as self, as it defends itself against the defence mechanisms of the nonself, the infected organism. When Jan Klein (1982) provided the subtitle *The science of*

Table 3.1. Three notions of the self

Notion	Biological Level	Explanatory level	Function	Instantiated by
Organismal self	Organism	Epistemology	Defence against nonself	Innate and adaptive system
Immunological self	Molecule	Ontology	Tolerance towards self	Adaptive system
Immune self	System	Metaphysics	Identity making	Adaptive system

self-nonself discrimination to his textbook *Immunology,* he utilised the concepts self and nonself in the organismal sense. In the nearly seven hundred pages exposition the terms self and nonself were indexed with one entry, to a single paragraph on page 5, where Klein stipulatively defined the self as "everything constituting an integral part of a given individual", and the nonself as "all the rest". He apparently held the meaning of the two terms to be self-evident and so never undertook a closer analysis of their meaning.

While the catchphrase provided in Klein's subtitle gives a seemingly precise overview of what immunology is all about, a closer look at the bearings of the dichotomy reveals some deep ontological and metaphysical conundrums. These were alluded to when Macfarlane Burnet (1969, p. vii) in his book *Self and not-self* neglected to index and define the title terms, but instead wrote in the preface that "I hope that the central theme that gives the title Self and Not-Self to the book is everywhere well to the fore. The need and the capacity to distinguish between what is acceptable as self and what must be rejected as alien is the evolutionary basis of immunology." So in a sense, Burnet left it up to the immune system itself to perform the defining of the self. It is likely that Burnet's reluctance to define the terms was owing to his sensitivity to the metaphysical commitments that the terms alluded to. These commitments, which are loosely implied when it comes to the immunological self, become manifest when the immune self is encountered.

The immunological self is the self alluded to when questions like "What is the molecular basis of the immunological self?" are being asked. Nossal (2001, p. 685) answered the question by insisting that "the immunological self is particular and idiosyncratic, reflecting an individual's MHC genes and the entire genome, as represented in the peptides situated in the MHC grooves." Since the immunological self is measured at the molecular level, and since infecting agents and their hosts have overlapping characteristics owing to common descent and molecular mimicry, there is a zone of overlap between the entities that make up the immunological self and the antigenic world.

The immunological self is identified as the organism's own gene-products that the adaptive immune system recognises, and it is thus not a reciprocal property. The term non-self becomes irrelevant in this conceptualisation, what matters is the self as seen by the immune system. But even though the immunological self can be given a fairly precise scientific interpretation, the conceptualisation runs into metaphysical conundrums because of the implicit answers to questions like – what if no antigen receptor recognises a molecule encoded by the genome, is the molecule non-self? – what if an antigen receptor is generated at a later time during ontogenesis, does the molecule then move from being non-self to becoming self? – should such ignored self-encoded determinants be termed "non-immunological self"?

As opposed to the organismal and immunological selves, the third self delineated in Table 3.1, *the immune self*, is metaphysical all the way. It is a relation that emerges as a consequence of the self-organising properties of the immune system, and should, as will become clearer in later sections, be addressed at the pre-scientific level. The immune self is, like virulence, a relation. But whereas virulence is a two-way relation between a host and its pathogen – between the organismal self and the other, the immune self is a relation that relates to itself.

A consequence of the tripartite conceptualisation of the self is that identical twins with identical organismal and immunological selves, may well differ radically when their immune selves are being compared. Hence, the observation that identical twins may be discordant in expression of autoimmune diseases is a manifestation that their immune selves have been differently connected to the environment, and that their immune systems therefore inhabit the world in different ways.

The endangered self

While stipulative definitions that divide the world into non-overlapping domains may serve an epistemological role, their use often obscures biological ontology by imposing unrealistic demarcations between entities. An unfortunate consequence of the self-nonself dichotomy is that it gives a simplified and unrealistic account of how the adaptive immune system performs its diverse functions. It gives the impression that immune reactivity is a deterministic and mechanistic activity, and fails to account for some well-known facts about immunity. Klein's definition does not account for the presence of physiological autoreactivity, and it misses out the important relation between the recognition and effector phases of immunity. The dichotomy is also problematic in view of the presence of naturally selected molecular mimicry of infecting agents and their hosts, as well as of the close genealogical relatedness between molecules of interacting species (Stebbins and Galan 2001).

The notion of the organismal self has been criticised on both conceptual and scientific grounds for not giving a proper accounting of immune reactivity, and pleas for the concept's abandonment have been raised. Upon reviewing the evidence favouring the self-nonself model of immune reactivity, Tauber argued that the "self has become an impediment to furthering the conceptual horizon of immunology" (Tauber 2000, p. 242). In this he resonated the recalcitrant views of Polly Matzinger (1994) who proclaimed the death of the organismal self and the ascension of the danger concept in her landmark paper" Tolerance, danger, and the extended family".

Being dissatisfied with the idea that the primary goal of the immune system is to distinguish self from non-self, she forcefully argued for the replacement of the self-nonself distinction with a danger/non-danger dichotomy embedded within a new theoretical framework, the danger model.

While the danger hypothesis was met with harsh criticism when first introduced, several experimental data now lend credence to the hypothesis. The defenders of the self-nonself conceptualisation do, nonetheless, not accept the bearings of the model. They either claim that the model is tautological, in that danger is defined as just about anything that can induce an immune response (Medzhitov and Janeway 2002), or that nothing can be explained by the danger model that can not be explained equally well or even better by the self-nonself model (Langman and Cohn 1996). Attempts at reconciling the two models have also surfaced. It has, for example, been argued that both models comply with the observations, but that one model explains one facet of the immune response and the other another facet. Accordingly, the models are to be viewed not as competing but as complementing (Heath and Carbone 2003).

The complementary bearings of the two models can be visualised by placing them within a two-dimensional framework (Figure 3.1). In this figure, the organismal self and the nonself are located at the very ends of the self-nonself axis. The immunological self is located along the entire axis, thus indicating that the immunological self can be overlapping with structures or functional characteristics of parasites. The immunological self is therefore better viewed as a continuum of gradation in which a variety of mixtures of the characteristics are possible. The immune self, being a metaphysical concept, has no place in the figure.

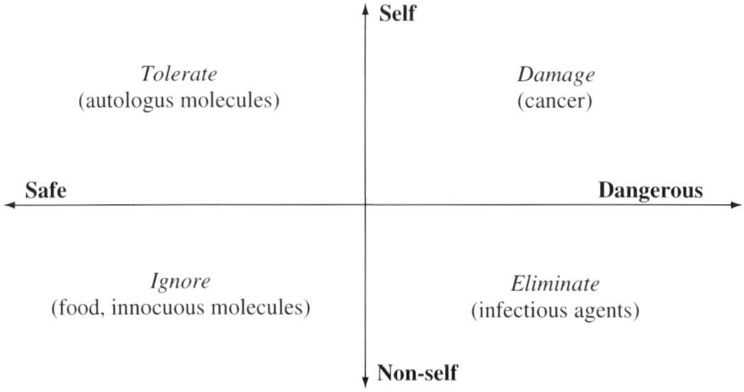

Figure 3.1. The response chart delineates response options as determined by gradations of danger and self. Some self is viewed as non-self in autoimmune disease, while some non-self is viewed as self when the individual is tolerant or ignorant of the antigen. Self-nonself discrimination is an all or none case when the organismal self is considered, but comes in degrees when the immunological self is considered. The safe/danger and the self-nonself discrimination are context-dependent discriminations. While the organismal and immunological selves are amenable to analysis along these lines, the immune self has no place in the chart

Like the immunological self, danger may be viewed as a continuum along a separate axis. Danger is a context-dependent property that varies in time and space. What is regarded as safe in one context, may be a danger signal in another context, and what is self in one context may be non-self in another. If we envision the cut-offs as being sensitive to the physiological context in which they are experienced, neither the self-nonself nor the safe-danger divides need to be understood as absolute in the sense that the cut-offs between them are fixed. This allows the immune system to be viewed as experiencing non-self and danger in a wider context, and hence envisions the immune system's response characteristics as being influenced by the physiological dynamics of the organism and its relation to the environment at a given time.

The immune self

Subject, object – both or neither?

Three arguments serve to clarify that the immune system is a true being-in-the-world. First, the immune system is shaped by its ecological surroundings. This is evidenced from the fact that most organisms in a given ecological niche have immunoglobulins and T cell receptors that reflect the microbes in the given environment. Second, the organism shapes its own immune system. This is evident form the fact that most organisms have few if any antigen receptors that react strongly with self-constituents. Third, that the immune system shapes the antigenic universe can be exemplified by the way certain infections, such as influenza and measles, are eliminated from the environment following epidemics (Ferguson et al. 2003).

Most attempts at identifying the immune self have been guided by a theoretical and empirical orientation, in which the investigators have taken the role of detached observers trying to circumscribe the essential characteristics of their study-object. As clarified by Alfred Tauber (1994), the results of these attempts have been disappointing. If anything, the studies demonstrate that the immune self is not a subject amenable for empirical investigations. That attempts at pinpointing the immune self have been disappointing suggests that our intuitions of the immune self are either wrongheaded, or that the strategies utilised have been inadequate to the task encountered. As detailed below, there are several arguments in favour of both construals.

In the concluding section of his exposition into the self's historical roots, Tauber (1994, p. 295) summarised the largely negative results of his investigation by proclaiming the view that the immune system makes its own practical definitions of what the immune self is, "the immune self is neither subject nor object, but is actualised in action". Although this may be looked upon as an insignificant elaboration of Burnet's (1969) view, Tauber's suggestion, that "we abandon attempts to pinpoint and define the self, allowing its definition by expression in its phenomenological address" (1994, p. 295) is actually an important methodological statement as it points the way towards a relational understanding of the immune self.

The emerging self

Within the *res extensa*, the immune system has sometimes been described as a system that recognises itself first and foremost, and only secondary does it recognize significant others (Stewart 1992). This view has been epitomised as "know thyself" (Greek *gnothi seauton*) (Avrameas 1991), the famous words of the Delphic oracle that Socrates set as his philosophical motto. The argument was set forth to make clear that one must know oneself to be wise. But, by implication, this dictum also implies that one must know everything else to know oneself. For how is a self to differentiate itself from other if it does not know what is other and what is not?

Whether or not there is a way out of the ensnarement is a question for further metaimmunological speculations, but in the context of the previous discussion we may point one way forward. This follows the path where Tauber left off, with his advice to address the immune self in its phenomenological expression. However, before so doing it will be necessary to give a brief discussion of Edmund Husserl's phenomenology as it provides a good explanatory contrast to Heidegger's existential phenomenology. While I see Husserl's phenomenology as misguided when it comes to explaining the immune self, Heidegger's phenomenology provides the panacea to the riddle.

While Husserl's treatment of the self is one of the most complicated topics in his phenomenology, it seems clear that he interpreted the self as residing within the subject-object epistemology that has burdened philosophy since the times of Descartes (Lübcke 1982). To Husserl, phenomenology meant the study of phenomena as they appear to the conscious subject. The phenomenologist should, accordingly, encounter the phenomena by "bracketing" all previous theoretical and other contextual knowledge, thus allowing the phenomena to be encountered as they truly are (Dreyfus 1991). His view was that the self is a punctual entity that creates a unity in the individual experiences, and that the conscious subject encounters self-knowledge with certainty precisely when the subject is isolated from the world. To Husserl, as to Descartes, the self was the given that the world could be erected upon.

A central idea of Husserl's phenomenology, and which Heidegger forcefully attacked, is the insistence that the conscious subject, the self, can perform its interpretative and constituting activity independent of the world, as an independent spectator so to say. Heidegger noticed that if the constituting self was bracketed away from the world, as maintained by Husserl, it would loose contact with itself. For how can something constitute the world if it is not itself being constituted? Heidegger's proximal target of criticism thus became Edmund Husserl's understanding of the self's constitution, and he argued, in opposition to Husserl, that the self develops as the organism distinguishes itself from the world. As the organism creates its self, it also creates its world.

Heidegger's main project was to lay out an explication of the self in its various aspects (Dreyfus 1991). His refusal to accept previous philosophers' understanding of the self led him to dismiss their terminology, and he thus invented a new word "Dasein" to emphasise this. To Heidegger (1927, p. 78), "Dasein is an entity which in each case I myself am." This I, the self, is "what maintains itself as something identical throughout changes in its Experiences and ways of behaviour, and which relates itself to this changing multiplicity in so doing." (p. 150). The self can not be found by "tracking down and inspecting a point called the Self" (p. 187), but involves "seizing upon the full disclosedness of Being-in-the-world throughout all the constitutive items which are essential to it" (p. 187). Hence, the self is truly a situated being.

In contrast to the Cartesian self, Heidegger's self is already thrown into the world, and this contingent fact determines the constitution of the self. The true self is, nonetheless, not immediately present to the organism. To the contrary, the self can only reach itself through the world of which it is already a part. Before the self starts its constituting activity, it is already imprinted by the world. Thus, the self is inseparable from the world. In his discussion of the immune self, Francisco Varela (1991) took over Heidegger's interpretation of the self. To Varela, as for Heidegger, the word *environment* denotes a theoretical construction made by scientists, whereas the word *world* denotes the surroundings as seen by the living system, the Umwelt. The living system must perform the contradictory project of distinguishing itself from the environment while at the same time maintaining the coupling. In this project, Varela insisted that there is an

> important distinction between the environment *of* the living system as it appears to an observer and without reference to the autonomous unity – which we shall call hereafter simply the *environment* – and the environment *for* the system which is defined in the same movement that gave rise to its identity and that only exists in that mutual definition – hereinafter the system's *world*. (Varela 1991, p. 85).

Thus, the immune self inhabits a world that is replete with meaning. What is important becomes significant, and what is unimportant goes unnoticed. In addition, when viewed through phenomenological lenses it emerges that empirical investigations that fail to account for the self-reflexivity of immune system activity misses out on a constituting aspect of immunity.

A relation which relates to itself

One problem that scientists have had in singling out the immune self is similar to a major problem in philosophy, that of understanding self-reference. The immune system stands in a communicative relation with the surroundings, and it is in such relations that the problem of self-reference emerges. While most philosophical discourse on the self has presupposed conscious beings, and Heidegger (1995, p. 233) explicitly stated that he reserved "the expression 'self' and selfhood to characterise the *specifically human peculiarity*", there are good reasons to depart from this restriction (Dreyfus 1991).

Heidegger was concerned with the practical dealings of organisms, and although these can be performed consciously, more often they are not. Heidegger's ambition became to describe how the self is constituted within the world, and how these constituting acts define the self. In this he leaned on insights that had been elaborated principally by Hegel and Søren Kierkegaard (1813–1855). In the *Phenomenology of Spirit* Hegel (1807) attempted to show that the self organises the experiences of the subject as it moves through history, and he famously claimed that "the self is the sameness and simplicity that relates itself to itself" (Hegel 1807, p. 12). Hegel's self-consciousness is, accordingly, being shaped by its encounters with other selves, and it is not, as Descartes' self or the organismal and immunological selves, punctual and self-sufficient.

Hegel's thoughts were taken over and expanded by Søren Kierkegaard (1849), who in the opening paragraphs of *The sickness unto death* struggled to capture the conscious self:

> But what is the self? The self is a relation which relates to itself, or that in the relation which is relating to itself. The self is not the relation but the relation's relating to itself. ... Such a relation, which relates to itself, a self, must either have established itself or been established by something else (Kierkegaard 1849, p. 43).

Although the above paragraph is difficult to grasp in its full complexity, the paragraph states that the human self is neither a substance nor an entity, but a relation. The self-interpretation that is being performed is not something external to the self. This relating to itself is fundamental to the self's being. Thus, while Husserl held that all things exist in the same way, Kierkegaard and Heidegger held a special kind of existence for the self. Unlike relations between entities, which like the self-nonself dichotomy are often of the subject-object type, the self is a relation that relates subjectively to itself.

In the same treatise Kierkegaard further claimed that the self is a synthesis of three polarities; the infinite and the finite, the temporal and the eternal, and possibility and necessity. And since the self is a synthesis between these three polarities, the self is always emerging. Whether or not the self develops in a healthy way depends upon whether the polarities are balanced or not. When out of balance, the individual experiences despair, which is "a consumption of the *self*, but an impotent self-consumption not capable of doing what it wants." (Kierkegaard 1849, p. 48). Someone in despair despairs over something, and this something is oneself; what one wants to get rid of is oneself. This is, however, a contradictory project. "He is eating himself up. For that is just what he despairs of doing, that is just what to his torment he cannot do, since with despair a fire takes hold in something that cannot burn, or cannot be burned up – the self." (p. 49). Thus, despair is a contradiction experienced by the self as it is torn between wishes to be both infinite and finite, temporal and eternal, and possibility and necessity.

Kierkegaard's enquiry, which was performed under the influence of Hegel's thinking, was published ten years prior to Darwin's publication of the theory of natural selection. His work was thus an enquiry into the abysses of the mind, and his solutions to problems were found in the Christian God. Yet, by reading Kierkegaard through the lenses of modernity, stripped of his reliance on consciousness and religiosity, his deliberations are strikingly close to what I understand as the immune self. This becomes especially evident if *The sickness unto death* is read within a Darwinian and Heideggerian framework. It then emerges that the relation between the finite and the infinite is analogous to the relation between the individual organism and its world; the relation between eternity and the temporal is analogous to the synthesis of the lineage's evolution with the organism's development; and the relation between possibility and necessity is analogous to the organism's freedom to act according to its inner disposition and the restrictions laid down during the evolution of the lineage. This is why Kierkegaard (1849, p. 59) could state that "The self is freedom. But freedom is the dialectical element in the categories of possibility and necessity".

Like the Kierkegaardian self, the immune self is a many-way relation, in that it relates to itself, to the organism in which it dwells, as well as to the infectious world. The being of the immune system, its self, is standing in relation to something. This "something" includes the immune system's own being as well as everything else that impinges on it[1]. A consequence of this is that entities that make up the immune system, which are the structural bearers of the immune self, can not discriminate precisely between own organism and non-self other. In this sense, the immune self becomes located in-between the subject and the world (Figure 3.2). Within the same Kierkegaardian framework, the despair experienced by the immune system takes the form of autoimmunity; the immune system's eating itself up – the fire within that cannot but still burns.

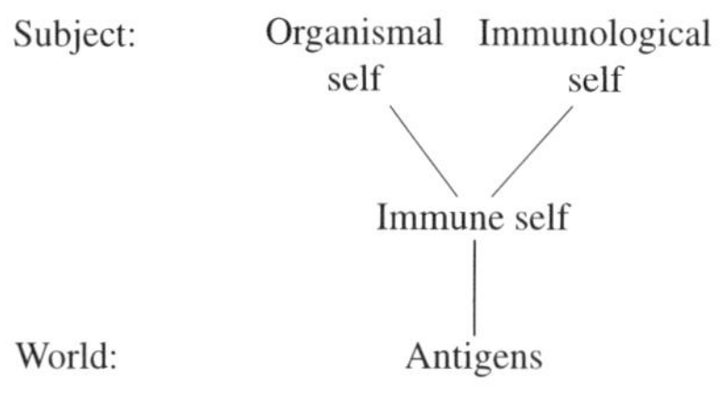

Figure 3.2. The immune self is not primarily a subject that relates to the organism's inner states. It is rather a being that relates itself to existence, and which can only exist insofar as it relates itself to everything else

[1] The near impossibility of performing the task properly becomes clearer when considering that a given number of entities (n) can take n(n-1)/2 numbers of relations.

The know thyself may thus be reframed into a question of whether the immune system is able to make itself into an object for itself or not; is it possible for the immune system to simultaneously be subject and object? I think not. The reason for this is that if the immune system makes itself into an object, if it accomplishes to envision itself as an objective fact, this fact can not include the fact that it is the immune system itself that constitutes the objectified self. The object being grasped must necessarily be what it was in the past, not what it presently is (Sartre 1943); the self can not be caught as a fact because it continuously transcends its past. And even if the constitution of the self is attempted another turn around, the same problem emerges anew; the self eludes objectification because it looses the present as it catches the past.

Attempts at demasking the self reveals masks all the way down. The self who attempts the objectification but who eludes any attempt to be objectified, is the transcendental self. It is invisible to itself because it would have to perform the impossible task of seeing itself through itself to become visible. Failure to identify one's own self is thus not some form of self-deception. It is not an epistemological but an ontological and metaphysical impossibility.

By imposing such a transcendental self we have to accept not only that it is impossible to objectify oneself, to be both participant and spectator at the same time, but also that dualism cannot be overcome, although it can be mediated to a certain extent.

3.2. Signalling behaviour

SIGNS MATTER

Laws and rules

Communication between biological entities, from interacting cells within an organism to interactions of organisms within and between populations, is based on sign operations. Broadly defined, a sign is any signal that is interpreted by a receiving system. Thus, a sign is anything that stands to somebody for something else, which has a meaning, and which is composed or generated by material structures that conform to regularities determined by physical laws. While physical laws restrict sign operations, signalling behaviour is not completely understandable or explainable by the same laws. Signs derive their meaning from functional interactions, and since signalling behaviour is the product of a history of evolution by natural selection, a sign's usefulness is restricted by spatial and temporal contingencies. Hence, a sign is not a universal structure and has no intrinsic meaning outside the context in which its signification evolved.

Signs are selected because of their fitness-related value, and only by virtue of the selected signalling aspects of material structures do the law-determined physical aspects of matter become functional. Or stated differently, while physical laws describe those events that control the life and potentialities of organisms, the process of natural selection allows lineages to increase their control of physically constrained events (Pattee 1995). Thus, experience can be divided into things that leave no alternatives and never change – the physical laws, and things that do change – the rules based on signs. While laws operate according to the principle of determinism, rules operate according to the principle of probabilism. A system that acts according to its inner dispositions, instead of reacting directly to input stimuli, is not determined by laws but chooses according to rules that have been selected. Since signs must have a material embodiment, physical laws and natural selection are complementary models, and both must be invoked when biological events are to be explained.

The naturally selected signs allow organisms to focus on important aspects of their Umwelt. To stay alive, an organism has to rely on its evolutionary selected measuring devices – its sensory organs, and it has to be able to measure something without having to measure everything (Pattee 1995). In an evolutionary context, the "somethings" that are being measured are those entities that have, in some way or other, provided the ancestors of the organism with increased fitness. For example, receptors of the innate immune system recognise specific molecular patterns of infecting agents while ignoring others, and the signs recognised are the ones that have proven useful in an evolutionary timeframe.

To capture the interplay between laws and rules, Peirce worked out a metaphysics that looked upon living entities as guided by triadic relations – between the experiencing organism, the experienced object, and a signifying entity of the object. While he saw efficient causal relations as being dyadic, and therefore incapable of transcending into novelty, he held that triadic relations were capable of transgressing mere necessity. This was a consequence of the ease by which triadic, as opposed to dyadic relations, could be coupled together in networks. Since repeated triadic relations can make up a network of relations, non-deterministic communication becomes possible.

The triadic relations, which make contact with each other through signs, are thought to be integrated within the self, which Peirce envisioned as a decision-maker. In his 1868 paper *Questions concerning certain faculties claimed for man* (Peirce 1992, p. 20) he stated: "In short, *error* appears, and it can be explained only by supposing a *self* which is fallible".

Although Peirce did not state clearly what characterises the self he alluded to, he was well aware of the ontological importance of relational properties. And since he took over the Hegelian belief that logic consists of triadic relations, in which the third is a mediator of a first and a second, his thoughts on the self may be reframed into an existentialist reading. There are therefore certain affinities between Peirce's pragmatism and Heidegger's existential phenomenology. For example, the experiencing system is termed the interpretant in Peirce's philosophy, owing to its relatedness to an interpreter. And since the interpretant refers to the property of being a unifier of impressions (Bergman 2003), it is in several respects closely associated with Heidegger's concept of the self. And like Dasein, the interpretant integrates other-reference and self-reference within the context in which it dwells. This is also the case with the immune self, which thus comes on par with Heidegger's Dasein and Peirce's interpretant.

To denote the behaviour that followed from the use of signs Peirce invented the word semiosis, which he described as "a cooperation of three subjects, such as a sign, its object, and its interpretant, this tri-relative influence not being in any way resolvable into actions between pairs." (Peirce 1998, p. 411). The triadic relation can be represented by a triangle (Figure 3.3 A). The interpretant in the primary triangle can act as a sign for a new interpretant, thus engaging a signalling network consisting of iterated triadic relations. An example of this is seen in Figure 3.3B where the antigen serves as a sign of the bacterium to the B lymphocyte, which again, in the context of danger, provides the immune system with vital information.

Peirce did not restrict semiosis to conscious experience, but instead claimed that "the definition does not require the logical interpretant ... to be a modification of consciousness" (1998, p. 411). Thus, semiosis can occur in any system that is capable of representing a sign. While mental representations of signs are characteristics of the higher animals, more primitive forms of semiosis can occur in all living entities, from single-cell bacteria to multicellular organisms, and even within self-organising systems like the immune system.

This idea can also be exemplified by the process of phagocytosis initiated by the secreted products of B cells, the immunoglobulins, as they encounter infectious agents. Antibodies are Y-shaped molecules that utilise the two upper parts of the Y, the Fab-part, to react with epitopes on a bacterial surface, while it uses the base, the Fc-part, to initiate the effector function. As shown in Figure 3.4, the Fc part of the antibody can bind to an Fc-receptor on the surface of a macrophage. Upon binding to the receptor, the macrophage becomes activated and starts to phagocytose the complex consisting of antibody and bacterium. In this case, the bacterium is the

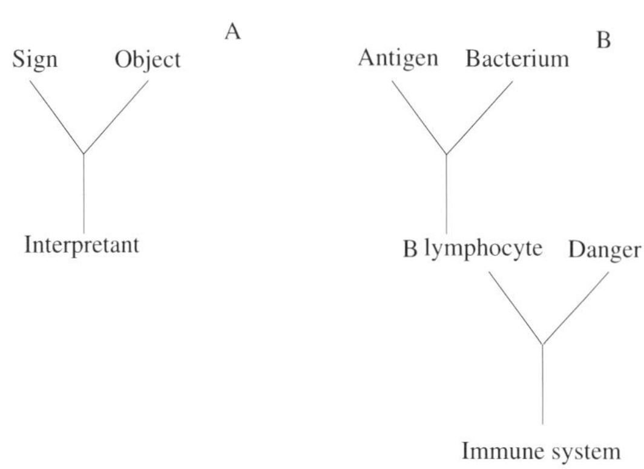

Figure 3.3. A: The semiotic triangle, in which the sign signifies the object to the interpretant. B: The B lymphocyte (interpretant) becomes activated by a bacterium (object) which is represented through the antigen (sign). The B lymphocyte thereafter serves as a sign to the immune system in the context of danger

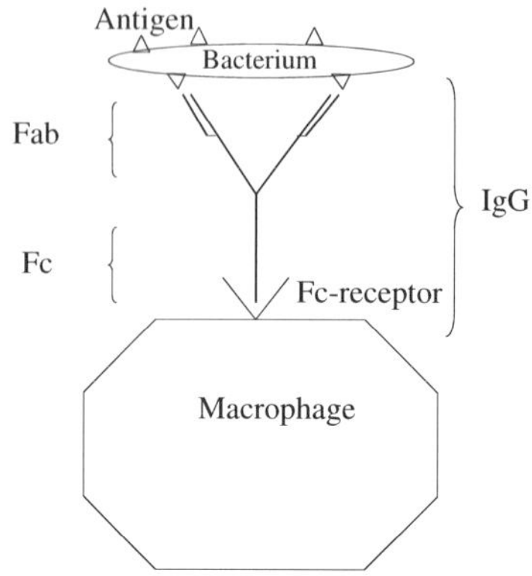

Figure 3.4. The immunoglobulin (IgG) serves as a mediator between the infecting agent, the bacterium, and the phagocytosing cell, the macrophage. The Fab-part (antigenbinding fragment) binds to the antigenic epitopes, while the Fc-part binds to the Fc-receptor on the surface of the macrophage. In semiotic terms, the object is the bacterium, the sign is the antigen, and the interpretant is the macrophage

object, the antigenic epitope the sign, and the macrophage the interpretant. A new triadic relation is thereby initiated, as the macrophage, in turn, releases a variety of signalling molecules which stimulates or inhibits other cells, including B cells, T cells and tissue cells.

Peirce was well aware that signs can be differentiated on the basis of their relations to their objects. He termed signs that are directly connected to the signified object as indexical. For example, an antigenic epitope on the surface of a bacterium is an index. Iconic signs, on the other hand, resemble or imitate the signified. In biological systems this may occur when for example an infecting agent evolves molecular characteristics that mimic the molecular characteristics of the host, so called molecular mimicry. The third mode, the symbolic, consists of signs that are arbitrarily connected to the object signified. Symbols are conventional markers and so their relationship to the object must be learned. While symbols are ever-present in human mentalistic semiosis, their role in non-conscious semiosis has been questioned (Heidegger 1995; Deacon 1997).

Since there is always an element of indeterminacy in semiosis, interpretants of both the conscious and unconscious type have to base their decisions upon insecure information. Hence, scientific as well as immune abductions are prone to errors. For example, since the sign is different from the object and so cannot substitute for the object, it follows that if a sign may be coupled to a variety of different objects, the true object may not be truly revealed by the interpretant unless the context reveals the true relation. Vaccinologists take advantage of this phenomenon when they immunise animals with purified antigens that are being dissolved in adjuvant. In this case, the adjuvant tricks the interpretant, or the immune system, to treat the immunisation as occurring in an infectious context.

Biosemiotics

Throughout man's intellectual history theories of communicative behaviour have focused on conscious interpretative relations between fellow humans. There are, however, some noticeable exceptions to this practice. The Hippocratic tradition emphasised signs of disease and their connection to invisible disease-generating mechanisms occurring within the body. The signs allowed inferences to be made about specific sorts of disturbances underlying the signs. The fact that these inferences were sometimes mistaken, that for example pneumonia was thought to be related to an excessive amount of phlegm instead of to the influence of bacterial infections, is typical of rule-based signalling behaviour.

But conscious humans are not unique in interpreting and acting out on signs of disease. For example, female birds are capable of reading signs of health and disease on males in their vicinity, and they use this instinctual knowledge to select healthy uninfected males for mates (Hamilton and Zuk 1982). Signalling behaviour is also characteristic of interactions between cells, for example between immune cells and their infectious world. Signalling behaviour amongst cells is performed through the use of signal transduction mechanisms. These may come in a multitude of forms, but they basically share a similar mechanistic logic. A molecular message, or ligand, binds to a receiver molecule, the receptor. The binding of ligand causes the receptor to change shape, or to cluster with other receptors. This sets off a cascade of interactions amongst molecules inside the cell, which either results in the activation/deactivation of enzymes, or the switching on/off of transcription factors. The activity of the enzymes serves to alter the molecules within the cells, while transcription factors regulate the expression of genes. It is the enzymes and transcription factors that serve to alter the behaviour of the cell in response to the signal initiated by the ligand.

Semiosis thus permeates all aspects of life, and in a sense life can be equated with the capability to perform semiotic interactions. To capture this notion, Jesper Hoffmeyer (1996) claimed that life is located within the semiosphere, the world of signification and meaning that encompasses life at all levels, and into which "every organism on Earth is tossed at birth" (Hoffmeyer 1996, p. vii). Semiosis that occurs through time, for example during parent-offspring signalling, has been termed vertical semiosis, while horizontal semiosis designates exchange of signs between biological entities within a spatial ecosystem. Horizontal semiosis can be divided into endosemiosis, which denotes the exchange of signs between cells and tissues within the organism, and exosemiosis, which designates signalling between organisms (Hoffmeyer 1996).

The idea that the organism has a perspective on the world and that signs play an important role in channelling decisions made by the organism, has been termed biosemiotics (Emmeche 1991; Hoffmeyer 1996; Kull 1999; Sharov 1999). Biosemiotics, which is derived from the Greek word for sign, *semeion*, literally means the study of living systems from a sign-theoretical perspective. As claimed by Hoffmeyer (2005, p. 425), biosemiotics "substitutes an outdated metaphysics, according to which life is nothing-but-chemistry and molecules, with a broader and more up-to-date philosophy of life in which the sign rather than the molecule is placed as the basic constituent of living beings."

Biosemiotics is founded on theoretical constructions erected by Charles Sanders Peirce and Jakob von Uexküll. While Peirce provided the terminology and semiotic insights, von Uexküll provided the insight that animals experience and respond individually to stimuli within their Umwelt (Kull 1999). Although Peirce was well acquainted with Darwin's ideas, he had a only a rudimentary understanding of the new and expanding biological fields of embryology, bacteriology and immunology (Braga 1999). And similarly, Uexküll had only a rudimentary understanding of semiotics (Kull 1999). So even though the necessary background knowledge had been available during almost the entire 20th century, biosemiotics was not recognised as a scientific field of its own until the 1990s.

Sign-phenomenology

The acceptance of biosemiosis leads us into seeing the biological world as an unruly activity that functions through decision processes made upon a contextual foundation. Or stated differently, the activity of biological systems appears to be non-deterministic and therefore unpredictable because every action "that consists of perception and operation imprints its meaning on the meaningless object and thereby makes it into a subject-related meaning-carrier in the respective Umwelt (subjective universe)" (Uexküll 1982, p. 31).

The entities encountered by biological recognition systems are not fully describable by the properties they exhibit; they are also influenced by characteristics of the recognition system. Entities have several possible ways to be, and these possibilities are revealed in different manners by different recognition systems. So when pollen becomes recognised as an allergen by one immune system and as an innocent molecule by another, or when the same streptococcus induces rheumatic heart disease in one patient but merely a sore throat in another, it needs not be something peculiar to the recognised entities that induce disease. Rather, the examples testify to the importance of studying entities as they appear to the recognition system during interactive encounters.

Semiosis is a phenomenon that can be studied by an appropriate phenomenological method. Husserl's phenomenology is obviously inadequate since it is too tightly connected to the subject-object epistemology, but also because Husserl held signs to function in causal relations (Dreyfus 1991). Heidegger's phenomenology is, in contrast, well adapted for semiotic studies. Heidegger held that signs not only point to other objects in the interpretative situation, they also function as signs because they point out the context of the situation. "Signs always indicate primarily "wherein" one lives, where one's concern dwells, what sort of involvement there is with something." (Heidegger 1927, p. 111). Individual signs do not signal in isolation from other signs, and the message conveyed by a signalling molecule is to a large degree influenced by the context in which it is expressed. Signs are thus important because they allow organisms to give flexible responses to changing environments.

To perform as intended, signs need to be reliable. But as any honest signaller knows, cheaters sometimes take advantage of honest signs and utilise them for their own advantage, without ever paying the costs that necessarily follow from being included in a signalling network. Infectious agents are masters in disguising within the organism and of confusing the internal signs to their own advantage. This is mainly owing to their having a short generation time and thereby a potential for rapid evolution. The evolution of the adaptive immune system within the jawed vertebrates can be seen as a highly successful countermeasure against this microbial strategy. However, as so often happens during coevolutionary interactions, the infectious agents have evolved counter-counterstrategies against the vertebrate counter-strategies, and several infectious agents have even evolved the capability to utilise the signalling network of the immune system itself to infect and mediate disease in their host organism.

The adaptive adjustments that take place between the microbes and the immune system are not driven by purely mechanical adjustments. As any other practical being in the world, the immune system responds concernfully to the signs encountered. The decisions are based on two sorts of information, the proximate signs and learned predispositions. The proximate signs are the ones that directly trigger receptors while the predisposition consists of previous knowledge that the immune system has accumulated, partly during the evolutionary process and partly during the individual's own lifetime.

The decision process can be given a formal representation by treating the proximate signs and predispositions as chance-occurrences, thus allowing us to estimate the plausibility that a given sign is derived from for example nonself. If we let the probability that nonself is encountered given a certain sign be $\mathrm{Pr}(\mathrm{nonself}|\mathrm{sign})$, the probability of encountering a sign if nonself is instantiated be $\mathrm{Pr}(\mathrm{sign}|\mathrm{nonself})$, the chances of encountering the sign in the environment be $\mathrm{Pr}(\mathrm{sign})$, and the prior probability of encountering the nonself be $\mathrm{Pr}(\mathrm{nonself})$, the relation between the chance occurrences can be defined according to Bayes' theorem as follows:

$$\mathrm{Pr}(\mathrm{nonself}|\mathrm{sign}) = \mathrm{Pr}(\mathrm{sign}|\mathrm{nonself})\,\mathrm{Pr}(\mathrm{nonself})/\,\mathrm{Pr}(\mathrm{sign})$$

The conditional probability $\mathrm{Pr}(\mathrm{nonself}|\mathrm{sign})$ is the posterior probability of nonself given the observed sign, and gives an estimate of how certain we can be that we observe nonself. The conditional probability $\mathrm{Pr}(\mathrm{sign}|\mathrm{nonself})$ is the likelihood of nonself, and answers the question: how probable is it that sign will be encountered if nonself is present? If the likelihood is small, it is improbable that the sign will be observed if nonself is encountered, whereas a high likelihood indicates that it is very probable that the sign will be encountered if nonself is present.

Predispositions to react in a certain manner enter the immune response through several routes, the most important being through developmental, environmental and genetic paths. The predisposition is baked into Bayes' theorem as the prior probability of encountering the infectious agent, Pr(nonself); the higher Pr(nonself) the stronger the predisposition. If Pr(nonself) is high enough, one would expect the organisms that risk infection to evolve receptors that directly target the nonself, whereas a low Pr(nonself) should likely be met with an induced or adaptively learned response.

The posterior probability can be easily calculated if the values on the right hand side of Bayes' formula are known. However, in real life situations this is seldom the case, and the probabilistic values may sometimes be hard to obtain. So instead of making decisions based upon posterior probabilities, decision-makers sometimes take a more pragmatic route. Instead of measuring the posterior probability of a hypothesis, they compare the plausibility of two likely hypotheses, and then choose the most likely. If one would like to know whether one hypothesis is more plausible than another hypotheses, for example whether the hypotheses Pr(nonself|sign) is more plausible than hypothesis Pr(self|sign), instantiation of the following inequality suggests the answer:

$$\Pr(\text{nonself}|\text{sign}) > \Pr(\text{self}|\text{sign})$$

This inequality can be rearranged to:

$$\Pr(\text{sign}|\text{nonself})\Pr(\text{nonself})/\Pr(\text{sign}) > \Pr(\text{sign}|\text{self})\Pr(\text{self})/\Pr(\text{sign})$$

Since the denominators in the two expressions are the same, we find that the hypotheses Pr(nonself|sign) is more plausible than hypothesis Pr(self|sign) precisely when:

$$\Pr(\text{sign}|\text{nonself})/\Pr(\text{sign}|\text{self}) > \Pr(\text{self})/\Pr(\text{nonself})$$

This inequality demonstrates that the comparison of two hypotheses involves two factors, the hypotheses' likelihood and their prior probabilities. The prior probabilities can not always be given precise interpretations, but in some biological situations they can be depicted as being fairly constant within certain spatiotemporal limits. The probability of encountering an epidemic infectious agent, like influenza, can for example be given a fairly precise probabilistic interpretation during epidemic outbreaks. If we, for our comparative purpose, accept that the prior probabilities Pr(nonself) and Pr(self) are relatively constant, the plausibility of the two hypotheses reduces to a relation between the two likelihoods.

A sign is something that guides the behavioural response of a biological system. By comparing the likelihood that a given sign is elicited by nonself versus it being elicited by self, a decision-maker can acquire information that enables him to respond in an adaptive manner. In the long run, the decision that is based upon the highest likelihood also has the highest fitness. The differential value of the two likelihoods can be calculated as the likelihood ratio (LR), defined as:

$$\text{LR} = \Pr(\text{sign}|\text{nonself})/\Pr(\text{sign}|\text{self})$$

In nature, decision making is often envisioned as a choice between alternatives. For example, biological fitness is a relative concept – what matters is whether an organism survives or reproduces better than its competitors. In a very real sense, the properties of a system are not defined as much by the system's characteristics as by the decisions it makes. Very seldom are scientific inferences or biological decisions uniquely determined by the observations. This is because observations are contextually embedded – the same observation may have a different meaning in a different context. Within the immune system, a given sign may be interpreted as self in one context and nonself in another.

Although the same probabilistic methodology is applied to explain decision-making by the scientific mind and the immune system, there is a marked difference between the types of "knowledge" that the scientific mind and the immune system strive to achieve. While the scientific mind efforts to reach theoretical knowledge that is true irrespective of time and place, the immune system is selected by evolution to be practically useful. Whether or not the decisions it makes are true in the theoretical meaning is non-essential, what matters is the practical usefulness of the "knowledge". In nature, truth is measured as survival and reproduction, whereas truth in science denotes correspondence with the world. In real situations it may sometimes pay to act on a false representation if this acting increases fitness relative to acting on a true representation of nature.

IMMUNOSEMIOSIS

Agents that have a perspective, that are related subjectively to the world encountered during their everyday coping activity, are situated in their world. The situatedness of the immune system and its self becomes evident when attempts are made to identify the entities that make up the non-nonself. In logic, the statement *A or non-A* identifies two non-inclusive groups of entities, and so the *non-non-A*, which is the negation of *non-A*, is identical with *A*. In immunology, however, the non-nonself is definitively not identical with the self. Hence, unlike *A* and *non-A*, self and nonself do not belong to exclusive categories. So when Jan Klein (1982) defined the nonself as "all the rest", he did not intend bees, cars or mountains, but implicitly alluded to those parts of the world that the immune system had been evolutionary selected to cope with. Hence, "all the rest" is defined in terms of the perspective taken by the self.

That someone has a perspective implies that something in the environment reverts to the background and something else becomes sign. But this does not have the further implication that the background is made into something like the non-A. The background can not consist of nothing; there is something there. This something can, for the situated observer, be termed nothingness owing to the fact that he does not directly recognize it. But in stark contrast to nothing, the nothingness of the non-self may be as important for the response as the sign is. The importance of nothingness for human coping activity has been forcefully argued by

Sartre (1943), and immunologists have recently begun to think along the same lines when explaining malfunctions of the immune response. In the context of immune reactivity nothingness is sometimes termed adjuvancy to indicate this active role. For example, the hygiene hypothesis, which explains allergy as being the consequence of a lack of infectious stimuli on the developing immune system, implicitly acknowledges the importance of nothingness as a major explanatory element.

When the immune system's interaction with the environment is seen in terms of the internal structure of the system, it becomes clear that environmental perturbations do not prescribe or determine changes in the system. Rather, the reaction induced in the system by the environment is guided by the internal structure of the system itself. It is the structure of the system as well as its previous history of perturbations that determines what reactions the new perturbation will induce. The immune system's perspective has been evolutionary shaped and is being developmentally defined through encounters with the antigenic world. It is made up according to what is significant within the environment and what is not, and environmental entities that make no fitness-related differences therefore do not become part of the immune system's world. Thus, the immune system, like any other discriminating agent, has a perspective of the world that is in a certain sense biased.

The immune system discriminates between the environment and the world by utilising signs. The signs employed are those entities that have been evolutionary serviceable and therefore have received a fitness-based value. The immune system either recognises these signs through innate receptors or through receptors that are adaptively shaped. So in a sense, the immune system brings forth its world by establishing a coupling to the significant entities encountered in the environment, or as more generally claimed by Peirce (1998, p. 305): "whenever one thing acts upon another it determines in that other a quality that would not otherwise have been there."

A major immunological task is that of understanding how the immune system represents the world. To do so would necessitate a scientific vocabulary that is appropriate to the task. While immunologists seem to accept semiosis at the ontological level, that the immune system is concerned with signalling behaviour, many immunologists are nonetheless reluctant to accept that semiotic language can provide much of interest to immunology at the epistemological level. This was, at least, a major outcome of a congregation of biologists, immunologists and semioticians who discussed *The semiotics of cellular communication in the immune system* (Sercarz et al. 1988).

There seemed to be an agreement amongst some of the participants that semiotics has little to offer immunology. The semiotician Umberto Eco (1988, p. 3) claimed that "I feel still unable to say whether semiotics can help immunology, but I discovered that immunology can help semiotics". And the physician Giergio Prodi (1988, p. 63) argued that "whether the concepts of semiotics can help the immunologist in his experimental work is, in my opinion, doubtful." The immunologist Edward Golub (1988) accepted the fact that immune system activity is semiosis, but argued, like Eco, that semiotics had little to offer immunology.

There were, however, also differences of opinion. The immunologist Tomio Tada (1988, p. 85), upon criticising the lack of rigour that many concepts of cellular communication are bestowed with, believed that "semiotics may provide important clues to correct and restate what we have been abused to use in describing immunocyte interactions." In his presentation he distinguished the immune response from cause-effect relations that occur in the responses of muscle fibres to electric signals, and so made clear that interactions between lymphocytes and antigen presenting cells could well be restated in semiotic terms. He thus defined immunosemiotics as "the study of general principles underlying the structure of sign systems perceived by different cells within the immune machinery" (Tada 1988, p. 85).

The perspectives offered during the discussions and presentations apparently thrilled the participants so much that the organisers were tempted to declare their delight and their looking "forward to the next episode in this sage." (Sercarz et al. 1988, p. viii). Still, as the tempting thrill tempered, immunologists showed little interest in continuing the sage. Although it is quite true that signalling has taken centre stage in contemporary immunology and that immunologists now conceptualise the immune system as a signalling system (Dustin and Chan 2000; Sen 2001; Sambrano et al. 2002), the signalling behaviour is not conceptualised in semiotic terms. This is probably owing to the fact that immunologists, who mostly perform their studies at molecular and cellular levels, already have a language that serves their science well. Given this serviceability, it is doubtful whether it is of much interest for them to translate their vocabulary into the semiotician's vocabulary without this translation providing added value to immunology.

Like many molecular biologists, immunologists are somewhat reluctant to expand their notion of causality to include formal, material and teleological causality. However, the fact that the immune system is situated within its world implies that immune reactivity can not be studied simply as an organ that operates according to stimulus-response patterns. The appropriate framework for studying situated agents is semiotic, in which the ecological context, the evolutionary and developmental history, and the perspective of the interpretant are considered. By taking a biosemiotic perspective, immunologists would probably achieve a better understanding of the system properties of the immune system, and it would also facilitate the integration of endo- with exoimmunosemiosis, by linking them within a unified explanatory framework.

The latter aspect, that of integrating endo- with exoimmunosemiosis, has gained an unexpected impetus from recent studies of microbial semiosis. It has now become apparent that bacteria have evolved intricate signalling networks that allow co-operation between individual cells, thus making the cells capable of organising into structured colonies with elevated environmental adaptability. To perform these feats bacteria utilise signalling networks that allow them to sense environmental cues, exchange chemical information, and thus make collective decisions. Such meaning-based communication is set in association with a sort of social intelligence (Ben Jacob et al. 2004).

Owing to the pathogenic potential of microbial social signalling, one would expect on theoretical grounds that organisms of various lineages would evolve countermeasures that served to inactivate or interfere with the microbial signalling. This expectation was recently demonstrated when epithelial cells were shown to have the capability to inactivate and exploit specific forms of bacterial semiosis (Chun et al. 2004).

THE RELIABILITY OF SIGNS

Selection would favour organisms that made adaptive decisions based on reliable signs. The reliability of signs is, however, seldom of an all or none type. Signs can be entirely truthful or entirely deceptive, but may also be located anywhere along a continuum between these two extremes. In many situations there will be conflicts of interest between signaller and receiver, and dishonest signs may thus evolve if they increase the fitness of its bearer. When for example infectious agents evolve molecular structures that mimic molecules of the host, the molecular mimicry is an example of dishonest signalling that serves to increase the fitness of the infectious agent.

The idea that life is semiosis explains why life is open to misinterpretations. In order to reduce the chances of misinterpretations, an important aspect of honesty involved in biological signalling has evolved. Honest signs are signs that come with a fitness-related cost, thus making them less susceptible for being exploited by cheating. Honest signals are costly precisely because this ensures their honesty. The idea of honest signals can be invoked at any level of the biological hierarchy, from cells to organisms and their groups, and some theorists have even set honesty as the foundational principle for signalling behaviour (Zahavi and Zahavi 1997). Other theorists have disputed the generality of the principle, but recognise its importance in selected cases (LaPorte 2002).

In 1892 Metchnikoff postulated that strong and healthy somatic cells would repel attack by phagocytes by secreting inhibiting substances. Sick or weakened cells would, in contrast, be unable to produce the same substances, thus leaving them susceptible for phagocytosis. In more recent years studies of the cells of developing organisms have demonstrated that the principles of honest signalling, which were alluded to by Metchnikoff, seem to be an important aspect of somatic selection. Since a signal has to be costly to be a reliable indicator of self-quality, cheaters that cannot afford to pay the costs of producing the signal are being revealed and eliminated. Thus, a low quality cell, which attempts to produce a signal indicating high quality, cannot do so either because it does not have the resources, or if it does produce the costly signal it is penalised by a large reduction in fitness. This reduction in fitness is much larger in low quality cells than the one that occurs in high quality cells. This idea has, for example, been used to model the development of stable phenotypes in which deficient or malignant cells are selected against (Krakauer and Pagel 1996), and costly signs have also been applied to explain signalling amongst cells of the immune system (McKean and Zuk 1995), much like the theory has been utilised to explain how animals coordinate their social behaviour, for example during sexual selection (Folstad and Karter 1992).

3.3. Signal, decision, response

TUNING THE THRESHOLD

A peculiar aspect of semiosis within the immune system is that signs exchanged between B cells, T cells and antigen-presenting cells take place during short-lived interactions that last from minutes to hours. In contrast, the contact-dependent exchanges of signs between epithelial and neuronal cells, which result from close membrane to membrane appositions, are stable for years. Short-lived and sequential interactions allow for a cumulative and sequential tuning of the various signs exchanged. But although such interactions allow for flexibility of responses, they also increase the risk that many of the sign-operations may turn into non-functional or even malfunctional responses.

Activation of lymphocytes involves simultaneous signalling through a multitude of ligands and receptors. Two signals are essential for specific activation of the individual lymphocyte. The first is the antigen-specific signal that is transmitted via the antigen receptor, and the second is the costimulatory signal. Costimulatory signals initiate a positive response in the responding lymphocyte independent of the signal conveyed by the antigen receptor (Frauwirth and Thompson 2002). Much research has been performed to understand the complex intracellular pathways that regulate the behaviour of lymphocytes upon stimulation. Although much work remains to be done, the results of the studies are compatible with the view that context is of essence for the specificity of the signalling response (Downward 2001).

An example of this occurs during the interaction between the T cell and the MHC-peptide complex. It has been demonstrated that T cells exhibit different sensitivity thresholds for activation at different points in the developmental cycle. By being able to tune the response in accordance with developmentally shaped thresholds, the integrity of the organism can be regulated also at the level of the sign and receptor. Such tuning minimises the risk of overreaction, which can lead to autoimmune disease, and underreaction, which can lead to increased susceptibility to infection (Dustin and Chan 2000).

In their "tunable activation threshold hypothesis", Grossman and Paul (2001) purport to explain why lymphocytes with degenerate antigen receptors are tolerant to self-antigens while still being reactive towards exogenous antigen. The key elements of their hypothesis are that lymphocyte activation is a threshold phenomenon, and that the threshold varies according to the stimulatory experience of the cell. Physiological signals received by the cell are not only effective because of the chemical make up of the signalling molecules. In addition, the intracellular signals of multiple interactions become integrated into excitation and de-excitation response patterns, which then results in metabolic perturbations in the cell. Excitation denotes the biochemical changes that cause gene activation, while de-excitation consists of changes that reverse or negate the effects of excitation. The cell becomes "tuned" to its environment according to the relative levels of excitation and de-excitation factors, and it responds fully only when a sufficiently rapid increase in excitation factors occurs, for example owing to disease caused by infectious agents. Hence, the same lymphocyte can be induced to express alternative responses, depending upon its activation threshold.

In addition to giving an integrated explanation of the accumulating data concerning lymphocyte activation, the tunable activation threshold hypothesis serves to modify the bearings of the clonal selection theory. Selection can no longer be envisioned as an automatic process. Rather, it is dependent upon the microenvironment of the responding lymphocyte, its previous history with the same or a similar antigen, and whether or not antigen and receptor bind to each other with a sufficient affinity and precision so that an effective defence response can be initiated.

An important facet of the tunable activation threshold hypothesis is its focus on the integration within a single cell of various environmental signs. The stimuli, which are delivered through receptors for antigens, cytokines, chemokines, and adhesion molecules, inform the cell about the physiological context within which the sign is encountered. The receptors may thus be thought of as individual diagnostic devices that inform the cell about aspects of the microenvironment.

THE SIGNAL DETECTION FRAMEWORK

The bearings of the tunable activation threshold hypothesis can be analysed on a more general basis by the use of signal detection theory, a wide-ranging method that is useful to investigate the adaptive value of different response strategies in complex situations (Swets 1996). Signal detection theory decomposes the responses of a cell, organism or system into two independent processes – one of decision and one of discrimination. Although the two processes are tightly integrated, the theoretical framework of signal detection theory allows them to be analysed and evaluated independently. A *decision* process determines how strong the evidence inherent in the sign must be in order to make a stimulus-adapted appropriate response, whereas a *discrimination* process assesses the degree to which the evidence in the observed sign favours the existence of a stimulus of category S1 relative to S2. S1 and S2, which signify two different states of the world, may exhibit distinct or overlapping characteristics.

It may seem peculiar that the signal detection framework so strongly emphasises the *response* that follows after discrimination and decision. The reason for this choice is purely practical. Since no biological recognition system is perfect, some propensity for error is inevitable, and the signs are therefore seldom capable of providing certain information about the state of the world. In this setting, the establishment of a biological response is the only way to mark a true recognition. A signal that does not elicit a response is thus not a sign.

Signal detection theory has incorporated uncertainty into its main body by looking upon diagnostic evidence as if it represents the state of the world in a probabilistic manner. In situations where there are only two states of the world and where the responding system has access to signs that truly categorises these two states, the system has little need for a complex apparatus to discriminate and decide. However, in cases where the signs are probabilistically related to one or the other state of the world, the sign becomes a fallible indicator of the worldly state. In even more complex cases, where the world may take on several different states and of which there are no definitive signs that allow the system to truly discriminate between them, the tasks imposed upon the responding system becomes formidable. In such scenarios the systems that are capable of mounting a plastic and stimulus-adapted response would have the highest fitness. Signal detection theory is a useful device to describe and model such plastic responses.

To understand the deliberations involved in determining the optimal response threshold, it is useful to start simple, in a world already dichotomised as S1 and S2. Let the stimulus be the presence (S1) or absence (S2) of an infectious agent, and the response be the initiation (R1) or the absence (R2) of an immune reaction. There are four possible stimulus-response outcomes. If an infectious microbe has invaded the organism, and the organism responds to microbial signs, this is a true positive reaction. If the organism does not respond, it is a false negative response. Similarly, if the organism responds in the absence of an infecting agent, this is a false positive reaction. If the organism, in the absence of an infecting agent, gives no response, this is a true negative response (Table 3.2).

Let us imagine the immune system as a measuring device that delivers test results to the organism. The signs encountered by the immune system are embedded within a context shaped by the innate immune system, and so the adaptive immune response becomes tailored to the accumulated evidence that reaches the lymphocytes. In this setting, the sign X may be thought of as the concentration of an infectious agent or its metabolic products. Infection ensures a high concentration of the sign, while low or undetectable concentrations occur without infection. When the sign occurs at intermediary levels it may, however, be especially problematic to interpret and to use as a predictive indicator.

In such problematic cases, the immune system can be modelled as if it was a signal detection device. The sign X can take on a variety of values, some of which are more likely to occur in S1 than in S2 (Figure 3.5). For some intermediate values

Table 3.2. The two-by-two table of environmental state and response. The payoffs are depicted as costs (c), which are associated with wrong responses, and benefits (b), which are associated with correct responses for the different states of the world

		State	
		S1	S2
Response	R1	True positive (b1)	False positive (c2)
	R2	False negative (c1)	True negative (b2)

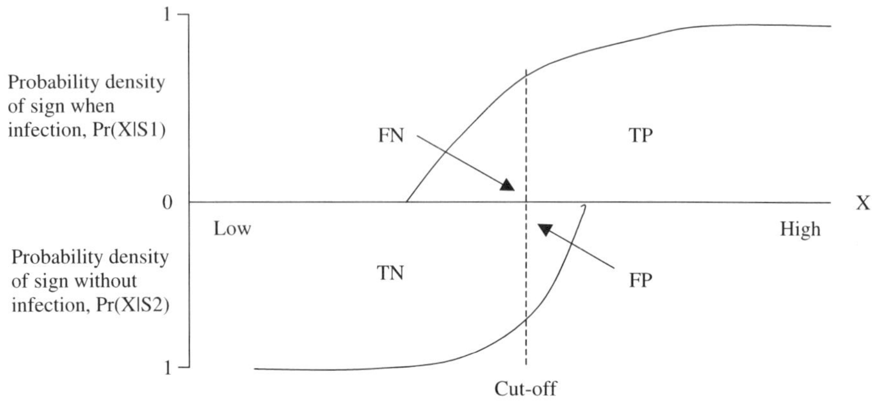

Figure 3.5. Probability density of test result (X) in organisms with (S1) and without (S2) infection. True positive (TP), true negative (TN), false positive (FP) and false negative (FN) responses are indicated. When the two likelihood functions have a large area of overlap, the immune system receives ambiguous inputs concerning the state of the world, while the system receives input of higher quality when the areas have a small overlap. Increasing specificity of receptors on the lymphoid cells, or combined utilisation of innate and adaptive receptors, can be thought of as a means by which the organism reduces the overlapping areas

of X, the sign does not discriminate in a reliable manner between the two states of the world, and so the immune system has to decide whether to respond or not based upon uncertain information. In order to make a stimulus-adapted immune response, the cells of the immune system have to establish a decision threshold that discerns, in a probabilistic manner, between the two states of the world.

The optimal cut-off depends upon the prior probability of occurrence of the event in question, as well as upon the costs and benefits involved in the decision. The prior probability that the world is in S1 can be depicted as P and the probability of S2 as $(1-P)$. The prior probabilities can be though of as response constraints that have been learned through the evolutionary experience of the lineage, during which the organisms that exhibited maladapted prior probabilities became selected against. The costs and benefits involved in making maladaptive or adaptive responses can be given fitness-based interpretations, and costs may be measured as damage, while benefit may be measured as avoidance of damage. As demonstrated by Swets (1996), the expected payoff to the decision-maker is optimised when a decision-criterion is chosen such that

$$\frac{Pr(x|S2)}{Pr(x|S1)} > \frac{P}{1-P} \cdot \frac{b1-c1}{b2-c2}$$

The term on the left hand side is the likelihood ratio, $Pr(x|S2)/Pr(x|S1)$, and measures the likelihood that a given value of x will be detected in S2 relative to the likelihood of the same value of x occurring in S1. The decision threshold, or cut-off, is a value of X that satisfies this inequality. The fraction on the right hand side is the ratio of the expected importance of the two states of the world. Expected importance is defined as prior probability multiplied by the sum of the costs and benefits. The formula tells us that location of the threshold is dependent on the relationship between the expected importance of the two states of the world. When the expected importance of S1 increases, the inequality can be satisfied by increasing Pr(x/S2) and reducing Pr(x/S1). This may be accomplished by moving the threshold to the left in Figure 3.5.

By lowering the response threshold to reduce the amount of false negatives, the system risks to generate an increasing amount of false positive responses. This is not always tolerable to the system, as it may lead to increasing risks of generating immunopathological or autoimmune diseases. An alternative strategy that would reduce the risk of disease would be to increase the discriminatory capabilities of the measuring device, or to alter the effector mechanisms that are coupled to the response. This is precisely the strategy that is utilised by the immune system during encounters with infectious agents. As the immune response develops, the specificity of the responding T and B lymphocytes increase markedly. This is demonstrated in immunisation procedures, in which the earliest antibodies that are being generated are of lower specificity and affinity than are the antibodies generated during the later phase. In addition, the immunoglobulins generated during the response change from IgM to IgG, IgA or IgE, which have other effector properties than IgM.

DISCRIMINATION

The process of discrimination depends on the capacity of the diagnostic system to differentiate between two stimuli, and is therefore a measure of the system's specificity. For example, a lymphocyte with antigen receptors that differentiates between two related substances to a higher degree than another lymphocyte receptor has a higher specificity and hence also a higher discriminatory capacity. The discriminatory capacity is an inherent characteristic of the diagnostic system, and is not like the cut-off decision dependent on prior probabilities of events or on the costs or benefits involved.

The discriminatory capacity of the immune system can be modelled by comparing it to the discriminatory capacity of a medical test. When for example a physician speculates as to whether a patient has an infectious disease or not, he may submit the patient's serum to a test which has some operating characteristics that are set independently of the serum being tested. The most important operating characteristic of the diagnostic test is its capability to reliably detect the disease in question. This ability is a consequence of the test's capability to deliver a high fraction of true positive results and a low fraction of false positive results.

When a diagnostician decides to validate the operating characteristics of a diagnostic test, he first collects a sample of sera from patients with and without the disease in question. The sera are then submitted for testing, and the results of the test are then displayed graphically and statistically in order to evaluate the test's performance characteristics (Ulvestad et al. 2000). The results of this procedure may turn out as the results depicted in panels A and B of Figure 3.6. The figure gives the results of a diagnostic test that was elaborated and adjusted to discriminate between patients with and without a certain infectious disease. In the test the cut-off was set at 50 units. But as the figure shows, several healthy individuals score higher than 50 units. These individuals are thus considered to be false positive for the disease in question. In the patient group, the true positives are those that score above 50 units. If we lower the threshold to 40 units, the frequency of false positives increases markedly while the true positives increase only marginally. Hence, the cut-off appears to be set at a nearly optimal level.

The information provided in panel A can be converted to probabilistic values and given a statistical interpretation as the area under the curve (AUC) in a receiver operating characteristic (ROC) plot (Figure 3.6 B). When the false positive proportion is plotted against the true positive proportion for all values of X, the line

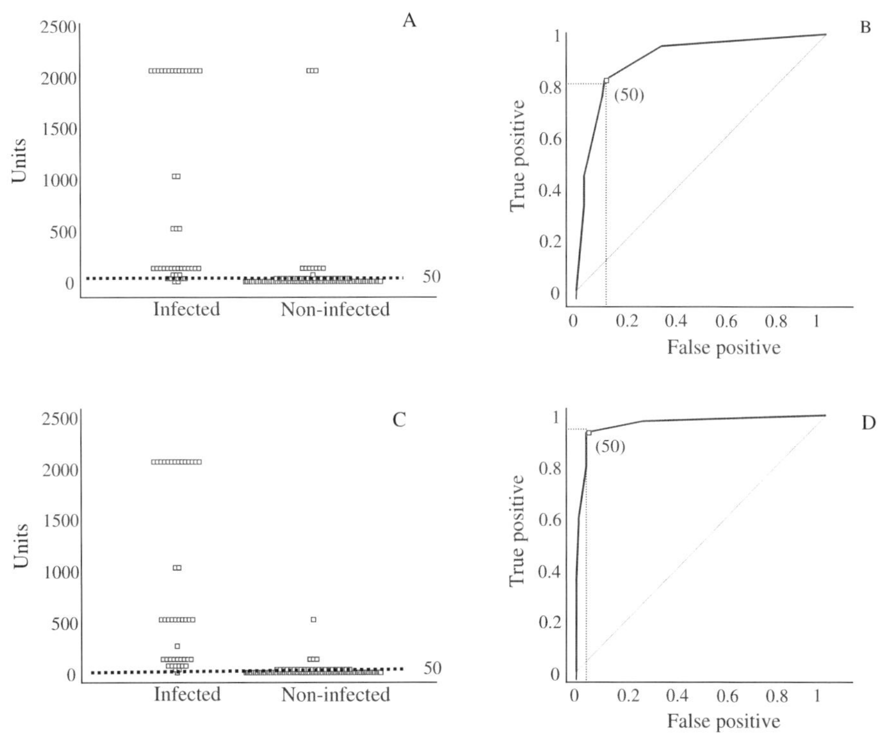

Figure 3.6. Two diagnostic tests were applied to sera from patients with and without a specific infectious disease. The concentrations of antibodies in the sera were measured by an old (A) and a new (C) test. The cut-off was set at a concentration of 50 units. Receiver operating characteristics (ROC) curves for two measuring devices are shown in B for the old test and D for the new test. The ROC curves display true positive results compared to false positive results at various cut-off levels. The fractions of true and false positives indicated at the cut-off value 50 are indicated. The true and false positives for all other cut-offs can be similarly read from the curves

drawn through the intercepting points gives the curve from which the AUC can be calculated. As demonstrated in Figure 3.6 B, the cut-off point of 50 gives a true positive frequency of 0.82 and a false positive frequency of 0.11. The likelihood ratio of the test at the given cut-off is thus $0.82/0.11 = 7.25$.

Two measuring devices that measure the same quantity may perform differently. The ROC plot is ideal to compare the results from two different diagnostic tests. Figure 3.6C displays the results from a new test that measures the same disease parameter as the test in Figure 3.6A. At cut-off 50 the new test gives a true positive frequency of 0.93 and a false positive frequency of 0.04. The likelihood ratio of the test at the given cut-off is thus $0.93/0.04 = 22.6$. The area under the ROC curve for the old test (panel B) is 0.903, while the area under the curve for the new test (panel D) is 0.967. If we envision that the optimal cut-off is 50 units, the results are compatible with a hypothesis stating that the new test is a better discriminatory test than the old test. The new test has a higher likelihood ratio at the given cut-off and also has a larger area under the curve when all the cut-off values are taken into consideration.

The described model can incorporate the responses of several different receptors. While one test device may deliver unreliable information, two independent test devices may be better, and so on for three and more test devices (Ulvestad 2003).

In an analogous manner, by utilising several tests to diagnose the environment before responding, a system may shape the reliability properties of the signs to which it responds. Thus, multiple testing enhances discrimination. In addition, because it takes into consideration a variety of signs, the context within which the sign is embedded is taken care of. The shaping of sign reliabilities by multiple testing ensures that the AUC increases since the true positive proportion increases while the false positive proportion decreases. Since there are costs involved in establishing and maintaining novel recognition systems, there will be diminishing returns to the organism of additional systems. Accordingly, the efforts at improving the detection systems will not be worth the energetic candle after a certain point.

CHAPTER 4

Natura naturans

A thorough understanding of vertebrate immunity requires knowledge of why the immune system responds to some but not all non-self that it encounters. To get a grip on this complexity it is useful to have an idea of how life originated, what the first entities endowed with life were, and how the variegated life forms that now populate the Earth came to be.

That there is life at all, and that it has arisen out of mere matter, calls for an explanation in terms of natural processes. It has not been possible to get reliable data on how life originated, but we can with a high degree of confidence claim that the organic system into which life was "originally breathed" was the single cell – not a multicellular creature. Since cells are the smallest units that exhibit many if not all of the properties that characterise extant living organisms, including communication with the environment, self-maintenance, spatial and temporal organisation, adaptation to external stimuli, and reproduction, it is commonly held that the cell is the smallest living system and as such the common currency of life.

At first, interactions between cells were probably little more than chance occurrences. But as the density of cells increased, it is conceivable that some cells came to prey on other cells. It is likely that this served as an impetus for the evolution of rudimentary defence mechanisms. At some point cells began to cooperate and evolved tighter relations with each other. This laid the foundation for the evolution of division of labour between the cells and eventually for their differentiation into separate cell-lineages within multicellular organisms.

The capacity to defend, which became a foundational characteristic of cellular adaptation to the surrounding environment, was retained in one form or other in the individual cells of multicellular organisms. But so was the capacity to exploit the environment for resources. The cells of multicellular animals thus needed to modulate their exploitative and defensive dispositions against the requirements of the higher level, the organism. If these trade-offs were violated the organism would succumb to attack and exploitation by the lower level. In extant organisms such occurrences are manifested as disease – exploitation as malignant cell growth and unrestricted attack as autoimmunity.

The study of evolution is often a study of trade-offs. While natural selection may improve the functioning of a trait, improvements in some respects are often coupled to becoming worse in other respects. No absolute claim can be advanced about what will happen when a selection process that occurs at one level is being opposed by a selection process that takes place at a higher or lower level. The direction taken will be influenced by characteristics of the trait's genetic system, by changes in the environment, or by changes in the relational properties between the organism and the environment.

With the deciphering of whole genomes of several species, including humans, mice, and rats, as well as a variety of bacteria and viruses, one might think that all that researchers would need to do to have flexibility fall into place was to link up the interacting genes of hosts and parasites. There are, however, a variety of obstacles of both ontological and epistemological kinds that would hinder such an achievement. For example, recent research has demonstrated mind-boggling complexities in the world of nucleic acids, and concepts that describe the intricacies involved have been judged as insufficient to the task. Furthermore, since the linkage between a string of DNA and cellular function is many-to-many, not one-to-one, the description of gene-interactions between hosts and their parasites are not likely to reveal the one true story. Despite expectations to the contrary, phenotypic investigations of host-parasite interactions will certainly continue to be paramount.

Current biology views life is a multileveled hierarchical phenomenon in which the various levels are physically entwined, simultaneously instantiated and highly interdependent. They interact in a complex regulatory network in which fitness at one level is traded off against fitness at the higher or lower levels. The set of various trade-offs represents the total fitness of the biological entity. Much research has been devoted to discern the developmental mechanisms that stably reproduce the organisms that currently populate the Earth. It is the purpose of this chapter to describe evolutionary mechanisms as they relate to the evolution and development of the vertebrate immune system and the parasites it encounters. Understanding the mechanisms by which hosts and parasites relate to each other will hopefully enable us to appreciate the sophisticated workings of the immune system in health and disease.

4.1. Situating life

LIFE'S NEBULOUS NATURE

The demarcation problem

Fuelled by sunlight's energy, the once sterile and intimidating planet Earth gradually transformed into a fruitful garden loaded with biogenic formations. In a period of 3.8 billion years living entities evolved and adapted to almost every imaginable environment – they spread from the poles to the equator, from the bottom of the deepest sea to several kilometres up in the air, and from freezing waters to thermal vents. Their metabolism produced atmospheric oxygen, decaying bodies fertilised the soil and constructive activities altered the ecosystems. Some lineages managed to evolve new adaptations to ever-changing challenges and became evolutionary winners, while other lineages succumbed and disappeared. The harsh turmoil naturally endowed life with a capability to defy and transcend its own constraints. But whilst this proficiency granted life with the blessings of immortality, it simultaneously condemned life's conveyors to a ceaseless race for all-embracing fitness.

The material and energy that life is made of is commonplace in the universe, and it is not unlikely that life has emerged at other locations than Earth. Nevertheless, as of today there are no certain observations of life beyond Earth, and astrobiology – which would like to study life in space, is so far a science without subject matter. In this respect, astrobiology is analogous to the science of artificial life, which attempts to investigate life-as-it-could-be instead of life-as-we-know-it. Attempts at identifying life in space and the creation of virtual life in computers have emphasised the need to clarify life's being in a general manner that does not conflict with biological life-as-we-know-it. This endeavour, which has been approached by researchers from a wide variety of disciplines, has been hampered by the lack of a coherent and agreed upon understanding of what life-as-we-know-it really is.

The things on Earth endowed with life have certain similarities and appear, at least to the layman, well demarcated from non-living entities. This is so elementary that a leading proponent of the intelligent design movement, William Dembski (1999), with confidence claims that he has designed a diagnostic test that fully discerns between life and non-life. As it turns out, Dembski's test is but another way to state that what is alive is alive. And being a tautology, it is of little scientific value. Still, Dembski has managed to explicate the fact that biologists and philosophers have repeatedly but unsuccessfully attempted to give a unified and general definition of life. They have proposed a variety of conditions supposed to be necessary and sufficient, but so far any attempts at definition, including Dembski's, have been unsuccessful.

While the proposed properties – the most frequent being reproductive invariance, autonomous morphogenesis, goal-directed behaviour and communication – capture essential aspects of life's being, they are either too broad in that they encompass non-living things, or too narrow, so that counter-examples that we intuitively feel to be alive are included among the non-living. Application of the conditions to concrete cases may therefore lead to counterintuitive results. Our everyday concept of life, which is vague and ambiguous, is not easily transformed into a scientific concept, and many scientists have therefore regarded defining life as a futile endeavour. Even though much effort has been put into clarifying the demarcations between life and non-life, the scientific understanding of how, when and where life began is still rudimentary. Perhaps Aristotle was right when he in *The history of animals* proclaimed that "Nature proceeds little by little from things lifeless to animal life in such a way that it is impossible to determine the exact line of demarcation, nor on which side thereof an intermediate form should lie" (Aristotle 1984, book VIII part 1).

Unearthing the common ancestor

Life is something acquired or lost according to species-specific rules. For a period of more than 2000 years these rules were thought to be of a transcendental nature and as such inaccessible to scientific investigation. The tradition known as vitalism held that living entities were animated by an immaterial life principle that demarcated them from non-life. The mechanism of spontaneous generation accounted for the transmutation of inanimate matter into living forms, and organismal death was explained by spiritual loss. Even though vitalism and the belief in spontaneous generation of animals received several injurious marks during the nineteenth century, the beliefs were not considered falsified until the twentieth century when it was finally established that living entities are being made up of the same basic ingredients as non-living things.

The realisation that life's phenomenon lies within the realm of the natural sciences allowed researchers to propose and test hypotheses about life's origin without invoking supernatural forces. These hypotheses are, nevertheless, all beset with uncertainties. There is, for example, a shortage of fossilised data from which hypotheses of life's origin can be abducted and tested. Owing to this shortage, researchers have been forced to extrapolate contingencies of extant life backward in time. Such reversal of the historical process is methodologically demanding since natural selection tends to do away with the variation on which it acts. The many attempts at synthesising life under controlled laboratory conditions have also been disappointing. While initial experimental results were promising, in the sense that many of life's macromolecules were formed when inanimate chemicals were subjected to conditions supposed to exist on ancient Earth, life was never an outcome of these experiments.

While it has not been possible to reconstruct how life emerged from non-life or to identify what the first forms of life were, it is generally thought that production of abiotic macromolecules on ancient Earth was a prerequisite for the emergence of life. It is further believed that these molecules, upon reaching a sufficient concentration, assembled to form intricate molecular machines with a capacity for replication. The encasement of these self-assembling machines within a semipermeable membrane marked the final step towards what was to become the biological cell.

There is currently much debate as to whether the first steps towards life were of a genetic or a metabolic kind. Proponents of the genetic start-up emphasise the role of replicating molecules (Elitzur 1994; Maynard-Smith and Szathmáry 1995) while proponents of the metabolic start emphasise the need for self-organising macromolecules and the establishment of a cell membrane. The genetic systems, they maintain, were introduced at a later stage (Kauffman 1995; Rose 1997). This view is close to the biosemiotic perspective, which reframes the origin-of-life problem into a question of how material structures that follow physical laws gradually were transformed into entities that function according to sign operations. How can sign processes arise out of something that signifies nothing? The most plausible answer would be that semiosis originated simultaneously with the origin of life (Sharov 1999).

In an influential analysis Richard Dawkins (1989) seized upon the fact that the things being replicated in the life cycle are the cells' replicators, the genes. Therefore, he argued, the fundamental entities of life must be the genes, and organisms are thus merely appendages to the living genes. "The replicators that survived were the ones that built *survival machines* for themselves to live in"

(1989, p. 19). These survival machines, which were mortal and therefore devoured after a specified time, stood in contrast to the replicators that "did not die out, for they are past masters of the survival arts." (1989, p. 19).

But even though it may turn out to be true that replicators have temporal priority to cells, this does not suffice as an argument for the claim that autonomous replicators were endowed with life. The scenario that may be called "genes first" is certainly a plausible one, but it is not empirically better founded than the "cell first" perspective. Proponents of the latter view hold that genes were incorporated as the cells' handmaidens, and that they should be retained only in so far as they fulfil a function that is necessary for the cell. That contemporary bacteria living in close contact with other organisms can obtain many compounds of intermediate metabolism from their host, and thus can discard biosynthetic pathways and genes, may serve as support of this view.

There is no consensus between the two sides as to the adequacy of contemporary analyses and experiments that can be done to resolve the controversy. Nevertheless, since there are no vestiges of life that are less complex than bacterial cells, and since bacteria can inhabit almost any niche on the surface of the earth, it is conceivable that bacterial cells retain clues to the chemistry of Earth's surface as it existed in the remote past. It is therefore plausible that investigations into microbial life, especially in extreme environments, will provide further clues to life's origin.

Tracking life

Transcendence

Besides discussing a number of observations that fit into his evolutionary view of nature, Darwin provided two distinct but interrelated conceptualisations; that life is connected in a pattern and that this pattern has been generated by the process of natural selection. The two ideas are foundational for the Darwinian metaphysics which holds that all living beings, directly or circuitously, have been connected since life's origin. This proposition does not exclude the possibility that there may have been multiple start-ups of life, but does indicate that other life forms either went extinct or became absorbed by the surviving. Upon closing his book, Darwin marvelled when realising that this way of framing life gave a totally new and mechanistic conception of biology.

> There is grandeur in this view of life, with its several powers, having been originally breathed into a few forms or into one; and that, whilst this planet has gone cycling on according to the fixed law of gravity, from so simple a beginning endless forms most beautiful and most wonderful have been, and are being, evolved. (Darwin 1859, pp. 459–60).

The Darwinian metaphysics envisions life as a relation between a living entity and its surroundings. And since any relation involves transcendence – a going beyond a given to something novel, life is essentially a dynamic process. Evolutionary individuals are thus relational beings that continuously outdo their own limits. The transcending activity, while being a prerequisite for continued living and the emergence of novel traits, simultaneously leads to intrusion and devouring of the boundaries of fellow individuals. It is this reciprocal activity between individuals that generates conflicts and thus serves as an impetus for the creation of defence mechanisms.

Since life in the Darwinian world is individualistic, semiautonomous, self-preserving and transcending, it would seem that life becomes the creator of life. Life forms and is formed by the interactions it makes with the world, leaving no room for immutable and universally stable organic elements. As the features of living entities become dependent upon the contexts in which they are shaped, probabilism and contextualism, not determinism, should be life's metaphysical foundations.

Process

The creative force of life is akin to a goal-directed process. But while natural selection accounts for the evolution of entities that are goal-directed, evolution in itself does not provide any goals. Stephen Jay Gould (2000) claimed that life's tendency towards increased complexity is owing to the statistical summations of circumstances and coincidences, and that life lurches from one moment in time to the next. It is therefore pure chance that evolution on Earth has produced the kinds of organisms that are alive today.

These ideas are in line with observations made in the late 1960s. New techniques in molecular biology then provided data that were inconsistent with the claim that natural selection was an all-encompassing force in evolution. The uncovering of an unexpectedly large amount of intraspecific variability at the molecular level, that appeared to be selectively neutral, prompted the geneticist Motoo Kimura to propose the neutral theory of molecular evolution. The theory proclaimed that most mutations that are not disadvantageous are not advantageous either – they are neutral. And since neutral mutations are far more common than advantageous mutations, the great majority of evolutionary changes at the molecular level should be caused not by Darwinian selection but by random drift (Kimura 1991).

Random genetic drift is the idea that alterations in gene frequencies, which occur through natural selection when populations are large and mating random, may be caused by sampling errors when the evolving population is small and mating non-random (Price 1996). When for example a population shrinks to a small size, the survivors may be a more or less randomly selected sample of the pre-extinction population. There is thus a chance that the genetic variation in the founders of the new population will be biased. The effects of this is especially dramatic when it comes to genes that were highly variable in the pre-extinction population, such as genes involved in immune responses (Klein and Takahate 2002), since most such genes would have to be generated anew in the post-bottleneck phase.

But even though chance is important, it is also clear that organisms faced with the same challenges repeatedly arrive at the same solutions. Gould and Lewontin (1979) forcefully argued that organismal form is constrained by previous adaptations and developmental constraints. Arguing from the study of evolutionary convergence, Simon Conway Morris (2003) presented a related view on the directedness of evolution. While acknowledging that the number of possible combinations of biological systems at all levels is gigantic, he claimed that physics and chemistry severely constrain biological possibilities, and that life has surprisingly few choices that can be realised.

Pattern

Since the 1930s it had been known that living organisms could be divided into two domains (Woese 1998a). Small single-celled microbes that lacked a nucleus (*karyon* in Greek) were termed prokaryotes (*pro*, Greek for before), while organisms that harboured a nucleus were termed eukaryotes (*eu*, Greek for good). While this distinction marked a decisive step forwards in classification of the microscopic lifeforms, the investigations of the relationship between bacteria, viruses and eukaryotes were long hampered by insufficient methodology.

For higher organisms morphological traits were sufficiently complex and numerous to allow a reasonably accurate phylogenetic classification, and the existence of a fossil record and the study of embryological development greatly aided this task. Microorganisms, on the other hand, are morphologically simple, and the traditional classification of microrganisms had to rely on phenotypic characteristics like differential staining with dyes, cell morphology, enzymatic activity, aerotolerance, growth patterns, and antigenic characteristics. These techniques are rather crude, and despite considerable effort, microbiologists scarcely managed to determine the phylogenetic relationship between microorganisms. This all changed in the 1960s with the introduction of new molecular techniques that allowed fine-grained structural comparisons between the nucleic acids of the microbes.

A major breakthrough in microbial phylogenetics was reached when the structure of ribosomal RNA (rRNA), a stable molecule present in all living organisms, was compared between microbial species. Such comparisons provided the first glimpses into the relationship between the prokaryotes and eukaryotes. Sequencing the nucleotides of the rRNA molecule even led to a totally unanticipated discovery that eventually did away with the two-domain world. It appeared that prokaryotes had to be divided into two domains, the Archae, which includes bacteria that live in extreme environments, and the Eubacteria, which contains the bacteria that are most familiar to us, such as those that cause disease. The third domain comprised the Eukaryotes, which contain the eukaryotic uni- and multicellular organisms. The data also revealed that eukaryotes and archaea are more closely related than either is to bacteria. A majority of microbiologists and biologists were initially reluctant to accept that there could be two types of prokaryotes, but the three domain world is now accepted as a proper description of life's domains (Olsen et al. 1994).

Molecular data have further allowed the classification of organisms into five kingdoms. The unicellular organisms belong either to the kingdom Monera, which comprise prokaryotic bacteria, or to Protista which are made up of a heterogeneous group of unicellular eukaryotic organisms. The multicellular eukaryotic organisms belong either to the kingdom Fungi, which is made up of molds, yeast and mushrooms, to the kingdom Plantae, which includes multicellular algae and plants, or to the kingdom Animalia, which comprise the animals ranging from sponges to vertebrates. Research in the latter half of the 20th century also revolutionised our knowledge of the physical nature of viruses, but comparable progress in understanding the origins and relationship of viruses to cellular life is conspicuously lacking (Hendrix et al. 2000).

The rRNA molecule has several properties that are valuable for phylogenetic studies: it is present in all organisms, its sequence is relatively conserved so that regions of homology between organisms can be established, and it is sufficiently large so that the information content is adequate for the phylogenetic reconstruction. The reclassification based on rRNA sequences was believed not to be liable to undergo much additional revision, yielding at least a robust, lasting classification of living organisms. Nevertheless, it has now become clear that some similarities of rRNA may be due to convergence (i.e. separate origin) and not to homology (i.e. common ancestry). Furthermore, since artefacts are created when modelling assumptions are introduced for creating phylogenetic trees, it has been proposed that rRNA studies need to be supplemented with whole genome studies (Philippe and Germot 2000).

In 1995 the first complete genome sequence for a free-living organism, *Haemophilus influenzae*, was published (Fleischmann et al. 1995). Since then, complete sequences of a multitude of organisms from all three domains have been published. This has given birth to a new field of research, phylogenomics, which uses whole genomes to infer phylogenetic relationships. While these investigations have revealed important facets of the evolutionary history of organisms and their genomes, they have revealed certain epistemological and ontological challenges that may be difficult to resolve. The problems are related by the fact that natural selection tends do conceal its own tracks, and hence that parts of the tree of life might prove difficult, if not impossible, to resolve with confidence (Delsuc et al. 2005).

That so many species had been delineated at the level of the genome was a tremendous benefit when the human genome was sequenced in 2001. By comparing the human genome to the genomes of other species, it appeared that most of our genes come from the evolutionary past (Baltimore 2001). Furthermore, it became clear that the most elementary cellular functions evolved just once, and that they stayed fixed in all organisms from the single-celled microbes onwards. The results also revealed that especially two types of genes differentiate vertebrates from plants, worms, flies and microbes. These are genes for multifarious systems, including neuronal complexity, blood clotting and the adaptive immune response, and genes that provide increased cellular connectivity such as genes for intra- and intercellular signalling (Aravind and Subramanian 1999; Baltimore 2001; Levine and Tjian 2003). The increases in such genes would be beneficial to the multicellular organism by allowing extensive cell differentiation, intra- and extracellular communication, and a tighter bond between cells.

While the molecular studies provided evidence supporting the view that all extant life derives from a common ancestor and that the evolution of any particular lineage is the product of a great number of historical contingencies of both genetic and ecological nature (Woese 1998b), they also revealed some previously unknown routes of inheritance, for example that lateral gene transfer between organisms may have been fairly common. Moreover, the remarkable genetic similarities between species along with the marked genetic variation within species substantiated the unitarian belief that traits that set organisms apart from each other are differences in grade. A major outcome of these studies is therefore that the genome of an individual should be viewed as a mosaic of discrete segments, each with its own unique history and relatedness to different contemporary and ancestral individuals (Paabo 2003). A practical consequence of this knowledge has been the further complexities it has introduced into prokaryotic taxonomy and thus for the assignment of microbial isolates to species membership (Gevers et al. 2005).

THE COMPOSITE ORGANISM

Conflict vs. integration

Organisms come equipped with a wide variety of forms and functions and are, like life itself, not easily determined or categorised by any set of defining conditions. The observed heterogeneity of living entities has been a rich source for speculation, and attempts to clarify conceptual and empirical issues related to the diversity have resulted in highly sophisticated theoretical frameworks. While these served well to organise the biological world, they also served as epistemological constraints that parsed investigations of the organism in distinct directions. This has been especially pronounced when it comes to understanding the importance of organismal development during natural selection.

The organism that develops has been conceptualised by two influential views, the Aristotelian and the Leibnizian. The Aristotelian metaphysics held the organism to be an individual composed of matter and form. The formal nature was thought to be causally responsible for the persistence of the living entity by integrating the material workings and thereby helping the organism to maintain itself (Wilson 1999; Lennox 2001). The idea of harmonious internal workings were elaborated further by Descartes in his mechanical model of the organism, and were given scientific credence by Claude Bernard and Walter Cannon with the ideas of a constant internal environment and the concept of homeostasis (Perlman 2000).

In the latter half of the 17th century Gottfried Wilhelm Leibniz (1646–1716) worked out a quite different view. His theory was formulated during the same epoch that microscopic organisms were discovered within macroorganisms by Leeuwenhoek, and the theory was probably made in an effort to accommodate these discoveries (Nachtomy et al. 2002). In his Monadology Leibniz (1714) proposed that organisms contain other organisms within themselves, and so radically broke with the Aristotelian position by claiming that an organism is typically made up of other organisms in a hierarchical structure, one nested within the other. To coordinate and unify these organisms he invoked a vital force, the *entelechy*, which served as a dominating force that would activate, suppress and organise the nested organisms. Although organisms were thought to be nested and dominated by other organisms, they were nevertheless thought to be complete organisms and not mere parts of other organisms. The Leibnizian metaphysics thus allowed for conflicting interactions between the lower-level units, and the harmonious internal workings was not something given but something to be negotiated.

Both the Aristotelian and the Leibnizian metaphysical views have contemporary supporters. On the one hand, there are those who subscribe to an integration perspective, in which cell adhesion and signalling are emphasised as being the pivotal processes for development, while competition and selection are given a minor role. Proponents of a competition perspective are, on the other hand, more concerned with conflicting interactions and emphasise that co-operation is a result of multilevel selection processes and mediations of conflicts between the levels of selection. In practice, though, most researchers pragmatically adhere to both perspectives, and apply the one or the other depending on the developmental process being investigated (Winther 2001).

Development

The now obvious idea that development is an integral part of evolutionary biology was largely ignored by evolutionary biologists from about 1900 to about 1980 (Hamburger 1980). The separation of the fields was so extensive that when Ernst Mayr (1961) in an influential paper discussed cause and effect in biology, he still distinguished between evolutionary and developmental biology as of two separate explanatory fields that differed in methods, explanatory projects and concepts.

The neglect of developmental constraints was in part justified by August Weismann's late nineteenth century claims that i) information passes from germ line to soma but not in the reverse, ii) that the germ line is separated from soma and iii) that only the germ line is continuous in evolution. These claims allowed geneticists and evolutionary biologists to treat development as a black box and thus to bypass it when trying to understand evolution (Buss 1987). But the two research fields were also to a large degree conceptually isolated. Developmental concepts were not easily integrated into the genetic and evolutionary perspective moulded during the making of the "modern synthesis" of evolution in the 1930s, in which Darwin's theory of natural selection was blended with the rediscovered Mendelian genetics (Hamburger 1980).

The modern synthesis portrayed evolution as being the interplay between mutation and selection, with the former providing a supply of variation and the latter acting as a fitness-based sieve. In the case of unicellular organisms, this representation is fairly accurate. But in the case of multicellular organisms where genes serve as modulators of biochemical and physiological parameters that in turn influence the growth of embryonic tissues, the effects of mutation on fitness are not directly accessible for selection. Since selection works on phenotypes and their functional characteristics, since development is a major determinant on the multicellular organism's phenotype, and since some ontogenetic trajectories are better for reproducing and survival than their competitors, development is an important determinant on the pathway taken by natural selection.

Development impinges on evolution because it ties the organism up in a system of references to other living and non-living entities in-between fertilisation and death. In accordance with the biosemiotic view, the individual organism may be envisioned as a node in a network with an internal configuration and autonomous form that refers to a world of significance. Hence, organismal life is not simply conforming to a predetermined trajectory but follows a variable path upon which developmental decisions are influential. Still, the plasticity is constrained by developmental agents of diverse kinds, including transcription factors and elaborate gene regulation systems (Levine and Tjian 2003) as well as environmental organic and physical influences (Gilbert 2001). Hence, owing to developmental constraints not every trajectory is attainable.

Modularity

Research in developmental genetics has revealed an astonishing conservation of many developmental processes in a wide range of multicellular animals. Many of these processes tend to be employed repeatedly by animals of various taxa in different contexts during development (Schlosser 2002).

Developmental modules, although lacking a rigorous definition, are generally characterised as networks of interacting elements behaving as relatively independent units of development. The immune system, which is surprisingly robust to variations in parameters that govern its up- and downregulation following environmental perturbations, is a typical developmental module. Modules, which can be relatively easy to dissociate and recombine or redeploy in new contexts, are expected to improve the capability of a lineage to evolve. Modular dissociability or quasi-independence, which denotes the propensity for independent variation and shuffling of phenotypic traits during development and evolution, has even been regarded as a prerequisite for both phylogenetic and adaptive plasticity and thus for the entire adaptionistic program (Lewontin 1978; Wagner and Altenberg 1996).

Different concepts of modularity exist in the various biological subdisciplines, including developmental biology, evolutionary biology, and molecular biology. The concept is usually used in two different ways. It either refers to the repeated use of a unit within a system, or it refers to the independence of several units from each other (Schlosser 2002). According to the first usage, bacterial colonies are modular organisms since they are composed of repeatedly used bacterial units (Andrews 1998). According to the second idea, which comprises the first usage, modularity refers to the fact that there may be relatively independent subprocesses or subunits within an integrated whole.

In their study of life-history evolution Jokela and Haukioja (2000) presented a simplified model in which they singled out developmental modules as consisting of a hierarchy of traits, tactics and strategies. There are modules at each level, one nested within another. At the highest level of the hierarchy they placed the organism's developmental *strategy*. This depicts the totality of plastic responses that the organism can perform during its interactions with environmental stimuli. At the bottom level of the hierarchy they placed the *traits*, which are the characteristics that directly interact with the environment. In-between the strategies and the traits are the *tactics*, which consist of interacting traits that coevolve as a response to the same selection pressures. The developmental tactics signify the various modes by which the organism exhibits tolerance to external and internal stressful perturbations.

The usefulness of the hierarchical view on modules can be exemplified by considering the function of organismal defence. The defence tactic is made up of the totality of defence traits that the organisms of a given species come with. Since the different traits within a tactic have more or less the same functional significance and respond to the same selection pressures, they may become integrated during evolution. This is, for example, the route taken by the various defence systems of vertebrate organisms, in which the adaptive immune system, the phagocytic system and the complement system have evolved into an interactive defence tactic.

The functioning of the traits is relatively autonomous of other physiological processes. But the traits are tightly connected to each other, and may therefore be considered as a higher level module – the defence tactic. But even though they are autonomous they are not isolated; rather, they are influenced by and influence other tactics. For example, stressful life events influence the capability of an organism to mount an adaptive immune response (Padgett and Glaser 2003). The defence tactic is thus influenced by the organism's stress-tolerance tactic. The defence tactic may also influence the reproductive tactic. For example, animals with a defective immune system are poorer at mating than immunocompetent organisms (Hamilton and Zuk 1982).

The variability in the hierarchy of developmental strategies, tactics and traits is not necessarily dependent upon variability in the genetic system of the organisms, although genetic variability would expand the variability. For example, the defence tactics of identical twins may differ as a result of environmental perturbations. The twins, while having genetically identical complement and phagocytic systems, will have different adaptive immune systems because of the epigenetic changes that take place during development of their immunoglobulins and T cell receptors.

SELF-MAINTENANCE

Embodied drive

A characteristic feature of life at all levels is directionality, of which autonomous morphogenesis of the unfolding embryo and stereotypic defence reactions are examples. For more than 2000 years goal-directed behaviour was explained by analogy with human purposive behaviour, in which organisms were thought to realise the potential of their species. René Descartes challenged this teleological explanation by claiming that nature always acts in accordance with the laws of physics, and that mechanistic categories alone are capable of explaining biological functions. Descartes' biology, which is a philosophy of the living machine (Hatfield 1992), was taken over by the defenders of the Darwinian theory.

But by treating the internal forces of the living as if they were non-existent, the mechanistic outlook simultaneously dismissed an important distinction between the living and the non-living. The need to explain life's dynamics was not eliminated by the falsification of vitalism; the problem remained but required an alternative explanation. And while Darwin provided a naturalistic explanation of organismal adaptations that virtually falsified the explanations from divine design, his explanation did not provide an alternative explanation for the dynamics of life. The sources of embodied drive, the urge an organism has to act upon signs of internal need and environmental perturbations, were left unexplained and to a certain degree ignored.

The biological problems to be explained in association with embodied drive are succinctly addressed, albeit not adequately explained, in the philosophical literature on self-preservation. These ideas are vehemently expressed in the Stoic writings. The Stoics transformed the Aristotelian notion of goal-directedness and the Greek philosopher Epicurus' (342–270 BC) teachings, that life's goal is to minimise pain and maximise pleasure, into the principle of self-preservation, and looked upon it as being the first principle of animate nature (Wolfson 1934). These ideas were expanded upon by Benedictus de Spinoza (1632–1677) who, in contrast to the Stoics, held that self-preservation is a necessary constituent of everything, from inanimate to animate matter. "The endeavour (*conatus*) wherewith a thing endeavours to persist in its being is nothing else than the actual essence of that thing." (Spinoza 1677, p. 90). The endeavour, which Spinoza also called appetite, was seen as an emotional state that served to direct the living entity outwards to fulfilment of its inner desires.

Like the Stoics, Spinoza held that self-destructive behaviours like suicidal and altruistic actions can not be caused by the organism itself but must be related to some external force. He thus claimed that things "are contrary by nature, that is, they cannot exist in the same subject, in so far as one can destroy the other" (Spinoza 1677, p. 89). These thoughts were taken over in an almost unmodified form by for example Charles Darwin and Paul Ehrlich. Darwin's claim (1859, p. 229) was that "natural selection will never produce in a being anything injurious to itself, for natural selection acts solely by and for the good of each", and Paul Ehrlich's statement of 1901 was that "the organism possesses certain contrivances by means of which the immunity reaction, so easily produced by all kinds of cells, is prevented from acting against the organism's own elements and so giving rise to autotoxins … so that we might be justified in speaking of a "horror autotoxicus" of the organism. These contrivances are naturally of the highest importance for the individual" (cited from Silverstein 1989, p. 162).

By attempting to give naturalistic explanations of natural events without recourse to explanations in terms of vitalism or teleology, Spinoza may be conceived of as a speculative forerunner of the scientific worldview. To him, Nature and God were but two words for the same phenomenon.[1] But in contrast to the God of the Genesis, Spinoza's God was an immanent creative force, not an external agent. And since God was immanent in all of Nature, Spinoza held all living entities to be creative. Nature is therefore to be looked upon as an active power that directs and governs life, the creating nature (*natura naturans*).[2] He dismissed final causality by claiming that "nature has no fixed aim in view" and therefore that "all final causes are simply fabrications of men" (Spinoza 1677, p. 33). Hence, living entities are self-driven by immanent efficient causes that are emotional in nature. But Spinoza also acknowledged that nature is found in the created things we see around, the passive reality of our daily experience, the created nature (*natura naturata*).[3] Hence, he saw a contrast between what is creating and what is created.

In his 1929/30 lecture-course Martin Heidegger (1995) opposed Spinoza's unifying idea of the living universe by explicitly stating that there is a difference between lifeless matter and life. He maintained that it is a characteristic feature of the living, as opposed to the dead, that the living entities have access to and respond upon environmental stimuli. For example, stones are lifeless because they are worldless, while animals, plants and microbes are living because they have access to their world and therefore can respond to perturbations in a functional manner. Furthermore, he held that neither mechanism nor vitalism is able to explain life's true being. Mechanism was disapproved of because machines are incapable of self-production, self-regulation and self-renewal. And even though vitalism correctly focuses on various forms of self-preservation as distinguishing characteristics for life, vitalism was denounced simply because it explains nothing. By invoking a vital agent, he claimed, the vitalists simply eliminate the problem to be explained.

Instead of mechanism, which does not allow the question of purposive behaviour to arise, and vitalism, which tries to solve the problem by eliminating it, Heidegger

---

[1] It is also interesting that Spinoza, in contrast to Descartes, believed that mind and matter are but two aspects of the same thing, and hence, that there is no need to invoke any mentalistic entity to explain organismal behaviour.

[2] *Natura* is the Latin word "nature." Spinoza added participle endings to that noun. *Naturans* is thus "nature" plus the active participle ending, which is "-ing" in English; so "*natura naturans*" is "nature naturing."

[3] *Naturata* is "nature" plus the past passive participle ending, which is "-ed" in English; so "*natura naturata*" is "nature natured."

sought recourse to the concept of drive. The driving forces, which arise from within the organism, are never simply mechanical. They are capacities that are propelled but not purposive, and any idea of consciousness or soul must be set aside. To Heidegger, the drive characteristic of the living is located within the self. It is the self that is preserved during self-preservation. But while he restricted the term self to human beings, he acknowledged that self-like structures are to be found within "every self-like being, every being that possesses the character of personality in the broadest sense (every personal being)" (Heidegger 1995, p. 233). Hence, animals and parts of animals such as modules and organs may contain a self, even though their selves differ fundamentally from the human self.

Self-preservation and the corollary denial of a desire for self-destruction are to a certain degree foundational for the understanding of biological entities and their complex biological processes, and several attempts have been made to naturalise teleology. One of the most utilised concepts, teleonomy, was invented by Pittendrigh in 1958 to explain goal-directed behaviour as a mechanistic process, while at the same time avoiding vitalism (Mayr 1988). According to Ernst Mayr (1988, p. 45), "All teleonomic behaviour is characterised by two components. It is guided by a "program," and it depends on the existence of some endpoint, goal, or terminus which is foreseen in the program that regulates the behaviour". The programs can be laid down in the genes of the organism or can be acquired through a learning process, or both, as a result of natural selection.

But even though the concept of teleonomy gives a naturalistic conception of how behavioural responses are initiated and regulated, it does not account for an organism's embodied drive. The two concepts explain different aspects of organismal behaviour. While teleonomic behaviour is the effect of ultimate causes shaped by natural selection, embodied drive is the proximate energetic that drives organismal behaviour. Hence, embodied drive makes natural selection and hence teleonomy possible.

This idea can be made clearer by analysing the role that fitness plays in natural selection. The concept of fitness, which is an indirect measure of physiological and ecological performance, is composed of both a relational and an intrinsic dispositional property. The relational property indicates the degree to which the organism fits into its environment, measured as survival, while the dispositional property is the organism's fertility, measured as expected number of offspring. The process of natural selection is driven by fitness differences between organisms, and those individuals with the lowest fitness are negatively selected. Natural selection can to a certain degree explain changes in survival and reproduction, but it can not explain why the organisms live and reproduce. That's where the concept of embodied drive makes its irreducible contribution.

The concept of embodied drive is active, in contrast to other drives in biology, including molecular drive (Dover et al. 1982), meiotic drive (Pomiankowski 1999) and developmental drive (Arthur 2001). Embodied drive operates in all living systems, also in sterile mules where reproductive fitness is zero. But if all living entities possess embodied drive, does this allow us to state that entities with embodied drive are alive? It is for example controversial whether viruses should be included among the living; they lack metabolism and can not survive outside their host cells for extended periods of time. But they do respond adaptively to external stimuli and actively preserve their own integrity through a self-maintaining process. It thus appears that they have embodied drive. So even though they are not living according to some definitions, the embodied drive concept would include viruses amongst the living.

Embodied drive occurs at the level of the smallest living system, the cell. But there are also cases in which embodied drive seems to operate at subcellular levels. In the process called meiotic drive, a driving gene evolves because it makes more copies of itself than does the gene against which it competes. The gene's evolution need not be for the sake of the organisms in which it occurs, the gene may have neutral or even detrimental effects on the organism and still increase in frequency (Pomiankowski 1999). But the seeming autonomy of the driving gene is only a statistical outcome of processes that take place at the cell level during the reproductive process. The driving gene of meiotic drive lacks agency, it is pseudodriving, and the effective driving agent is the cell.

The idea that organisms embody fundamental drives has recently been supported by experimental and modelling work on emotions. Emotions, which have been investigated in a wide variety of animals – from molluscs and flies up to the higher vertebrates, are equated with internal driving forces that motivate behavioural decisions. They are responsible for drives such as hunger and thirst, and allow organisms to cope successfully with situations that are dangerous or advantageous (Damasio 2001). Some biologists have recently taken the emotional drive of individuals seriously by noticing that optimality models of life can be made more realistic if an organism's internal motivation, its hedonic tone, is included as a model parameter (Giske et al. 2003).

The genesis of life's embodied drive, the active urge to transcend limits, goes all the way back to life's origin. Self-maintaining cells are actively engaged in exploring environmental resources; without this activity life could not be. The concept of embodied drive is teleological. But in contrast to vitalism, in which living entities received their teleological characteristics from an external goal-conceiving agent, embodied drive signifies an immanent goal-directed property. Embodied drive, which is found only in the region of living entities, is a naturalistic, not a super-naturalistic, concept, and is in many ways closely related to the concept autopoiesis.

Autopoiesis

Humberto Maturana and Francisco Varela, who in the 1970s developed a theory that considered life to be an organised and dynamic phenomenon, asked themselves what it is that distinguishes life's dynamics from any other set of molecular interactions and transformations that occurs in nature, and their answer was autopoiesis[4] (Maturana and Varela 1987). Autopoiesis theory attempts to define, beyond the diversity of all living organisms, a common denominator that allows for the discrimination of the living from the non-living. It is based on direct observation of how extant cells work. Cells are seen as the minimal autopoietic units. They are capable of sustaining themselves owing to an inner network of reactions that regenerate all the system's components. Autopoietic machines are autonomous in that their main activity is the maintenance of their own organisation – they produce themselves from within. Autopoietic machines are different from allopoietic machines in that the latter produce something different from themselves (Varela 1979).

The metabolic processes that occur within the cell are enclosed by a cell membrane, through which information is exchanged with the environment. The cell membrane, which is semipermeable, ensures that nutrients are transported to the inside while waste products become dispelled to the outside. In this way, the cell creates a boundary between its self and others. The autopoietic view holds that the difference between a living and a dead cell is not so much to be found in their constituent biochemical components as in their metabolic functioning. Since viruses are unable to reproduce unless they enter cells of other living organisms, viruses are mere mixtures of chemicals. Hence, the traditional conviction that bacteria are alive while viruses are not can be defended by the fact that bacteria but not viruses are autopoietic.

All living systems share a common organisation, and it is this organisation that is living. The components, which are the main focus of molecular biology, constitute the structure of the living entity. The process of life is multiple realisable in that life's organisation is independent of the properties of its structural components – many different kinds of components can realise life. But life is also an emergent property. An emergent property entails that qualitative novelty occurs at the level of the whole. Biology is thus the study of an emergent phenomenon that cannot be reduced to physics, and an exclusive focus on structure misses out the systems-properties associated with life.

Varela (1979) held that all that is alive must be autopoietic, and conversely, that anything that displays autopoiesis is living. Autopoiesis is thus held to be both necessary and sufficient to characterise life. This characteristic is probably too ambitious. For example, self-reproducing micelles can be considered as simple autopoietic systems, but hardly deserve the characteristic of living. Hence, that a unit is autopoietic is not sufficient for it to be alive (Luisi 2003). Furthermore, while

[4] The meaning of this neologism is captured by combining Aristotle's concepts of *praxis* and *poiesis*. To Aristotle life was *praxis* in that it is an activity that fulfils its end while being done. Procreation, feeding and all other life activities of animals, including immune reactivity, are instances of *praxis* (Lobkowicz 1967). *Poiesis*, in contrast, is an activity that reaches its completion by stopping and leaving behind something different from itself. Hence, the making of artefacts by artists or the making of the immune system during the lifetime of an individual instantiates *poiesis*. The development of blood cells in the bone marrow is aptly termed haematopoiesis.

the extensive knowledge of cellular dynamics is compatible with autopoiesis theory, there is not enough information about the dynamics of multicellular organisms to decide whether they are autopoietic units or not (Maturana and Varela 1987).

The autopoietic system is referred to itself, and its actions consist in establishing and maintaining the dynamic process of staying alive. But while autopoiesis explains viability, the same process cannot account for the reproduction of cells. In fact, both cells and organisms can be alive without having the ability to reproduce, and reproduction is therefore not a constitutive feature of what it means to be living. Thus, reproduction must be a function derived secondary to life. Extant cells that reproduce do this as part of their life cycle. This was probably not the case with the first replicating units. Maturana and Varela (1987) hypothesised that reproduction initially occurred as a result of autopoietic cells being accidentally fragmented. In the course of the historical process,

> some odd cells underwent reproductive fracture as a result of their internal dynamics. These variants possessed a dividing mechanism from which derived a lineage or stable historical succession. It is not clear how this occurred. These origins are probably forever lost to us. But this does not invalidate the fact that cell division is a special case of reproduction that we can legitimately call self-reproduction. (Maturana and Varela 1987, p. 66).

These ideas have been supported by theoretical and experimental investigations. The division of a single cell into two occurs as a result of the internal dynamics of the cell as it expands, and is thus not something that needs a selective explanation (Maynard-Smith and Szathmáry 1995). The experimental results of Luisi and co-workers lend further support to this idea. They managed to combine a replicating molecule within a membrane system, in which the microspheres divided while the genetic material replicated, retaining the relative proportions of the two components (Oberholzer et al. 1995; Zepik et al. 2001; Luisi 2003).

Apoptosis

A phenomenon that is not captured by the concept of autopoiesis, and which stands out against it, is the phenomenon of programmed cell death. During this process there is no rupturing of the dying cell's membrane, and the cell dies by a non-inflammatory process that occurs by one of two mechanisms – autophagy or apoptosis. While autophagy occurs when clusters of cells or whole organs die during development, by chewing themselves up with little or no help from phagocytes, apoptosis is a process that occurs in isolated cells and in which the dying cells are being ingested by phagocytes (Baehrecke 2002). Programmed cell death is thus distinct from necrosis, the process in which severely damaged cells die by cell breakage and accompanied inflammation.

Programmed cell death was initially considered to be a phenomenon confined to multicellular animals, but the phenomenon has now been described also in unicellular organisms like bacteria (Lewis 2000) and yeast (Madeo et al. 2002). It is as yet not clear whether cells of the domain Arhchaea undergo programmed cell death, but cells from the two other domains of life, Eubacteria and Eukaryota, do have this mechanism to control life (Lewis 2000). Since apoptosis as defined in multicellular eukaryotes involves organellar disassembly, and since organelles are not found in prokaryotes, the term apoptosis cannot properly be used to describe the programmed cell death in prokaryotes.

The processes of autophagy as well as regulations of cell proliferation, differentiation, and apoptosis are dependent upon protein synthesis. When for example yeast are treated with cycloheximide which inhibits protein synthesis, induction of apoptosis is hindered, indicating an active role of the cell in the death process (Madeo et al. 2002). A similar inhibition of apoptosis has been described also in bacteria and cells from multicellular animals (Lewis 2000). Hence, organisms, both uni- and multicellular, must continuously inhibit self-destruction to stay alive. There are thus three developmental alternatives for a cell – it can divide, differentiate or die.

In multicellular organisms apoptosis has a normal, formative role during development, and the developmental stability of the body is maintained by signals that control the life and death of single cells. Without apoptosis the immune system, with its continuous renewal of cells, would become a monstrous module; it has been estimated that an 80-year old person without cell death would have 2 tons of bone marrow and lymph nodes to carry around (Melino 2001).

The processes of life and death have evolved together to become tightly integrated at the cellular level, and the same cellular machinery functions in both processes. This is made evident during the process of autophagy, in which a cell that is deprived of nutritious substances slowly degrades some of its own constituents, including entire organelles, to stay alive (Levine and Klionsky 2004). Another example would be the protein tumour necrosis factor (TNF), which can signal either death or survival to cells in the immune system. How a cell responds to TNF depends on the physiological context in which the cell is located. The physiological tuning of the cell influences the way in which intracellular multiprotein complexes co-ordinate cellular physiology to decide whether a cell should die or not (Tang et al. 2001). Since embodied drive and apoptosis are opposing forces, there is apparently more to fitness than survival and reproduction.

The phenomenon of apoptosis can hardly be understood solely from the autopoietic-situated perspective. Apoptosis in unicellular organisms reduces survival and reproductive potential, and would naturally be selected against in an environment in which apoptotic cells competed with cells having compromised apoptotic mechanisms. The explanatory dilemma, that cell death must have evolved by natural selection even though it is fitness-reducing, has been termed the kamikaze conundrum (LeGrand 1997). The dilemma can hardly be solved if natural selection is taken to be a process that only works on the individual. If, on the other hand, the theory is expanded to include selection at multiple levels, evolution of programmed cell death may be explained as having evolved through social interactions between cells.

Apoptosis may well have started out as a defence mechanism that evolved following competitive interactions between bacteria. Several bacterial species are able to produce toxic metabolites that target the membranes of other bacterial species, and at least some of these metabolites are able to induce membrane permeabilisation in mitochondria as well as bacteria (Blanke 2005). Thus, the apoptosis-inducing mechanism may have been a weapon used by bacteria to control their environment, a mechanism that the once free-living protomitochondrion later used to gain control of the eukaryotic cell that engulfed it (Blackstone and Green 1999). And in their scenario for the evolution of apoptosis in unicellular eukaryotic yeast, Madeo et al. (2002) claimed that apoptosis provides an evolutionary advantage for the kin of the dying cell. When damaged cells die during nutrition-shortage they simultaneously promote their own genome's survival by allowing their kin to survive.

4.2. Social evolution

CO-OPERATION AND CONFLICT

Social behaviour

That sociality is common in biology is a truism; social life is found everywhere and includes phenomena ranging from cellular cooperation within multicellular organisms to interacting social groups of pro- and eukaryotic organisms. To explain how sociality came into being is, nevertheless, not by any means a platitude. This is because the emergence of social life runs into a self-devastating dilemma as it becomes manifest; it derives its reimbursement from coordinated action and shared benefits that, in turn, are vulnerable to cheating by selfish individuals. Thus, the evolution of sociality requires a solution to the cheating problem.

It is useful to depict the evolution of social behaviour as an iterated game between an actor and a recipient (Table 4.1). Four basic types of behaviours may emerge: the behaviour of the actor may either benefit both players, in which case the behaviour is mutualistic; it may be costly to the actor but beneficial to the recipient, in which case it is an act of altruism; it may be selfish as when it is beneficial to the actor but costly to the recipient; or it may be costly to both as in spiteful behaviour. While selfishness seems to flow directly from the principle of natural selection, the three other options are in need of selective explanations that do not violate the Darwinian explanatory framework for conflicting interactions.

Table 4.1. Classification of different types of social behaviour

		Effect on recipient	
		Beneficial	Costly
Effect on actor	Beneficial	Mutualism	Selfishness
	Costly	Altruism	Spite

Darwin was the first to frame the problem of co-operation in the context of natural selection, and his proposed explanations as to how competitive interactions became transformed into co-operative interactions is strikingly close to recently elaborated quantitative models. A foundational principle in Darwin's theory of natural selection was the competition and conflicts that occurred among the living. The struggle for existence included both the individual's struggle for preservation of its own life as well as its success in leaving progeny.

> Hence, as more individuals are produced than can possibly survive, there must in every case be a struggle for existence, either one individual with another of the same species, or with the individuals of distinct species, or with the physical conditions of life. (Darwin 1859, p. 117).

The result of this struggle would be a gradual increase in the perfection of the individuals of the evolving lineage. Darwin did not envision this to be global perfection, from which all small changes would lead to deterioration. Rather, individual perfection was local and relative to the perfection of competing individuals. Furthermore, natural selection was envisioned as an utterly individualistic process.

But being an astute naturalist, Darwin was well aware of the extensive co-operation and self-sacrifice performed by organisms in their natural surroundings. In the context of his treatise he discussed both the sting of the wasp, which causes its own death by tearing out its viscera, and the evolution of sterile female workers in insect colonies. Since suicidal stinging and female sterility dramatically reduce the fitness of the altruistic organisms relative to that of selfish individuals, the competitive process of natural selection could hardly account for the evolution of the traits if selection worked only in the interest of the individual organism. Darwin's solution to this dilemma was to look upon how parts and wholes are related to each other.

Darwin realised that co-operation represents the benefit of group living. Though altruistic individuals do worse than selfish ones, groups with a high proportion of altruists might do better than groups with a lower proportion. Like individual organs co-operate for the wellbeing of the whole organism, individual organisms may co-operate for the wellbeing of the group. Darwin thus recognised that selection could take place at several levels in the biological hierarchy. But this would imply that natural selection, which breeds on conflict, must somehow manage to weather the conflict and to make the participating cells or organisms co-operate.

> we can perhaps understand how it is that the use of the sting should so often cause the insect's own death: for if on the whole the power of stinging be useful to the community, it will fulfil all the requirements of natural selection, though it may cause the death of some few members. (Darwin 1859, p. 230).

These ideas, while long dormant, were reconsidered during the 1960s with publications of Hamilton (1964a–b) and Williams (1966), in which they discussed the explanatory value of group selection. They argued that group selection was biologically unlikely, and that many of the cases in which group-level selection had been utilised as the explanatory model could be explained as well or better by invoking selection at the individual level. In addition, mathematical models were developed that showed that group-level selection was likely to occur under limited and strict conditions. The kinds of explanations dismissed were such that proclaimed that selection acts for the good of the species or the group, when in fact individual level selection would explain the phenomenon better. That traits are good for the species was seen as uncontroversial; it was the mechanism that generated the species-advantageous traits that was at issue.

A good example of what the debate was all about can be found in explanations provided for the emergence of genes encoded in the MHC complex. Since different allelic products bind different arrays of peptides, it will be advantageous to have a high MHC variability at the population level because such a population has a higher probability of not becoming extinct than has a population with low MHC variability. This has led to the belief that MHC diversity is maintained because it "provides broad immunological protection for the species as a whole (Roy et al. 1989, p, 574) and because diversity is involved in "a strategy to keep parasites from spreading through the entire population" (Klein and O'Huigin 1994, p. 355), and hence the statement by Hull, Langman and Glenn (2001, p. 520) that "the polymorphism of the MHC locus provides a particularly clear example of selection on alleles of genes that must occur at the population level while still being executed at the level of the individual organism". However, it is also advantageous for an individual to be heterozygous at MHC loci. This is because a heterozygote will be able to present a broader array of antigens and thus resist a broader array of pathogens. Thus, there should be no need to invoke species or group level selection for the diversity of MHC genes. Diversity of MHC genes may fore example be maintained because of balancing selection,[5] which refers to forms of natural selection that work to maintain polymorphisms within a population. The outcome may, for example, be owing to heterozygote advantage or frequency dependent selection. And since there is no conflict between the group and the individual, the selection of MHC diversity can be explained as individual-level selection, thus implying that the benefits to the group arise as statistical summations of the individual adaptations.

Altruism and spitefulness

A solution to the most difficult behaviours to explain, behaviours that are costly to the actor, was worked out by Hamilton in two papers published in 1964. In what has come to be known as Hamilton's rule he showed that co-operation will spread if

[5] The term balancing selection stands in contrast to directional selection, which occurs when selection favors a single allele. Directional selection is sometimes referred to as positive selection.

$rb > c$, where b is the benefit conferred to every other individual in the population, c is the cost to the benefactor or altruist, and r a measure of relatedness.[6] For the cells of a single individual or the clonal descendants of an organism $r = 1$. For clonally derived organisms, altruism evolves precisely when the benefit to recipient exceeds the cost to donor, $b > c$. When groups are composed of relatives, altruism can evolve because relatives resemble each other. This implies that altruism evolves if the beneficiaries of the altruistic behaviour are bearers of the same altruistic trait as the altruistic individual. For example, a bird that cries out that predators are under way, and that does so at a great risk to itself, promotes the self-sacrificing gene in the group of related birds. And so does a yeast cell that undergoes programmed cell death, thereby promoting the survival of close kin.

Nothing in this mathematical formulation alludes to how animals are able to recognize kin. Nevertheless, such recognition is important because otherwise kin would not be visible for selection. As genes become accessible for evaluation only through their phenotypic effects, kin recognition should involve phenotypic correlates of genes. The polymorphic MHC system has been given a role in kin recognition in vertebrates, as it allows animals to perceive the genetic similarity of other animals and to take action upon these percepts (Brown and Eklund 1994; Ziegler et al. 2005). Recognition of differences in MHC associated odours have been demonstrated in both mouse and man (Penn and Potts 1999), and women are able to detect differences of one MHC allele among male donors with different MHC genotypes (Jacob et al. 2002). Hence, kin selection in vertebrates may in part be served by signals derived from the MHC molecules.

In his original formulation of kin selection, Hamilton (1964a–b) described relatedness in terms of genes identical by descent. But in a later formulation he showed that what matters is not common ancestry but statistical associations between the genotypes of donor and recipient (Hamilton 1970). In this reformulation, the coefficient r became a measure of the extent to which actors are either more similar to recipients than to the average members of the population (positive r) or less similar (negative r). An interesting result of this reformulation is that relatedness can be negative, thus allowing evolution of spiteful behaviour. This evolves when it is adaptive for animals to harm themselves in order to harm others more. Spite can be favoured because the product of negative relatedness and a negative benefit to a recipient (harm) is positive. Benefit multiplied by relatedness can outweigh the cost, and it may pay for an animal to take the life of a "negatively related" neighbour even at some cost to itself. If available resources limit group size, attacking the offspring of others will increase the survival chances of one's own offspring.

A defence mechanism that is harmful to the cell that defends but helps its relatives may also be termed spiteful. A bacterium may die to kill negatively related conspecifics to save its kin (Gardner et al. 2004), or a phagocyte may harm itself by attacking a bacterium, but in so doing helping the organism of which it is

[6] It remains conceptually unclear how to estimate b and c, which both depend on ecological conditions, and researchers have therefore tended to estimate r and to focus on reproductive dynamics.

part. Spiteful behaviour can hardly be explained by the organism's striving for self-preservation, and also seems to contradict Darwin's claim that natural selection will never produce in a being anything injurious to itself. Nevertheless, when natural selection is envisioned as a process that occurs at multiple levels, such behaviour is seen as one of several possible outcomes of the selective process.

Multi-level selection

When Hamilton (1970) revised his theory he utilised a mathematical formulation in which evolutionary change was described as the covariance between a trait and its fitness (Price 1970; Price 1972; Price 1995). Briefly, the Price equation states that in order to asses the fortunes of a particular trait one must know whether having that trait correlates with having more children, and how likely it is that those children will actually inherit the trait. The equation, which formalises Lewontin's (1970) idea that a trait is selected if it has heritable variation in fitness, offers a method for the hierarchical analysis of natural selection and can be used to assess the effects of selection at different levels. For example, when groups differ in the frequency of individuals with a trait, say altruism, the trait will be selected for if groups with many altruists perform better than groups with few altruists and many selfish individuals.

The Price equation has had a major impact on the level-of-selection discussion, and there is now a general agreement that natural selection can occur at all levels in the biological hierarchy (Sober and Wilson 1998; Keller 1999; Michod 1999). The consensus view is that selfish individuals are generally favoured by *within-group* selection since they receive benefits that are generated because of the actions of the co-operators, but pay none of the costs. As opposed to within-group selection, *between-group* selection favours co-operation if groups with more co-operators do better than other groups. The relative strength of within- and between-group selection then determines whether co-operation evolves in any particular situation. Multi-level selection theory is thus able to explain the evolution of altruism, spitefulness and even apoptosis.

Since biological entities, be they organisms or groups, evolve under natural selection when there is heritable variation in fitness, and since variance between groups is usually smaller than variance between organisms, selection among groups is usually less effective than selection among organisms (Leigh 1999). Moreover, since groups tend to live longer than their component organisms, selection at the level of the group is slower than selection at the lower levels.

SHAPING THE MULTICELLULAR ORGANISM

Communal unicellular organisms

The two essential ingredients of sociality, co-operation and conflict, have traditionally been studied in insects and higher animals. But the sorts of mechanisms that lead to sociality and its disruption have repeatedly been discovered also in unicellular organisms. Many recent discoveries in microbial ecology and evolution have shown that social phenomena that are already well known in vertebrates and invertebrates also occur in communities of microorganisms. Like multicellular organisms, social microorganisms co-operate to build shelters, forage, reproduce, and disperse (Crespi 2001). They even perform altruistic suicide to preserve the integrity of kin. Such observations lend support to the generality of evolutionary theory, and further support the view that a deeper understanding of the co-operative mechanisms in microorganisms and the ways in which selfish individuals are dealt with could lend new insight into how complex organisms develop and evolve.

The interactions between microbes, like those between cells in eukaryotic multicellular organisms, are regulated by chemical signals that serve to co-ordinate social activities. Many species of bacteria are, for example, able to sense and respond to variation in local bacterium density by the release and uptake of signalling molecules. This sort of cross-talk between microbes, which may occur between members of the same as well as between cells of different prokaryotic species, and even between prokaryotes and eukaryotes (Joint et al. 2002), has been termed quorum sensing[7] (Fuqua et al. 1994). Some bacteria have multiple quorum-sensing systems, and it appears that certain bacteria are able to recognise and respond to multiple threshold points, each tied to a distinct biological response (Bassler 1999).

Quorum sensing has recently become associated with the pathogenicity that occurs during infectious disease. An illustrious example of this is the pathogenic relationship between the bacterium *Erwinia carotovora* and the soft rot that it causes in a variety of plants (de Kievit and Iglewski 2000). When the bacterium is present at a low concentration on a plant, the genes associated with pathogenicity are quiescent and the defence mechanisms of the plant are kept silenced. However, upon reaching a sufficient bacterial concentration, the quorum signals reach a threshold concentration and two distinct sets of genes become expressed. One set of genes produces enzymes involved in the maceration of plant tissue and thus liberates nutrients from the host, while another set of genes synthesises the antibiotic carbapenem. The production of the antibiotic ensures that the nutrients released are available only to *E. carotovora* cells because bacteria of other species are killed.

It appears that quorum signals and their receptors originated early in bacterial evolution and that such systems evolved independently in many bacteria (Gray and Garey 2001). Many of the functional pairs of signal and receptor are apparently inherited as a co-evolving regulatory module, but in some cases there is evidence of horizontal transfer between species. Attempts to model the evolution of such systems have seized upon the observation that singular bacteria bear a cost while the group in which they live receives the benefits, thus suggesting that the selection for quorum sensing occurred as a group selective process (Brookfield 1998; Brown

[7] The word "quorum" refers to the number of group-members required to be present at a meeting in order to legitimise a given decision. The term quorum sensing was coined in 1994 by Fuqua et al., and the rapidly developing field of bacterial cell-to-cell signalling has crystallised around this term.

and Johnstone 2001). This may have taken place through a two-step mechanism, in which the first step was evolution of a diffusion sensing system that measured the stability of the organismal environment. Diffusion sensing would allow bacteria to secrete degradative enzymes and antibiotics only under conditions in which losses due to environmental fluctuations were minimised (Redfield 2002). During the second step, groups with intact quorum-systems may have survived better than their competitors. Hence, lower-level selection lead to the evolution of signal-response systems that benefited the individual organisms, while higher-level selection led to the evolution of quorum sensing that benefited the group.

Quorum signalling offers an important advantage of surprise to bacterial invaders by synchronising the production of pathogenicity factors. The dynamic nature of living entities has, not unexpectedly, led to the evolution of counter measures against quorum sensing. For example, *Bacillus subtilis*, like *E. carotovora* a common soil bacterium, produces an enzyme that inactivates the quorum signals of *E. carotovora*. This quorum quenching effectively renders the *E. carotovora* bacteria a-pathogenic (Dong et al. 2001). The counter measures against quorum signals may have evolved for several reasons, but one plausible hypothesis would be that they evolved as a counter measure against the bacteriolytic effects of carbapenem. Counter measures against bacterial quorum signals have also been observed in plants (Fray 2002). In vertebrate animals, quorum signals have immunomodulatory activity, affecting the stimulation of lymphocytes and production of antibodies (Ritchie et al. 2003), but no direct counter mechanism that interferes with quorum signalling has so far been described.

Transitions in individuality

The rules they are achanging

The history of life on Earth is the history of how new entities were created from previous forms; it is the history of how the eukaryotic cell evolved from the prokaryotic cell, and of how the multicellular organism evolved from the single cell through a developmental process. It is thus the history of how natural selection allowed the transition of life from one level to another, and of how this shaped the variegated forms of life that have inhabited the Earth. While lower self-replicating units in their free-living state were selected directly by the environment, they became selected by traits expressed by the higher level unit after the transition. Hence, the rules of natural selection changed at each transition.

In an attempt to integrate evolutionary and developmental biology, Leo Buss (1987) set out to explain the rules of such evolutionary transitions. Like Lewontin (1970) Buss envisioned living entities as being composed of interdependent structural units that could be targets of selection provided that they exhibited heritable variation in fitness. But while Lewontin's argument was set forth to underscore that natural selection operates at multiple levels, Buss (1987) and later on Maynard

Smith and Szathmáry (1995) set forth arguments to emphasise that the levels on which selection acts have themselves been generated by natural selection.

The latter transitions were, nevertheless, not necessarily more sophisticated than their priors. Prokaryotes and viruses were then as now the chief molecular innovators of the biosphere, and their ranges of metabolic complexities and ecological niches far exceeded those found in the eukaryotic world. While the transitions involved novel solutions to common problems, they did not necessarily involve novel solutions built from scratch. More often than not, the higher levels came to reutilise what was already made by the prokaryotes and viruses. Such knowledge has falsified the until recently held belief that unicellular organisms are nothing but undifferentiated cells with genetic programs focused only on individual growth and reproduction (Shapiro 1998; Kroos and Maddock 2003).

Vulnerability to cheating

Several unicellular organisms form multicellular structures with clear-cut divisions of labour for divergently differentiated cell types. These include the prokaryotic bacterium *Myxococcus xanthus* and the eukaryotic slime mold *Dictyostelium discoideum*. Both types of organisms live in soil where they feed upon bacteria – by the process of phagocytosis in *D. discoideum* and by secretion of extracellular digestive enzymes by *M. xanthus*. Upon starvation the microbes glide towards aggregation centres where they form multicellular fruiting bodies from mixtures of genetically different clones of the species. During aggregation *D. discoideum* sacrifices 20% of the cells to form a supporting stalk, while up to 90% of the cells die during the formation of the *M. xanthus* fruiting body.

Some strains of cells are over-represented among the spores in mixed-strain multicellular organisms (Strassmann et al. 2000; Velicer et al. 2000), thus allowing them a higher probability of reproduction than their competitors. This phenomenon demonstrates a characteristic feature of co-operative groups; their vulnerability to cheaters that reap the benefits of co-operation without paying the costs. Vulnerability to cheating is especially pronounced in organisms that form from mixtures of genetically different clones of the species. But because natural selection can act at multiple levels, all kinds of biological societies are under threat from selfish evolutionary individuals. Hence, the phenomenon of cheating is seen also in complex multicellular organisms.

One of the best studied examples of cheating in higher organisms is the one that occurs when two colonies of the ascidian *Botryllus schlosseri* grow into contact with each other.[8] Upon contact they either fuse, which includes establishment of vascular continuity and genetic chimerism, or they develop an inflammatory rejection that

[8] *Botryllus schlosseri* is a colonial urochordate that reproduces asexually but with an occasional sexual generation. The zygotes of *B. schlosseri* mature into tadpole larvae with a chordate body plan. Metamorphosis results in the loss of the chordate body plan and the development of a sessile oozooid with its own tissues and a circulation.

permanently separates the colonies (Stoner and Weissman 1996). When genetically disparate colonies fuse, their somatic and germ cells are allowed to move freely between the two previously disparate organisms, thus setting the stage for somatic as well as gonadic chimerism and thus for the parasitism of one colony by the other. The result of this is that some colonies become disproportionally represented in the gametes, and that gametes of one colony can totally outperform gametes of the other (Stoner et al. 1999). Thus, underneath the fusion event, which superficially resembles co-operative behaviour between genetically distinct organisms, is an act of competition at the cell level.

Conflict modification

The formation of co-operative interactions among lower-level units is a necessary step in any evolutionary transition. There is, nevertheless, always a "temptation" to cheat, and an important aspect of evolutionary transitions to new levels is therefore the harmonising of interests between the levels. Theoretical investigations have made clear that whether multicellularity develops or not depends upon the strengths of the control mechanisms involved in harmonising the competing individuals. Moreover, the way in which conflicts associated with transitions in individuality have been mediated can influence the potential for further evolution of the newly emerged evolutionary individual.

Richard Michod (1999) has stated that the basic problem in evolutionary transitions is to understand how previously independent and competing entities become a new kind of entity, possessing the properties of heritable variation in fitness at the new level of organisation. In the evolutionary transition from groups of cells to a multicellular organism there are, for example, two manifest levels – the cell and the organism, at which selection may occur. Multicellularity requires that the group of cells manages to restrict variability between its component parts so that there will be little option for selection at the lower level. To reduce the risk of intraorganismal conflicts among cells that vary, several modifier mechanisms that enforce cooperation have evolved (Frank 1995; Michod 1999).

That cheaters are constrained in *M. xanthus* and *D. discoideum* as they form multicellular fruiting bodies from mixtures of genetically different clones is a long time observation (Dao et al. 2000). In an attempt to explain the phenomenon, Wilson and Sober (1989) suggested that the random distribution and congregation of cells might allow the build-up of heterogeneous groups with variable fitness. Since random segregation of traits into groups is a mechanism which may lead to trait-group selection (Wilson 1979), they argued that the social behaviour is explicable by the process of trait-group selection on altruistic behaviour (Wilson and Sober 1989; Velicer 2003).

While this proposal gave a fairly good account of the observations, it did not allude to the mechanisms involved. Such a mechanistic explanation was provided by Foster and co-workers in 2004 when they described a gene, *dimA*, which has two contrasting effects on the cooperative interactions of *D. discoideum*. While cells that express the gene are apt to develop into prestalk cells, cells that do not express the gene behave like cheaters in chimeric slugs by increasing their representation in the prespore population. The *dimA* negative cells do, nevertheless, not benefit from this cheating because the lack of *dimA* excludes them from spore-formation later in development. The pleiotropic linkage between *dimA* of stalk and spore formation thus limits the potential for cheating (Foster et al. 2004).

But of course, the punishing of defectors is itself an altruistic action. It is costly to the punisher while being beneficial to the organism. The existence of a control mechanism can therefore not be taken for granted – it is necessary to explain how it evolved. By calculating the costs and benefits to cells and organisms in a multi-level selection model, Michod (1999) demonstrated that a modifier trait might evolve if the cost of policing is kept within defined limits. But the introduction of modifier mechanisms is apparently not sufficient to restrict cheating; there is still a drive for the individual cells to strive for own success against their neighbours. This drive may breach through as malignancy or as other types of unrestricted cell growth.

Evolvability

Most descriptions of evolution have assumed that mutations are completely random with respect to their potential effects on fitness. Since "selection lacks foresight" the argument goes, existence of mechanisms that may promote evolution are not conceivable (Dickinson and Seger 1999). Nevertheless, some empirical evidence favours the idea that genomes may be selected for evolutionary potential. The potentiality for useful mutations in some surface proteins of bacteria and in the immune system of vertebrates are, for example, so probable that they can be viewed as being encoded implicitly in the genome (Caporale 2000; Caporale 2003).

"The genome's ability to produce adaptive variants when acted on by the genetic system" (Wagner and Altenberg 1996) has been termed evolvability. But evolvability has also been defined as "the capacity to generate heritable, selectable phenotypic variation" (Kirschner and Gerhart 1998). Taken to its logical conclusion, this latter definition would allow evolution to occur without mutational change in the genetic system. And this is exactly what may happen in quasi-independent modular systems. In accordance with Kirschner and Gerhart's (1998) view, modular systems that are constrained by exacting and numerous requirements for their functioning are less likely to evolve than modular systems that exhibit i) broad-specificity proteins, ii) a capability to function in multiple contexts, iii) a propensity for self-organisation and somatic selection and iv) compartmentation into relatively independent subsystems. Such modules are weaker constrained than those that do not exhibit the same functions, and would, according to Kirschner and Gerhart's hypothesis, release the developmental constraints on the evolution of novel phenotypic traits.

Evolvability is a fitness-enhancing trait that manifests most vehemently at the level of the lineage – the greater the variability within a species, the better it can respond to selection as the conditions of life change. Interestingly, Dickinson and Seger (1999) claimed that the process of evolvability had to be dismissed because it could evolve only as a group level adaptation. This they claimed, citing Williams (1966), occurs only under very restrictive conditions, and is therefore not plausible. But even though it is true that species-level selection has been criticised for being implausible, multi-level selection theory does not argue against it being possible (Lloyd and Gould 1993; Gould and Lloyd 1999). In their discussion of the evolution of evolvability, Kirschner and Gerhart took this perspective when claiming that evolvability is selected both at the organismal level, because plastic modules contributes directly to organismal adaptedness, and at the level of the lineages, because lineages containing organisms with larger amounts of modularity leave more offspring – more lineages – than those with smaller degrees of modularity do.

Since variability can evolve only through systematic association with features directly under selection, selection for evolvability of the immune system might have been selected not only at the organismal but also at the species level; this may have come through if higher fitness was bestowed to the jawed vertebrates as compared to the lineages with which they competed. The rapid evolution of the adaptive immune system, in about 50 million years, indicates that there must have been a strong selective pressure for it. This may have come about if different species differed in the frequency of individuals with adaptive immunity, and if species with many organisms with adaptive immune systems performed better than species with few organisms with adaptive immune systems. If species with adaptive immunity reproduce and generate more new species than species without adaptive immunity, it could even have been the case that species selection for an adaptive immune system could have occurred, for example owing to herd immunity, despite adaptive immunity being fitness-decreasing at the organismal level.

CONSTRAINTS ON SOCIAL EVOLUTION

The self-other divide

The origin of life also marked the origin of environment. The two entities, cell and other, were inseparable – like yin and yang. Since the environment provided resources but also threats, the emergence of an enclosed self necessitated cautious relating to the environment. Thus came the defence systems that demarcated the self from some non-self. However, as the cell needed to obtain resources from the environment as well, it could not put up defences at all non-self. Hence, the defence systems needed to discriminate between various kinds of non-self. Because noxious and nutritious non-self may be very similar, the first living entities had to evolve defence systems that were sufficiently specific to discriminate between signs and to decide when to defend.

The rules of the game changed markedly with the emergence of the first multicellular organisms. The individual cells could no longer use the defence mechanisms against cells of its own group. Those that did dramatically reduced their fitness, and so their descendants are no longer around. To be successful, the multicellular creature had to allow some of its cells to differentiate into specialised cells. Some of the cells transformed into committed germ cells, others may have had a function in nutrition, while still others may have transformed into specialised cells of defence. After the successful transition to multicellularity the discrimination of self against non-self had to incorporate tolerance to self as well as defence against non-self. Tolerance to self was necessary because not all cells of the organism would be identical.

Evolutionary transitions from unicellular to multicellular forms also involve the de-coupling of cell division from reproduction (Michod and Nedelcu 2003). Whereas the fitness components of unicellular organisms, which involve cell reproduction and survival, are coupled at the cell level, the same fitness components have been diverted to two different types of cell populations in multicellular organisms. These include the somatic cells, which perform the survival behaviour of the organism at the cost of having a suppressed reproductive potential, and the germ cells, which perform the reproductive role but who have suppressed their potential for performing survival-related functions.

The costs of self-defence

Expression and use of defence traits, which are beneficial in terms of reduced pathology and increased survival, are usually thought to come costly to the organism. Nevertheless, numerous examples in several taxa have demonstrated that natural selection can minimize the initial costs of defence by rendering the susceptible organisms naturally resistant (Rigby et al. 2002). For example, humans achieve a degree of cost-free resistance to malaria through a mutation in the chemokine receptor Duffy. The mutation abolishes expression of the receptor in erythrocytes and thus the binding and invasion of malarial parasites (Pogo and Chaudhuri 2000). Another example of the same phenomenon is the natural and cost-free resistance to HIV in humans with a specific mutation in the chemokine receptor CCR5. Individuals homozygous for a 32-base pair deletion in the gene for CCR5 display a strong, albeit not complete, resistance to HIV-infection. The deletion, which is associated with the absence of CCR5 from cell surfaces and thus with the absence of a port of entry for the virus, was until recently believed to have no apparent immunologic or biochemical costs (Paxton and Kang 1998). Some recent observations do, however, suggest that CCR5 is important for resistance to the damaging effects of West Nile virus, a pathogen that causes encephalitis (Glass et al. 2006).

As there are additional cost-free defence mechanisms associated with behaviour, life-history and anatomical structure, the cost of resistance hypothesis is not universally valid. Still, there are numerous examples of costly defences to consider. The costs are manifested by the need to re-allocate limited resources from one fitness-related trait to another, for example from reproduction to survival. A significant increase in investment on defence mechanisms to enhance survival is for example likely to compromise the organism's reproductive fitness. The amount of resources devoted to defence should therefore be balanced against a possible alternative use of the same resources (Zuk and Stoehr 2002). For organisms of some species the investment in defence mechanisms against infectious agents may well be the most important determinant of their lifetime reproductive success (Lochmiller and Deerenberg 2000).

The freshwater pond snail *Lymnaea stagnalis* provides an illustrious example of the trade-offs associated with life history decisions (Rigby and Jokela 2000). The most common predator-avoidance behaviour in snails is to retreat deep within the shell when attacked. To do this, the snail must expel blood. After the attack the snail replaces the blood with surrounding water, which can contain pathogenic microorganisms that must be neutralised by the defence system of the snail. The expulsion of blood is simultaneously a depletion of immunocompetence, at a time when increased immunocompetence is required. Rigby and Jokela (2000) showed that predator avoidance reduced the ability of the snails to defend against infectious agents and that it had clear costs in terms of reduced reproduction and survival.

While this particular form of predator avoidance is clearly special, the results signify that behavioural and various ecological factors may modify the expression and costs of defence mechanisms against infectious agents. The coupling of defences against predators and parasites may be a more general phenomenon, though. Species with a high injury risk are supposedly more susceptible to wound infections than species with lower injury risk, and would therefore benefit from a more efficient defence system. There are indications that variations in the immune systems across primate species has been influenced by the risk of injury (Semple et al. 2002).

The fitness costs associated with defence mechanisms have been analysed in a variety of ways and in a variety of species. Most studies have measured costs of defence by measuring alterations in reproductive potential (Lochmiller and Deerenberg 2000), but survival has also been measured, for example in bumblebee workers where reproduction does not occur (Moret and Schmid-Hempel 2000). The trade-offs have also been investigated experimentally, for example by first restricting energy input, then challenging the organism with a stimulating agent, and finally measuring the capacity of the defence systems to mount a response (Demas et al. 2003); or by first stimulating the defence system and then measuring the effects on reproductive effort (Bonneaud et al. 2003). While the methods used to measure life-history trade-offs in association with defence mechanisms have been criticised (Norris and Evans 2000), the general result from the studies, that mounting defence responses are costly, is uncontroversial. Cost-benefit analyses of immune function, embedded within a life history framework, may thus be of use to model variations in the evolution of defence among different species, as well as of variations associated with pregnancy, age, exercise, sex and season of the year (Read and Allen 2000).

As for the innate immune system, the maintenance of the vertebrate adaptive immune system is costly in terms of metabolic energy. For example, the IgG that is secreted into plasma accounts for approximately 0.3% of the total protein synthesis of healthy humans (Thornton et al. 1996). In a similar way, the wasteful production of lymphocytes, of which a high proportion die before they become immunologically competent, is believed to be costly in terms of metabolic resources.

As compared to maintenance costs, the mounting of an immune response towards an infectious agent is even more costly in terms of metabolic energy. To give an example, the stimulation of mice with lipopolysaccharide mobilises about 7% of its genes (Yoo and Desiderio 2003). And while defence is generally thought to increase the chances to survive, the mounting of an immune response under stressful circumstances may actually lead to reduced survival, as has been demonstrated in ducks (Hanssen et al. 2004). There may also be a direct relation between immune activity and reproductive output. The energetic costs of utilising the cell-mediated immune response in house sparrows, *Passer domesticus*, has been estimated to be of the same amount as the costs of producing half of an egg (Martin et al. 2003). Hence, if reproductive activity were to coincide with infectious disease in these birds, one of the two fitness-related functions, survival or reproduction, would have to be reduced if the birds could not increase their daily energy intake.

Since defence must be balanced against the associated costs of hyperreactivity and autoreactivity, the optimal immune response is not necessarily the maximal response. There is much evidence that immune-reactivity becomes suppressed during stressful events. This phenomenon may have two adaptive explanations. One is the resource-trade-off hypothesis of life-history theory, in which limited energy must be redistributed between functions of vital interest (Sheldon and Verhulst 1996). The other idea is that immune-reactivity needs to be balanced against the risk of generating immunopathology (Råberg et al. 1998).

In accordance with the latter idea, it is conceivable that during the evolution of defence mechanisms, most encounters between hosts and pathogens were localised to a small area of the host body, and that the pathogen inoculum was of a limited size. This would select for defence mechanisms that would provide a strong but limited attack on the microbial invaders. While a localised, immediate and strong immune response makes sense when the inoculum is small, large inoculums and overwhelming disseminated infections would turn the same defence reaction into a disaster for the responding organism, as happens when patients succumb to septic shock.

In a series of papers Emi Shudo and Yoh Iwasa (2001, 2002, 2004) modelled the optimal defence response of a host under various conditions of defence and infection. They identified the optimal defence strategy of the host as the one that minimises the total cost, and defined it as the weighted sum of damage caused by the infectious agent and the cost of using and maintaining defence responses. Their results are compatible with the previously described experimental results, thus suggesting generality of the models.

Evolution of defence mechanisms

Defences have probably been with us since soon after life's inception. The evolution of defence mechanisms was probably a gradual process, during which more recent mechanisms were built upon the scaffold of older mechanisms. Within multicellular organisms the evolution of specialised defence traits by specialised cells marked that fitness at the level of the cell had been transferred to fitness at the organismal level. Since the costs were carried by the individual cell while the benefits were accrued at the level of the organism, the costly defences were seemingly altruistic for the defending cell.

The defence mechanisms that evolved were of two principled functional types – constitutive and inducible, of which the innate and adaptive defence mechanisms are subtypes. While constitutive defence mechanisms utilise an unconditional strategy, inducible defences utilise a defence strategy that is conditioned upon the way the world is. Inducible defence mechanisms allow the bearers of the mechanisms to respond flexibly to a variety of infectious agents, and in some cases to learn and remember the previous encounters. However, since it may sometimes be useful to meet a complex world with a simple strategy while a more complex response may be useful in other situations, the accrued fitness benefits associated with bearing an inducible versus, or in addition to, a constitutive defence response, are not straight forward to ascertain.

The utility of a defence mechanism depends upon the prevalence of infection as well as on the severity of harm that infectious agents generate in undefended organisms. And for costly defences, the utility needs to be traded off against the fitness-related costs involved as well. These costs are of two types. In the absence of infection, allocation of resources to maintenance of defence mechanisms carries fitness costs in terms of survival and reproduction. And during infection, there are fitness-related costs associated with the response. These include costs of self-harm and costs involved in performing the wrong response.

The requirements for the evolution of defence systems can be likened to the requirements involved in the evolution of phenotypic stable versus plastic strategies (Moran 1992), or to the evolution of innate versus learning strategies. Basic questions that call for principled answers include:

i) Under what conditions is it better to meet a complex world with an inducible as compared to a constitutive response?
ii) Under what conditions is it better to meet a complex world with an adaptive as compared to an innate response?

Making the production and utilisation of defence mechanisms conditional upon environmental characteristics may reduce the costs, allow increased allocation of energy to growth and reproduction, and thereby reduce the total fitness loss to the organism. The inducible strategy should thus be favoured over the constitutive response when the reduction in maintenance costs is higher than the expected cost of harm induced by the infectious agent; or stated differently, inducible defences should be favoured over constitutive defences when the organism benefits more than it loses in terms of a higher initial cost of attack.

While an innate defence mechanism may evolve if there is sufficient matching between the environmental sign and the fitness-enhancing response, an adaptive mechanism has the further requirement that the responding organism must be able to modify its behaviour upon receiving the sign. In most real world situations signs are distorted by noise. Hence, the various states of the world can seldom be discriminated by a 100% reliable sign. And as there are no deterministic linkages between signs and responses, selection for adaptive traits should be correlated with the likelihood that an organism will perform a certain response given a certain state of the world.

The question "When will natural selection favour adaptiveness over constitutiveness?" is not easily answered. Whether the defence mechanism will be adaptive or constitutive depends upon the costs, on the probability of encountering the environmental agents, on the reliability of the signs encountered, on the capacity of the system to discriminate between states of the world, and on the performance of a stimulus-adapted appropriate response. The parameters may be combined in multiple ways. In line with the parlance of the predictive accuracy framework, the parameters are adjustable. When the parameters have been adjusted based upon how the world is, it may well turn out that constitutiveness should be better in some environments whereas inducibility should be better in other environments. Thus, what evolves as a constitutive response in one species may be solved by an inducible response in another species.

But even though various defence systems become selected according to their long-term costs and benefits, this does not imply that the fitter response will replace the less fit. The expected importance of defence systems is high in nature, as evidenced by the fact that the defence systems of vertebrate organisms are mosaics of invertebrate defence systems (Salzet 2001). It thus appears that phylogenetically elder defence systems are not obsolete, but rather that new defence systems become integrated with the older systems, and that they function together in a modular fashion.

4.3. Coevolutionary dynamics

ENDOSYMBIOSIS

Symbiogenesis

Many evolutionary transitions have involved the assimilation and merging of previously independent living entities into an adaptive entity at a higher level. This phenomenon, in which organisms of different species live together in close association and which results in a raised level of fitness for one or more of the organisms, is termed symbiosis. The evolution of symbiosis relies on intimate relationships over extended periods of time to allow symbionts to co-evolve and become dependent upon each other.

When the symbiotic partners combine and unify their metabolic apparatus to form an integrated endosymbiotic unit, they partake in the process of symbiogenesis. The term endosymbiosis is used for the condition in which one organism lives inside another and where formerly independent individuals come to live and reproduce as a whole. The evolution of endosymbiosis may result in mutualism if it is of advantage to both partners, or in parasitism if it is of disadvantage to one of them.

While theoretical work has suggested that mutualism and parasitism may be dynamic alternatives to the partners in host-microbe interactions (Frank 1997), the generality of this idea has been questioned by recent observations of genomic structures. These studies show that establishment of permanent parasitic or symbiotic relationships usually lead to massive gene loss and to a high stability of the genome that remains after the loss (Tamas et al. 2002), thus precluding shifts between parasitism and mutualism.

The merged organisms sometimes become so entangled that their separate origins become obscured. The difficulties in tracing the blending of two organisms have been likened to the tracing of the grin of the Cheshire cat. As Alice in Wonderland watched the cat dissolve, the grin remained some time after the rest of the cat had gone. In accordance with this simile, "There are a number of objects in a cell like the grin of the Cheshire cat. For those who try to trace their origin, the grin is challenging and truly enigmatic". (Smith 1979, p. 128).

Symbiogenesis has resulted in the formation of chimeric organisms that exhibit new morphologies and physiologies that are evolutionarily more advanced than those of their constituents. This projected life to greater levels of complexity, posing new potentials, but also risks. The establishment of co-operation lead to the emergence of new functions and gave us the eukaryotic cell, cytoplasmic organelles and multicellular organisms (Buss 1987; Maynard-Smith and Szathmáry 1995). To explain the observation that the eukaryotic genome is chimeric, consisting of genes from both archae and bacteria (Katz 1998), the serial endosymbiosis theory has been proposed (Gray et al. 1999). According to this theory, eukaryotic cells evolved when prokaryotic cells merged into a symbiotic relationship (Goksøyr 1967; Gupta and Golding 1996; Margulis et al. 2000), and in a similar manner eukaryotic cells obtained organelles, like mitochondria and plastids, when they merged with bacteria (Gray et al. 1999).

Although no intermediates of these transitions have survived or left fossils to provide direct clues, genome sequencing and phylogenetic analysis single out endosymbiosis and co-evolution as the most fundamental processes leading to viable eukaryotes. In the context of the endosymbiotic theory of the eukaryotic cell, both the nucleus and the mitochondrion have characteristics that are closely related to bacterial cells. But it is also evident that genes have entered the vertebrate genome by lateral transfer of bacterial DNA to germ cells. The project that sequenced the human genome claimed that at least 113 genes had entered the vertebrate lineage from bacteria (Lander et al. 2001), but this number has been criticised on methodological grounds, and the relevance of bacteria-to-vertebrate transfer is still an unresolved issue (Genereux and Logsdon 2003).

The mitochondrion

The power generators of the eukaryotic cell, the mitochondria, have a genome of their own and divide within the cytoplasm, often uncoupled from the cell's division cycle (Osteryoung and Nunnari 2003). Their reproductive dynamics are restricted by their being inherited almost exclusively along the maternal lineage. This ensures limited competition between maternal and paternal mitochondria. But there are still opportunities for competition and conflict at several levels; within each mitochondrion there are multiple mitochondrial DNAs, and each eukaryotic cell harbours about 500–1000 mitochondria in its cytoplasm. The organelles are thus, in a sense, autonomous units with a complicated population dynamics of their own.

Recent whole genome studies have provided some insight into the mitochondrion's "grin". Evolutionary trees based on rRNA sequencing originally pinpointed the rickettsial subdivision of the so-called proteobacteria as the closest contemporary bacterial relatives of mitochondria. These results were confirmed by comparative studies of the complete genome sequence of mitochondria and *Rickettsia prowazekii*, an obligate intracellular bacterium that causes typhus, thus indicating that the DNAs of *Rickettsia* and mitochondria are close relatives. In addition, the functional profiles of *Rickettsia* and mitochondria are strikingly similar (Andersson and Kurland 1998).

There has been an extensive coevolution between the nuclear and mitochondrial genes, and mitochondria are examples of permanent mergers as they have lost their ability to live outside cells (Rand et al. 2004). The merging involved transfer of mitochondrial genes to the nucleus, thus making mitochondria dependent upon the host to encode the majority of their proteins. The challenges involved must have been overwhelming as the transfer of genes from the mitochondria to the nucleus required that the mitochondrial nucleic acid incorporated into a chromosomal location, that the gene acquired transcription signals to drive its transcription, and that the protein acquired targeting sequences that ensured that it moved to the mitochondria once translated.

Owing to the biochemical processes in the mitochondrion, mitochondrial DNA undergoes frequent mutations. Such mutations may out-compete unmutated DNA by selfish replication in mixed cytoplasms, and could thus lead to "mutational meltdown" and extinction of mitochondrial lineages (Lynch and Blanchard 1998). Still, meltdown rarely occurs, probably because dynamic selection and drift occurring at multiple levels purge the deleterious mutations (Rand 2001). For example, a selfish mutant that out-competes the unmutated DNA by extended replication within the cytoplasm of a cell may be selected against at the level of the cells. This occurs if cells with mutant parasitic elements reproduce or grow slower than cells with unmutated elements. If, however, cells with mutated elements are positively selected, there is still a chance that they may be selected against at the organismal level. This may occur if mitochondrial mutations lead to somatic mosaicism, in which different tissues become populated with different mitochondria. If these organisms exhibit lower fitness, they, and consequently their mitochondria, will be selected against.

Transposable elements

The genomes of vertebrates are rich in retroviral elements that have been integrated into host DNA (Ostertag and Kazazian 2001), and it is currently assumed that at some time during the course of evolution, exogenous progenitors of endogenous retroviruses inserted themselves into the cells of the germ line where they have been replicating along with the host's genes. One of the most striking findings that came from the sequencing of the human genome is that approximately 45% of the genome is composed of retroelements including retroviral-like sequences (Li et al. 2001). These elements generally move within the genome by a "copy and paste" mechanism, and are thus called transposable elements.

Transposable elements are repetitive mobile sequences dispersed throughout the genome. They come in two broad classes, DNA transposons and retrotransposons. The repetitive structure of transposable elements within the genome, their lack of phenotypic effects, and their autonomous self-reproductive potential, prompted several geneticists and evolutionary biologists to regard transposable elements as prototypic ultimate parasites (Doolittle and Sapienza 1980; Orgel and Crick 1980; Dawkins 1982; Dawkins 1989). The idea of selfish DNA as being utterly selfish has, however, been contested. Much evidence indicates that transposable elements play an important role in host evolution and development, and the relationship between the host and the transposable element is therefore better viewed as encompassing a continuum from parasitism to symbiotic mutualism (Kidwell and Lisch 2000).

Although most transposable elements are inactive, some have retained the ability to transpose (Ostertag and Kazazian 2001). At least some of these elements are negatively selected in humans, thus implying that they constitute a genetic burden for their bearers (Boissinot et al. 2006). Transposons mobilise in both germ and somatic cells, and it has been suggested that vertebrate organisms may be somatic mosaics with respect to mobile element insertions (Deininger et al. 2003). Transposable elements have played a major role in shaping mammalian genomes, and may have been involved in catalysing the process of duplication of genes involved in immunological defence (Ventura-Holman and Lobb 2002; van de Lagemaat et al. 2003). They have also been given a role as regulators of the class switching process of immunoglobulin heavy chains (Magor et al. 1999). More generally, transposable elements are of importance for the genome's response to environmental challenges (Deragon and Capy 2000). For example, transposable elements influence the expression of genes associated with immunity as well as other genes involved in response to external stimuli, while genes involved in highly conserved activities such as metabolism and development are little affected (van de Lagemaat et al. 2003).

A spectacular case of "useful parasites" was discovered in the vertebrate immune system in the mid-1990s when the RAG1 and RAG2 enzymes that mediate the joining of the V, D and J segments in maturing lymphocytes were shown to be the products of "domesticated" transposable elements (van Gent et al. 1996; Agrawal et al. 1998). Thus, a fundamental component of the vertebrate immune system evolved from a transposable element whose capacity for DNA rearrangement was exploited to produce rapid somatic variability in specific host cells (Lewis 1999). The RAG genes have been detected in the descendants of the oldest jawed vertebrates, the shark, but despite much effort homologues of RAG genes have not been detected in jawless vertebrates (lampreys and hagfishes). The genetic structure of the RAG genes makes it conceivable that they arose from a transposable element introduced to the ancestor of the jawed vertebrates by horizontal transfer from a prokaryotic transposable element (Schluter and Marchalonis 2003).

The fact that deleterious or beneficial transposable elements can spread in host genomes can be explained as an example of natural selection acting in opposite directions at different levels of the biological organisation. While the replicative advantage of transposable elements over non-mobile elements allows their proliferation in genomes, this proliferation may be offset by negative selection at the level of the organism in so far as organismal fitness becomes reduced (Kidwell and Lisch 2000). On the other hand, if transposable elements increase the fitness of their hosts they may be positively selected also at the organismal level and may become integrated in the host genome, as did the genes encoding the RAG enzymes.

ATTACK AND DEFENCE

The works of Sisyphus

Current models of pathogenicity have set the dynamics of the host-pathogen relationship at the centre of attention. The competition between hosts and their parasites is an arms race in which the selective pressures are always multiple in space and changing through time.[9] During infection, microbes are actively engaged in sensing and responding to intra- and extracellular host signals, in modulation of cell surface structures that are targets for immune responses and in exploiting their microenvironments to allow replication and propagation (Guiney 1997). Thus, the genotypes of the vertebrate lineage must change rapidly – they must keep running – to stay in place in the ecosystem and avoid extinction. The possibility to continue, either as staying alive or via offspring, is the only external pay-off in the game. The most vigorous evolution of this type occurs between bacteria that rapidly change their surface proteins and bacterial viruses, so called bacteriophages, that evolve exceedingly rapid to keep pace with the dynamic surface structure of their host bacterium (Doulatov et al. 2004).

---

[9] Organisms are never perfectly adapted to their environments. If they were, no parasite would ever breach the host's defences and, likewise, no potential host would ever escape infection. These imperfections, which are the impetuses that keep the struggle between hosts and parasites going, have been well described by van Valen (1973) in his Red Queen hypothesis. The Red Queen hypothesis has it's name from Lewis Carroll's (1965) book *Through the Looking Glass* in which Alice and the Red Queen suddenly begin running across the chessboard, but without getting anywhere. According to the Red Queen, it takes all the running one can do just to keep in the same place, and if one wants to get somewhere else one must run at least twice as fast as that.

An analogous coevolution, albeit slower, occurs between multicellular organisms and their infectious agents. This can be illustrated by the evolution of ligands and their receptors. The two must be tightly coupled, otherwise they would evolve into a non-functional unit. This idea, that changes in signalling systems must coevolve, has been called sensory drive (Endler 1992). Sensory drive, which alludes to the capacity of sensory systems to "drive" evolution in particular directions, is well exemplified by the coevolution that occurs between hosts and parasites. There is a continuous drive on infectious agents to evolve novel molecules that may help them to either attach to specific cells of the infected organism or to avoid the immune defences of the host, and analogously, hosts are continuously driven to overcome the adaptive strategies of their infectious agents.

But competition also maintains variability at the level of the species, sometimes owing to localised ecological interactions between populations of different genetic makeup. Benjamin Kerr and co-workers (2002) allowed three populations of the bacterium *Eschericia coli* to interact, either in a homogenous environment in which the populations were well mixed or in a heterogeneous environment in which dispersal and interactions were local. The populations either produced i) the toxin colicin and its antidote, ii) antidote to colicin, or iii) neither toxin nor antidote. In the heterogeneous environment the three populations were engaged in a rock-paper-scissors[10] competition in which i) outgrew iii), ii) outgrew i) and iii) outgrew ii). In the homogenous environment, however, type ii), which produced antidote, eliminated the two other populations. Thus, localised interactions of the rock-paper-scissors type can turn a "one winner" outcome into a dynamic coexistence of all three populations, the boundaries between them shifting back and forth.

Such interactions, which are a result of the genetic makeup of the bacteria as well as of their local interactions, may also be relevant for communities of cells like *D. discoideum* and *M. xanthus* in their free living pre-aggregation stage. If so, the altruistic characters could well be maintained in the environment without being selected for or against at the group level. It would thus appear that the level of selection that is effectuated, whether group or organismal, depends on the ecological context as well as the functional and genetic characteristics of the interacting microbes.

Virulence

The concepts of virulence and pathogenicity have been used to describe a range of diverging phenomena. Such divergent use reflects the fact that the definitions for these terms have been altered during the years. The capacity of a microbe or of a

[10] The game "rock-paper-scissors" is classically played by two children that upon a given sign use their hands to simultaneously signal either rock (fist), paper (flat hand) or scissors (two spread fingers). Rock smashes scissors, scissors cuts paper, and paper wraps rock. Hence, the display of a flat hand wins over a fist, two fingers win over a flat hand, and a fist wins over two fingers.

microbial product to cause disease in a susceptible host is, for example, variously termed pathogenicity or virulence, while the microbial molecules responsible for inducing the disease have been termed pathogenicity or virulence factors. In their attempt to provide precise definitions Casadevall and Pirofski (1999) seized upon host damage as the trait to be explained, and proposed that pathogenicity should be defined as the capacity of a microbe to cause damage, while virulence should be defined as the relative strength of the capacity to induce damage.

While this conceptualisation of virulence is in agreement with current medical usage, it is insufficient when it comes to explaining the evolution of host-pathogen interactions.[11] To capture this aspect, evolutionary biologists have come to define microbial virulence as the microbe's effects on host reproductive fitness (Combes 2001). Accordingly, a microbe with low pathogenicity can be highly virulent if it causes little physical damage to the host but severely affects host reproductive potential, while a microbe with high pathogenicity can be of low virulence if it does not affect the host's reproductive potential. Thus, pathogenicity and virulence need not co-vary, although they often do.

A central question in the field of virulence is why infecting agents harm their hosts, especially if a live and healthy host is beneficial to their transmission. There are three major frameworks that purport to explain this phenomenon; virulence as a co-incidental by-product, as an adaptation of the parasite, and as a co-evolutionary race (Ebert 1999). *The coincidental by-product hypothesis* asserts that infecting agents harm their hosts if genes responsible for virulence evolved under conditions other than those in which they are currently considered; even though the toxin produced by *Clostridium tetanii* kills people, this is probably not the reason why the bacterium produces the tetanus toxin.

The adaptedness hypothesis is based on the idea that virulence evolves to the degree that it benefits or costs the infecting agent, and that reduction in host fitness is an unavoidable consequence of the reproduction of the infectious agent within the host. Hosts are considered to be invariant and their evolution to be so slow that it can be neglected. *The coevolution hypothesis* extends the adaptedness hypothesis by viewing virulence as a long-term result of the host-parasite arms race. With reciprocal selection, virulence and defence are expected to be ever changing traits that are balanced by the antagonistic evolution of hosts and parasites. Accordingly, the virulence of an infecting agent can increase or decrease depending on the selective pressures involved.

The mechanisms by which the immune system shapes and is being shaped by the variations and evolution of HIV provides a striking example of host-parasite coevolution. Current evidence derived from phylogenetic studies of nucleotide sequences

[11] Evolutionary ecologists, who have made extensive investigations into the conceptual aspects of virulence, see the phenomena to be explained by the concept as being the products of complex interactions among evolutionary and ecological processes. The duration of infectiousness, transmission patterns, host immune response and natural longevity of the host all influence the importance of within-host to between-host dynamics of virulence evolution (Bull 1994; Frank 1996; Poulin and Combes 1999; Day 2002).

of various genes in different HIV isolates indicates that HIV entered the human population through multiple zoonotic infections from simian immunodeficiency virus (SIV)-infected nonhuman primates in Africa sometimes around 1931 (Korber et al. 2000). The most marked effect of HIV on the immune system is the induction of immunodeficiency, which is owing to the destructive activities of the virus on the CD4 T lymphocytes. Nevertheless, during both the acute and chronic phases of HIV infection, CD8 cytotoxic T lymphocytes exert a strong inhibitory effect on HIV growth and replication, thus inducing a strong selective pressure for escape mutations of HIV. Hence, it appears that the highly polymorphic human leukocyte antigens that present antigen to cytotoxic CD8 T lymphocytes could be responsible for the driving and shaping of HIV evolution (Moore et al. 2002).

Much theoretical work supports the assumption that, all else being equal, reduction in host fitness (virulence) should be balanced by transmission efficacy so that transmission success is optimised over the lifetime of the infection. Accordingly, vertical transmission (from mother to foetus) should reduce virulence relative to horizontal transmission (from host to host) because vertical transmission depends on host survival and reproduction. Therefore, reducing the efficiency of transmission could select for reduced virulence (Ewald 1995).

Lifetime optimisation does not necessarily imply that parasite virulence should evolve to a maximum level. Rather, when taking coevolution of the host-parasite relation into consideration, an intermediate level is to be expected. While observational data have supported this idea, experimental work was, until recently, not available. This all changed when experiments with the host *Daphnia magna* and the bacterium *Pasteuria ramose* provided direct support for the view that optimal virulence may indeed be established at an intermediate level (Jensen et al. 2006).

The evolution towards an intermediate optimum of virulence is the expected outcome of infections with genetically uniform agents. But when hosts are infected with genotypically diverse agents, the outcome of the evolutionary process may become even less predictable as the agents may then compete with each other as well as with the immune response. If, for example, the immune response tailored to one agent generalises badly towards the other, as occurs during the original antigenic sin of influenza infections, the evolution of virulence should depend upon the immunogenicity of the two agents. If the agent with higher virulence is the most immunogenic, the immune system would attack it more fiercely, thus increasing the relative fitness of the less virulent strain. This may select for lowered virulence. And vice versa, if the low-virulence strain is the most immunogenic, selection may promote the evolution of higher virulence.

If on the other hand the immune response is equally efficient against the diverse agents, the direction of selection should depend as much upon relations between the competing agents as upon their relations to the immune system. As in the previous example, the evolution of virulence could go in both directions, dependent upon the mechanistic interactions involved. Such interactions have not been much studied, but an experimental study in which mice were subjected to mixed infections with virulent and avirulent strains of the rodent malaria parasite *Plasmodium chabaudi* recently provided evidence that the mouse immune system inhibited the avirulent strain more than the virulent strain (Råberg et al. 2006). In this system, therefore, the avirulent strain has the lower fitness and there should thus be selection for increased virulence.

The coevolution hypothesis has provided a theoretical framework for the belief that microbes should evolve towards benign coexistence with their hosts and thus towards lowered virulence. Supporting evidence for this view has come from cases where pathogens are relatively harmless to those hosts with which they have had a long-established relationship, while the same agents have a serious impact on new populations. In their book *Natural history of infectious disease* Burnet and White (1972, pp. 20–21) described infectious disease as a coevolutionary conflict between the host and its parasites, and argued that for the evolutionary race to result in a balanced condition "long periods of interaction and selection between the two species must have elapsed, and any protective adaptations on the part of the host must often have been countered by adaptive change in the parasite" (Burnet and White 1972, p. 29).

But although benign coexistence is a possible outcome of the evolutionary process, it is not an obligate outcome. As claimed by Paul Ewald (1994), virulence may increase, stabilise, or decrease dependent upon the evolutionary process involved. Natural selection favours transmission, and hence if "more rapid replication of a virus inside of a person leads to a greater passing on of the genes that code for that rapid replication, then replication rate will increase even if the more rapid growth of the virus population within a person causes the person to be severely ill, or leads to an overall decrease in the number of virus among people, or hastens the eventual extinction of the virus." (Ewald 1994, p. 4).

Proponents of the theory of evolution to benign coexistence argue as if natural selection primarily occurs at the species level, and they sometimes even claim that it is in the long time interest of the pathogen species to become less, not more, virulent with time. While evolutionary biologists have acknowledged the possibility of multiple outcomes of host-pathogen interactions, the erroneous idea that interacting species always evolve to mutual tolerance is still expounded in medical journals (Dixon 2003). The naivety of this idea is equal to the related belief held in the 1970s that infectious disease, at least in the developed world, had been overcome by better hygiene, better nutrition, better living conditions, effective vaccines and broad-level antibiotics (Cohen 2000).

Knowledge of the mechanisms by which evolution of virulence proceeds is of uttermost practical importance from a public health standpoint. For example, by use of mathematical modelling Gandon and co-workers (2001) showed that some pathogens might evolve increased virulence in response to the selection pressure imposed by a vaccination programme. The increased virulence may have wide ranging public health implications since it may result in more severe disease in unvaccinated individuals. Fortunately, vaccines in current use have been effective in protecting and even eradicating infectious agents without selecting for resistant pathogens that are adapted to circumvent this evolutionary pressure.

Multi-level dynamics

The importance of a multilevel framework for the understanding of pathogenicity, virulence and defence was made clear during "one of the most interesting ecological experiments that has ever been undertaken" (Burnet 1959:22), the myxomavirus epidemic in Australian rabbits in the 1950s. Myxomavirus, which occurs naturally as a mild infection of South American rabbits, causes a lethal infection in the European rabbit. This capacity made it suitable as a biological weapon against the European rabbits that had been introduced into Australia in 1859 for sporting purposes, but who rapidly spread to become a major animal pest of the agricultural industries. When myxomavirus was introduced into the Australian rabbit population in 1950 the virus caused over 99% case-fatality rates, but by 1953 attenuated mutants of the virus and resistant rabbits had emerged throughout Australia. During the following years a dynamic equilibrium between relatively resistant rabbits and less virulent viruses was established (Fenner and White 1976).

Myxomavirus belongs to the poxviruses, a group of large DNA viruses with complex life cycles. In contrast to RNA viruses, whose high mutation rate effectively subverts host defences, the lower mutation rate of poxviruses has necessitated the evolution of other subversion mechanisms. Studies of the effects of wild type and gene-manipulated myxomaviruses on rabbits have uncovered a wide variety of virally encoded immunomodulatory molecules that contribute to the virulence of the virus. These molecules interfere with apoptosis of lymphocytes, recruitment of inflammatory cells to the infected area, and with the adaptive immune response to the virus. It is conceivable that mutational changes in these virally encoded molecules were responsible for the alterations of myxomavirus virulence in Australian rabbits (Zuniga 2002).

At least three models have been put forward to explain the evolution of lowered pathogenicity and virulence. These include Burnet's and Lewontin's group level selection models, Sober and Wilson's trait-group selection model, and Alexander and Borgia's individual selection model. Burnet (1959) explained the evolution of lowered virulence by claiming that rabbits infected with the lower virulence virus, who were not simultaneously co-infected with the higher virulence virus, survived longer than rabbits infected with the higher virulence virus. If such long-lived rabbits serve as "a source for infection for a dozen rabbits which are not simultaneously exposed to infection by the virulent type, the new strain will be in a position to compete effectively for survival with the original." (Burnet 1959, p. 26).

Richard Lewontin (1970) gave a similar but extended and more formal explanation of the myxomavirus story. He explained the co-evolution of increased rabbit resistance and lowered myxomavirus virulence as occurring through individual and group level selection, respectively. The rabbits evolved increased resistance, as would be expected from organismal selection in which the more resistant outcompete the less resistant. The myxomavirus, on the other hand, evolved through a group selection mechanism. As envisioned by Burnet, each rabbit is a group from the standpoint of the virus. When rabbits die, the groups of viruses become extinct since they cannot propagate from dead rabbits. Thus, while there is a high rate of group extinction amongst the virulent viruses, the groups consisting of virus strains with lower virulence are left extant. This mechanism increasingly selects for strains with lower virulence, despite the lack of selective advantage for the avirulent strain within groups consisting of both virulent and avirulent strains.

But as envisioned by Sober and Wilson (1998), selection for lowered resistance may also take place in rabbits infected with both the high- and low-virulence strains. If this group-level property is associated with transmission bottlenecks in which only a few viruses are transmitted from one infected host to another, there may be a stochastic loss of the most virulent types. In this case, lower virulence will evolve analogously to the evolution of altruistic behaviour in trait-group selection models, the requirement being that groups with higher frequency of altruists (low-virulence) are more fit than groups with lower frequencies of altruists. Hence, parasite virulence becomes determined by selective forces that act on fitness variation between hosts and fitness variations within hosts.

The phenomenon of lowered virulence has also been explained as an individual level adaptation. In their 1978 paper Alexander and Borgia claimed that reduced virulence evolved as a consequence of individual selection because a virus that kills its host is less fit than a virus that does not. And since lower virulence reduces the chance of killing the host, lower virulence is an individualistic adaptation. The viral units are thus not altruistic but selfish; they follow their own reproductive interest.

In his review of the myxoma case, Robert Wilson (2004) evaluated the various arguments for selection of lowered virulence, and concluded that the case is likely to remain subject to multiple alternative explanations. There are two reasons for this, one ontological and the other epistemological. First, there are insufficient data to resolve the case. And second, some theorists believe that the different theoretical models are equivalent; that they merely produce alternative pictures of the same selective process, and thus that choice among the various models is a pragmatic matter that partly stems from the metaphysical viewpoints of the researchers (Reeve 2000).

While disagreements on the evolution of lowered virulence in myxomavirus abound, there seems to be agreement that increased resistance in rabbits occurred as an individual level adaptation. In their attempt to investigate the mechanistic basis for the evolving resistance of rabbits to myxomavirus, Best and co-workers (2000) proposed that differences in the inflammatory responses may well be owing to a change from a predominantly Th2 cytokine response in susceptible rabbits to a Th1 response in resistant rabbits. The case is, however, still not sufficiently investigated, and more data are needed before firm conclusions can be drawn.

One way forward might be to study the MHC distribution in resistant as compared to susceptible rabbits. In their study Best and co-workers utilised the MHC antigens as membrane-markers to identify immunologically important cells, but did not investigate the variability of MHC genes between the two populations of rabbits. Because MHC genes are highly polymorphic and therefore particularly well suited to study whether a population has gone through a bottleneck (Klein and Takahate

2002), such an investigation could have given an idea of whether any loss of certain MHC lineages had occurred, and whether this loss was related to the infecting agent. Such an investigation of MHC diversity has proved useful in several other species, including cheetahs and chimpanzees. The MHC variation among cheetahs is, for example, almost non-existent. This was first inferred by their high acceptance of skin grafts from unrelated animals, and was later verified by analysis of the MHC genes (O'Brien et al. 1985; O'Brien and Yuhki 1999). The data are suggestive that the ancestral variability of MHC genes was lost through a population bottleneck in a near extinction event. The MHC uniformity in cheetahs has resulted in a high susceptibility to infectious disease, and it has been speculated as to whether the continuous survival of cheetahs is associated with their solitary territoriality which limits the spread of infectious agents (O'Brien and Yuhki 1999).

While it is not clear what caused the loss of certain MHC lineages in ancestral cheetahs, a similar but less dramatic contraction of the MHC class I gene repertoire of chimpanzees may give an idea of the causative incident. Several lines of evidence suggest that this reduction may have occurred as a consequence of a severe viral infection in the ancestral chimpanzee population (de Groot et al. 2002). The infectious agent was probably a retrovirus related to HIV, and chimpanzees may thus have conquered AIDS already. This interpretation is supported by the observation that HIV is relatively apathogenic in primates, largely because primate cells express innate cellular factors that restrict retroviral infections (Towers et al. 2003).

THE "LUCKY SPLIT" HYPOTHESIS AND THE EVOLUTION OF ANTIGEN BINDING RECEPTORS

The immunological big bang

Thorough studies of the nearest phylogenetic relatives of the jawed vertebrates, including urochordates (tunicates), cephalochordates (amphioxus) and agnatheans (hagfish and lamprey), have identified a wide variety of genes involved in innate immune responses. These comprise amongst others complement components, Toll-like receptors, and cytokines. The pivotal genes for adaptive immunity which include the immunoglobulins, T cell receptors, MHC class I and II molecules, as well as the RAG-genes involved in somatic generation of diversity have, however, not been identified in any species beyond the jawed vertebrates (Du Pasquier and Litman 2000; Secombes and Pilstrom 2000; Azumi et al. 2003). Such data provide evidence to the hypothesis stating that the adaptive immune system arose in a 50 million year interval in-between the divergence of lineages leading to agnatheans on the one side and the jawed vertebrates on the other, some 470 million years ago (Schluter et al. 1999; Laird et al. 2000).

It is obvious that the ancestor of the jawed vertebrates had defence mechanisms that were effective against the most prevalent infecting agents, and that the selective pressure to evolve an adaptive immune response upon an already effective innate defence response would depend upon the costs and benefits associated with the defence system as well as on the probability of encountering the environmental infectious agents and the reliability of the signs encountered. It is difficult to envisage the costs and benefits involved as the adaptive immune system evolved, but it is conceivable that the costs were as high as the benefits, at least before a full fledged immune system with regulatory properties had been established.

While it is unlikely that the types of infecting agents changed dramatically during the 50 million years it took to evolve the immune system, the risk of encountering infectious agents may have increased. Jawless fishes were microphagous suspension feeders, and the appearance of the jaw marked a major transitory event in the history of the vertebrates (Purnell 2002). It has been speculated that the rapid evolution of the adaptive immune system came about when jawed fishes began to eat animals with hard skeletons, which might have enhanced the need for protection due to the increased risk of microbial infections (Matsunaga and Rahman 1998). Hence, according to this hypothesis the jaw and the immune system co-evolved.

However, some sort of adaptive immune responses also take place in jawless vertebrates. These responses, which do not involve immunoglobulins or T cell receptors, have been attributed to cells with a lymphoid morphology, and preliminary evidence suggests that each lamprey lymphocyte bears a variable lymphocyte receptor with the capability of somatic diversification (Pancer et al. 2004). The genes encoding the variable lymphocyte receptor are extremely diverse, and it appears that the diversity-generating mechanism is of the same general principle as that of the jawed vertebrates. Hence, a flexible defence system has evolved also in the jawless vertebrates, and the hypothesis that the evolution of the jaw was the decisive event in the evolution of the adaptive immune system is thus weakened.

The idea that the adaptive immune system evolved abruptly, in a manner sometimes referred to as the immunological "big bang" (Schluter et al. 1999) does not imply that the immune system was created *de novo* during this interval. Rather, there has been a growing realisation that precursors of immune cells and molecules that had been established in the common ancestor of the jawed vertebrates were redeployed for alternative use in the evolving adaptive immune system. As claimed by Dreyer et al. (1967, p. 360), "The evolutionary process is a slow one, especially with regard to complex protein molecules, and it is pertinent to ask what type of molecules might have been the ancestral precursors of antibodies, and what function they might have had in their ancient cellular environment".

The sequencing of immunoglobulin molecules in the 1960s provided the first evidence as to the structure, the developmental assembly and the evolutionary path that generated the immunoglobulins. Extensive studies of immunoglobulin sequences allowed the construction of an evolutionary tree of the phylogenetic relationship between the immunoglobulin molecules, and it was demonstrated a remarkable relationship among the amino acid sequences from mice and humans (Dreyer et al. 1967). The evolutionary tree suggested that the immunoglobulins had evolved from a primordial gene pool coding for molecules likely involved in cell surface signalling, and that the similarities between the molecules probably resulted from gene duplications. To explain the diversity of the immunoglobulins, Dreyer and co-workers further postulated that a diversifying mechanism had evolved at the root of the tree. "The remarkable fact that the thymus, lymphocytes, and the immune response appear so late on the evolutionary tree would lead us to suppose that chromosomal programming of the copy-splice mechanism evolved for functions having little or no relationship to the vertebrate immune response" (Dreyer et al. 1967, p. 364).

RAG-insertion

Several of the proposals stated in the paper by Dreyer et al. (1967) have been confirmed and extended by more recent DNA studies (Tauber and Podolsky 1997). For example, the copy-splice mechanism that enabled the adaptive immune system to evolve so quickly was probably the insertion of a transposon into a previously undisrupted exon of an immunoglobulin-like gene. The genetic structure of the RAG genes makes it conceivable that they arose from a transposable element introduced to the ancestor of the jawed vertebrates by horizontal transfer from a prokaryotic transposable element. The integration of the transposon was most likely a unique chance occurrence, and hence, the immune system may owe its existence to a "lucky split".

The genes that encode the T cell receptors and immunoglobulins display highly similar patterns of organisation and rearrangement during transcription and translation. T cell receptors and immunoglobulins are closely related members of a large family of molecules called the immunoglobulin superfamily. The members of this superfamily evolved prior to the evolution of vertebrates (Mendoza and Faye 1999), and it is currently believed that the RAG transposon was inserted into an immunoglobulin-like gene that later became the variable parts of the T cell receptors and immunoglobulins. It is, however, debatable whether the RAG insertion stimulated the emergence of the adaptive immune system as suggested by Dreyer et al. (1967). Rather, following a comparative and evolutionary analysis of the available molecular data Martin Flajnik (2002) hypothesised that the adaptive immune system might have existed in a rudimentary form already prior to the "lucky split". According to this idea, the RAG insertion was not the decisive factor for the "big bang" although it allowed a further refinement of the system. The refinement must have been advantageous to the early vertebrates since all extant jawed vertebrates use the same RAG mediated mechanism.

The observation of multigene families encoding diversified immunoglobulin-like proteins in amphioxus indicates that precursor genes of the T cell receptors and immunoglobulins were in existence before the lineages separated (Cannon et al. 2002). Two multigene families of vertebrate innate immune receptors that bear a close structural resemblance to antigen receptors may also be predecessors of the immune receptors (van den Berg et al. 2004). Phylogenetic analysis of the constant and variable regions of antigen-receptors indicates that the most recent antigen binding receptor to evolve was the $\alpha\beta$ T cell receptor, with $\gamma\delta$ T cell receptors and immunoglobulins being the most ancient (Richards and Nelson 2000). This evolutionary pattern indicates that direct antigen recognition was the ancestral condition, and that indirect antigen recognition by the αβ T cell receptor is a derived property.

It is conceivable that RAG-like genes were introduced into the genomes of many organisms in the past, but that only the common ancestor of the jawed vertebrates exploited them successfully. This might have been because the insertion was acquired and expressed in the right context. Somatic rearrangement could have been deleterious to the organisms, but when it became sequestered into lymphocytes, the dangerous cells with autoreactive potential could be somatically selected and deleted. Precursors of lymphocytes were probably available to the ancestor of the jawed vertebrates, as lymphocyte-like cells have been observed in lamprey (Shintani et al. 2000). Since morphology may be a deceitful marker when considering evolutionary generated structures, the identity of the lamprey cells was ascertained by the identification of molecular markers of lymphoid developmental pathways as well as molecules essential for lymphocyte function (Mayer et al. 2002; Uinuk-Ool et al. 2002).

While it seems clear that the MHC class I and class II genes are derived from each other, the ancestor molecules that served as the predecessor for these two MHC molecules has not been ascertained (Klein and Sato 1998), although putative precursors of MHC molecules have been described in urochordates (Azumi et al. 2003). Neither the ancestral function of MHC molecules nor their timely appearance related to the αβ T cell receptor have been elucidated, but some evidence indicates that MHC molecules evolved after the αβ T cell receptor. Since variation necessarily precedes selection in evolution, it is likely that the insertion of a transposon into the primordial receptor genes caused variational binding properties of the T cell receptors. As the randomly generated diversity led to the generation of self-destructive structures, the MHC molecules may have evolved as a result of their being selected to restrict the reactivity of the αβ T cell receptor.

While the cells, tissues and molecules that make up the adaptive immune system have been retained in all extant jawed vertebrates, several adaptive changes of the primordial system have evolved. These alterations are especially pronounced for the immunoglobulins where different species have evolved distinct mechanisms of generating antibody diversity. The structure and function of T cell receptors and MHC molecules have been more conserved through evolution, and the development of T cells in the thymus is quite similar among the jawed vertebrates (Litman et al. 1999; Flajnik 2002).

The received repertoire

The birth and death model

It is generally accepted that infectious agents have served as a selective force on the immune system. It has, nevertheless, proved difficult to test the hypothesis by means of conventional population studies. This is because most organisms in outbred species are heterozygous at loci of large multigene families, and it would be necessary to survey a very large fraction of the population in order to get a large enough sample of homozygote individuals to compare their fitness with that of heterozygote individuals. This limitation is especially pronounced when the selective advantage possessed by certain alleles or allele combinations are small (Hughes 1999).

An approach to investigate the selection of molecular structures, and which takes advantage of the large amount of immune molecules already sequenced, is to investigate the nucleotide substitutions that have occurred in their genes. The rationale for this method is that there are two major ways that changes at the DNA level can change the function of proteins; either by altering the pattern of gene expression or by altering the amino acid sequence of the protein.

When two sequences of the same gene from two individuals are aligned, there may be differences between the nucleotide sequences. Some of these changes will lead to differences in the amino acids of the encoded protein. These changes are termed nonsynonymous changes. Other changes, because of the degeneracy of the genetic code, leave the protein unchanged. These changes, termed synonymous changes, record the background mutation rate of the gene. Since synonymous sites evolve much more rapidly than do nonsynonymous sites in most coding sequences, the comparison of synonymous to nonsynonymous nucleotide substitutions can be a powerful tool in the study of adaptive evolution (Hughes 1999; Hurst 2002).

According to their effects on the fitness of their bearers, mutations can be divided into three categories. First, neutral mutations, which are the most prevalent, have no effect on the fitness of their bearers. Although not under selective pressure, neutral mutations can become fixed in the population by random genetic drift. Second, disadvantageous mutations have a negative effect because they lower the bearer's chances of survival or reproduction relative to that of other organisms, and are usually eliminated from the gene pool by negative (purifying) selection. Since the majority of nonsynonymous substitutions are deleterious and hence eliminated by natural selection, it is to be expected that there should be more synonymous than nonsynonymous changes in most proteins. The third category of mutations consists of advantageous mutations. These mutations have a positive effect on their bearers. Hence, they have a greater probability of being selected by natural selection to be represented in the gene pool of the next generation. In such cases, an increased rate of nonsynonymous to synonymous changes is to be expected.

When variable regions of immunoglobulins, T cell receptors and MHC molecules are being investigated by comparing the observed molecular pattern with what would be expected in the absence of natural selection, that is with the expectation of the neutral theory of molecular evolution, one would expect the largest change in the regions of the molecules that come in contact with the antigen. This expectation has indeed been verified in extensive studies on genes encoding immunoglobulins, T cell receptors and MHC molecules performed by Masatoshi Nei et al. (1997) and Austin L. Hughes (1999).

Their results provide strong evidence that natural selection has acted on these molecules, but also that the nature of selection works differently on the immunoglobulins and MHC genes. Natural selection on non-neutral mutations works in two ways, either through balancing selection or through directional selection. The MHC genes are mainly selected according to the process of balancing or overdominant selection, which acts to maintain polymorphism, although directional selection may also be at work (Lohm et al. 2002). Immunoglobulin variable regions are, in contrast, selected in accordance with directional or diversifying selection. This type of selection occurs when advantageous mutations that occur in a population become fixed as a result of natural selection. Similar selection acts on the V region gene segments of T cell receptors, although the evidence is less clear-cut than in the case of immunoglobulins (Allen et al. 1996; Hughes 2002).

While nucleotide substitution mutation has been a strong evolutionary force in the generation of the immune system, the origination of new protein families and functions has not been achieved by mutation alone. Recombination of genetic elements at the molecular level, for example the combining of a gene with a new promoter or the combining of two or more nonproductive genes into a functional protein, has played an important evolutionary role (Kinoshita and Honjo 2001). There is also strong evidence that genes belonging to multigene families, like those of the immune system, have evolved as a result of repeated events of gene duplications (Prince and Pickett 2002). For example, the immunoglobulin and T cell receptor genes are the fastest duplicating genes in the rat genome (Gibbs et al. 2004).

The most likely fate of a duplicated gene is that one of the pair will degenerate to a pseudogene or be lost from the genome by purifying selection. A more infrequently expected outcome is that the duplicated gene might undergo mutation in its regulatory region so that it comes to be expressed in different tissues or at different developmental stages, or the mutations may affect the gene and thereby alter its functional characteristics. In the immune system, gene duplications have been demonstrated in the variable regions of both immunoglobulins and T cell receptors, as well as among the MHC molecules. The duplicated genes have subsequently accumulated nonsynonymous mutations in their antigen-binding regions, thereby exposing the new gene to natural selection. The non-functional genes have become psudogenes while the functional genes have become members of the antigen specific repertoire of the organism. According to this model, new genes are created by repeated gene duplications (birth) while some genes are deleted or made nonfunctional by deleterious mutations (death), hence the term the birth-and-death model (Nei et al. 1997).

Selection for variability

In his attempt to explain the evolution of antigen binding receptors, Melvin Cohn (1972) claimed that since immunoglobulins are made up of a constant and a variable part, in which effector functions are linked largely to the constant part while recognition is linked to the variable part, natural selection would work differently on the two parts. The evolution of the constant part would be uniquely germline, while the variable part would have a somatic as well as a germline selection component. The variable part, which would evolve by natural selection in the germline, would diversify in the individual organism by somatic selection. This two-stage selection model allowed him to differentiate the levels of selection. While the germ-line was selected at the level of the organism, the somatic modifications were selected at the level of the lymphocyte cell.[12]

In order to be both functional and evolutionary selectable, Cohn held that the repertoire of the immune system had to be formed in two stages. The stage I (germline) repertoire resulted from evolutionary selection on the genome, while the stage II repertoire resulted from somatic diversification of the stage I repertoire. To be selectable by evolution, the stage I repertoire had to be functional, which meant that the repertoire had to be small. But since antigen-binding receptors are efficient only above a certain concentration, the repertoire had to exhibit a high copy number. This functional stage I repertoire, the minimal unit that would protect the animal, was called the protecton. The smallest vertebrates would have one protecton unit, while the protectons had to be iterated in proportion to the size of the animal (Cohn 1972; Cohn and Langman 1990). Accordingly, the immune system is modular in construct, and the modules or protectons, are equivalent in function. Furthermore, the protecton is the unit upon which evolution selects.

Unlike the repertoire of the B cells, which has a germline-selected specificity for antigen, the relation between T cell receptor diversity and the MHC/peptide complex appears to be more indirect. For example, a single class II MHC/peptide complex selects T cell receptors with various binding specificities (Chmielowski et al. 1999), and a similar mechanism appears to operate for class I MHC/peptide complexes (Messaoudi et al. 2002). The results thus indicate that MHC molecules have a diversifying effect on the T cell repertoire. On the other hand, it is also clear that a single T cell receptor can recognise a number of different peptides that do not necessarily show high sequence homology (Mason 1998). Thus, the relation between the MHC and the T cell receptors is a many-to-many relation, not one-to-one. Given such a many-to-many relation it is difficult to conceive how infectious agents could select directly for variability of the T cell receptor genes.

It is possible that balancing selection, which leads to diversification of the MHC molecules, is responsible for diversification of the T cell receptor variable

---

[12] There are formal analogies between natural selection and developmental selection in that both processes occur by a two-step mechanism. Generative mechanisms create a population of alternative trajectories, while selective filters choose, from the excess of possibilities, those trajectories that satisfy specific criteria. Both processes result in adaptive systems in which the selected entities vary, have continuity, and display differential fitness. For the immune system the two selective processes are connected by the fact that the developmental process that generates the adaptive immune system has been evolutionary shaped by the process of natural selection.

genes as well. The contact between T cell receptors and MHC occurs through the CDR1 and CDR2 regions of the T cell receptor (Garcia et al. 1999). In contrast to the CDR3 region, which is generated through recombination, the CDR1 and 2 regions are germline encoded. The coupling between the allele specific determinants on the MHC molecules and the T cell receptor CDR1 and 2 regions must therefore be germline determined. Thus, the evolution of the CDR1 and 2 regions are constrained by the evolution of the MHC. A similar evolutionary scenario has also been described for the evolution of receptors for MHC molecules on NK cells, the killer immunoglobulin-like receptors (KIR). In a comparative study on the evolution of KIR in chimpanzees and humans, the results were compatible with an indirect selection on KIR. The hypothesis states that innovation in MHC molecules, driven by a selective pressure imposed by infectious agents, would in turn impose pressure upon KIR to follow up (Khakoo et al. 2000).

There is also a possibility that infectious agents may shape the T cell receptor repertoire during ontogenesis. Mice infected with the mouse mammary tumour virus (MMTV) in neonatal life experience a gradual deletion of T cells that express one specific T cell receptor β chain. This effect is caused by the presence of a gene encoding a superantigen in the MMTV genome (Li et al. 1999). Superantigens are molecules that bind to both MHC class II molecules and to T cell receptor β chains and thereby stimulate a larger number of T cells to proliferate than do conventional antigens. The responding T cells can either proliferate, become nonresponsive, or die. Superantigens bind to MHC molecules outside their peptide binding groove, and to the T cell receptor β chain via amino acid residues encoded by the gene encoding the variable region, but outside the hypervariable antigen detecting region of the T cell receptor.

Superantigens are not only derived from exogenous sources. Endogenous retroelements also produce molecules with superantigenic properties (Stauffer et al. 2001; Sutkowski et al. 2001). The induction of endogenous superantigens can occur either directly through infection with an exogenous virus such as the Epstein Barr virus (Sutkowski et al. 2001), or indirectly through stimulation with interferon α (Stauffer et al. 2001). The consequences of increased production of endogenous superantigens may be enhanced susceptibility to infectious disease or induction of autoreactive T cells that cause autoimmune disease. The hypothesis that endogenous superantigens may modify the expressed T cell repertoire and, furthermore, that the same superantigens may in fact be involved in positive or purifying evolutionary selection of the T cell receptor repertoire, warrants further exploration.

Coping with autoreactive receptors

The birth-and-death model presupposes a functional link between microbial structures and the host gene-products that binds to microbial antigens – the genes retained in the host are those that provide enhanced protection against infectious agents. But while some gene products directly determine susceptibility or resistance to specific infectious agents (Hill 1999), such links have only rarely been detected between a specific infectious agent and germline encoded T cell receptors and immunoglobulins (Hillson et al. 1993; Roost et al. 1995; Harada et al. 2003). Rather, and apparently in contradiction to the predictions of the birth-and-death model, a high proportion of germline encoded antibodies seem to be naturally reactive with self antigens (Baccala et al. 1989; Ailus and Palosuo 1995; Miller and Rodriguez 1995). These self-antigens include both immune-related molecules, like the immunoglobulins themselves, as well as non-immune molecules such as DNA and erythrocyte surface antigens.

The observation that the immune system harbours germline encoded autoreactive receptors, along with the proposition that molecules of the immunoglobulin superfamily originally evolved to be involved in cell-to-cell interactions, prompted John Stewart (1992) to hypothesise that the immune system originally served to integrate the internal molecular environment of the organism. Accordingly, he concluded that the defence function of the immune system is a derived feature that owes its existence to a redeployment of an already existing cell-to-cell communicative system. This in turn led him to postulate the existence of two types of immune systems – the central immune system, which, in line with Jerne's (1984) contention is concerned with self-reactivity and is regulated through an idiotypic network, and the peripheral immune system, which is non-reactive with self but reactive with infectious non-self.

The double-faceted immune system, in which one system is regulatory and the other destructive, but where both systems utilise the same molecular mechanisms, has been criticised from an evolutionary point of view; it could not have evolved by natural selection as postulated. The major weakness in the hypothesis, and which has been pointed out by Melvin Cohn and Rodney Langman on several occasions (Cohn and Langman 1990; Langman 1992), is the incapability of the hypothesis to explain the differential response to self and non-self. This is especially problematic since several of the effector mechanisms used by the adaptive immune system had already evolved prior to the emergence of the adaptive immune system (Kimbrell and Beutler 2001; Nonaka 2001). Hence, an auto-destructive catastrophe elicited by the immune system would occur if Stewart were correct.

Stewart's mistaken conclusion came from his insistence on explaining the autoreactive features of the immune system. If instead the focus is shifted towards explaining the defensive properties of the immune system, it will become clear that the autoreactive properties of the immune system, while occurring as a result of natural selection, need no independent selective explanation. Since the molecules of the immunoglobulin superfamily were originally involved in cell-to-cell communication, they were by definition self-reactive. The utilisation of these self-reactive molecules by the evolving immune system did not eliminate their self-reactivity. But following mutations and structural modification natural selection gradually moulded the self-reactive molecules to react also with infectious non-self. Repeated gene duplications and mutations gradually altered the antigenic reactivates of the antigen receptors, but since the molecules were moulded on self-reactivity, this reactivity remained to a certain degree in all antigen receptors. Hence, autoreactivity needs no selective explanation, although coping with autoreactivity does.

The inherent properties of the antigen receptors are no longer believed to be a one-to-one reactivity towards antigen. Rather, accumulating evidence favours the view that antigen recognition by immunoglobulins and T cell receptors is mediated by flexible interactions in which the combining site of the receptor undergoes a substantial structural change in the course of binding (Foote and Milstein 1994; Mason 1998; Manivel et al. 2002). Such interactions imply that autoreactivity cannot be selected against by natural selection working on the germline. Purging the germline of autoreactive receptors would purge non-self reactivity as well. But since self-reactivity is potentially dangerous, natural selection has generated somatic mechanisms that delete autoreactive receptors.

Immunoglobulins encoded by the VH4-34 gene are an example of the incapability of natural selection to purge autoreactive cells from the repertoire. These immunoglobulins, which are intrinsically autoreactive without requiring somatic mutation, are sometimes referred to as cold agglutinins since they bind to the erythrocyte carbohydrate antigens I/i and are capable of agglutinating erythrocytes with increasing affinity at decreasing body temperatures. Cold agglutinins are present in healthy individuals and patients with various viral and mycoplasmal infections. Their clinical effects are, however, most apparent in patients with the lymphoproliferative disorder chronic cold agglutinin disease (Berentsen et al. 1997; Ulvestad et al. 1999). As the temperature in the peripheral circulation of skin decreases, cold agglutinins bind to erythrocytes and thereby induce acrocyanosis and other clinical symptoms of intravascular agglutination. Due to a concomitant activation of complement during cold agglutinin binding, almost all patients also experience symptoms of anaemia. The anaemia is of haemolytic type and is caused by intravascular complement-mediated haemolysis and/or removal from the circulation of complement opsonized erythrocytes by phagocytic cells in the liver. The cold agglutinins in patients with cold agglutinin syndrome consist of monoclonal IgM antibodies with kappa light chains and anti-I reactivity (Harboe et al. 1965). The variable part of anti-I cold agglutinins are encoded by the VH4-34 immunoglobulin gene segment (Pascual et al. 1992), of which the framework region 1 is essential for antibody binding to I/i red cell antigens (Li et al. 1996).[13]

It might have been expected that a gene with pathogenic potential should have been deleted over evolutionary time. Nevertheless, VH4-34 is universally present in the human genome, without any polymorphism (Sanz et al. 1989). In addition, the VH4-34 gene segment is the most frequently rearranged VH4 gene family segment in immature B cells (Rao et al. 1999), and is found at a high frequency in peripheral

[13] The immunoglobulin heavy chain (H) variable (V) region is about 110 amino acids in length and is composed of four framework regions (FR) and three complementarity determining regions (CDR). The first three FR regions and the first two CDR regions are encoded by VH segments, the CDR3 is encoded by contributions from variable, diversity and joining segments, and FR4 is encoded by joining segments.

blood B cells (Kraj et al. 1995). Such lymphocytes are, however, rarely used during induced immune responses against infectious agents (Ohlin and Borrebaeck 1996). Furthermore, VH4-34 positive B cells are excluded from the germinal centres in the lymph nodes and do not differentiate into plasma cells (Pugh-Bernard et al. 2001). These properties suggest that somatic processes regulate the production and secretion of immunoglobulins by the VH4-34 encoding B cells. Such data are in accordance with the idea that natural selection cannot purge the germline of autoreactive receptors, and that somatic mechanisms therefore evolved to police the autoreactive receptors.

CHAPTER 5

Disabled defences

The adaptive immune system derives its functional characteristics from the self-organising properties of the developing lymphoid cells as they relate to the world. It is especially sensitive to initial conditions, and the imprintings being made during early life are constricting for later reactivity. The process of assembling the integrated and well-functioning immune system is a complex undertaking, and several developmental trajectories may lead the system towards a well-functioning state. The immune system is thus multiply realisable.

But while there are multiple ways to build a functional system, even more paths are available for the assembly of a malfunctional system. Since each organism self-organises its immune system in the light of a unique immune history, the malfunctional immune system may differ amongst individuals with similar phenotypic diseases. This idea suggests that immune system diseases are not natural kind entities; there are no unique ways to classify them, there are many plausible and defensible ways of doing so, and the best way will depend on both the purposes of the classification and the peculiarities of the disease in question.

Despite being fitness reducing, immune system malfunctions increase in prevalence. This is partly owing to the increasing prevalence of elderly people in the Western world. But there are also malfunctions of the immune system that increasingly occur in younger individuals. It is the purpose of this chapter to explain immunosenescence and diseases of the immune system by first giving an explication of how the concepts of function and malfunction apply to the issues at hand, and thereafter to disentangle these concepts as they relate to the immune system's being-in-the-world. It will emerge that immune system aberrations occur owing to evolutionary contingencies, and that they are proximally caused by spatio-temporal mismatches between the immune system and the world.

5.1. Failure to perform

GOAL-DIRECTEDNESS

With the demise of vitalism and the emergence of Darwinism, the Aristotelian concept of teleology was replaced by naturalistic explanations of life's directedness. But instead of purging the notion of goal-directedness from biology, as Newton did for physics, Darwin provided an alternative explanation that preserved goal-directedness. His evolutionary metaphysics effectively split the teleological explanation into two varieties – the functional and the intentional. The two can be differentiated on the basis of characteristics related to the agents that perform the behaviour to be explained, but there are no sharp distinctions, or so I maintain, between their explanatory scopes.

The Darwinian concept of function has attracted a great deal of interest. Several attempts have been made to delineate what a functional explanation is, but so far no unified meaning has emerged. The concept is currently being used by biologists and philosophers alike to designate at least two explanatory projects, one of which is a historical and another a real-time notion (Godfrey-Smith 1993). While the two concepts of function differ in explanatory ambitions, they are related by the fact that the real-time functional attributions are dependent on evolutionary biology's functional attributions. They are thus closely related to Mayr's (1988) distinction between ultimate and proximate causes.

The evolutionary view of functions holds that the function of something is the effect it has and so explains why it is there (Wright 1973). Functions in the evolutionary sense are based on origin and are as such closely related to the concept of adaptation. Hence, the reason why a trait has been retained and modified throughout evolutionary time is explained by the adaptive role the trait had in organisms that possessed it. The seemingly goal directed behaviour of the trait is explained as being caused by the process of natural selection, a strictly *a posteriori* process that never sets up future goals but, in contrast, rewards past events.

The second usage of function, which depicts a proximate real-time explanatory concept, appeals to the workings of the component parts of the system (Cummins 1975). When biochemists, physiologists or immunologists explain how a complex system manages to perform some capacity, they often explain the capacity as being the function of the system. By appealing to the causal capacities of the component parts, biologists need not refer to overall systemic goals or purposes when providing a functional explanation (Amundson and Lauder 1994). When immunologists claim that the function of the immune system is to eradicate antigen, they do not claim that the immune system anticipates the future. Rather, their explanation is wholly mechanistic. The immune system is, like any other biological system, regulated by causal mechanisms, and derives its goal-directedness from a set of subsystems whose non-teleological behaviour simulates overall goal-directedness. Antigen eradication is explained by the fact that antigen triggers receptors of the innate and adaptive immune systems that in turn activate a plethora of molecular mechanisms. These evolutionary selected mechanisms are active as long as the triggering antigen is sensed by the receptors. After eradication of the antigenic stimulus there is no further activation going on and, hence, the immune response is brought to an end.

While functional explanations pertain to all living entities, intentional explanations have traditionally been reserved for rational man (Elster 1979). The explanatory scopes of the two types of explanations are often contrasted by claiming that intentional agents are capable of using indirect strategies to realise their goals while non-intentional functional agents, who do not set themselves goals, are not. Nevertheless, the demarcation between the intentional and the functional is not always easy to draw in practice. Results from research on adaptive systems demonstrate that discriminatory assessment of environmental signs as well as proper responding to the signs can be performed without mental operations being invoked. Such meaning-making is, for example, characteristic for the adaptive immune system, which performs a differentiated response to a given sign on the basis of context (Neuman 2004). Even the smallest organisms, the bacteria, have developed intricate channels of communication to co-operatively self-organise into highly structured adaptively plastic colonies. Some authors even go so far as to equate this behaviour with social intelligence and to claim that adaptive plasticity in bacteria simulates intentional behaviour (Ben Jacob et al. 2004).

Despite there being a fuzzy border between intentional and functional explanations, Mohan Matthen and Edwin Levy (1984, p. 353) were definitively on the wrong track when they claimed that the immune system is goal-directed, and that "some of the things that the immune system does can be described and understood as being errors made in trying to achieve these goals". Although they acknowledged that the utilisation of teleological explanations is a precarious and anthropomorphic undertaking, they still claimed that "the immune system is capable of arriving at justified false beliefs" (1984, p. 365). Since errors can only be meaningfully ascribed to a system if the system is directed towards a goal that is known by the system beforehand, their explanation endows the immune system with intentionality. Invocation of "justified false beliefs", which is on par with Plato's maintenance in *Theaetetus* (1989, 201d) that knowledge is "justified true belief",[1] make perfect sense when truth and falsity are measured in terms of a correspondence between a statement and the world to which it applies. But where scientific explanations are concerned with correspondence and theoretical truth, the immune system has evolved to care for survival, reproduction and practical truth. Whether the immune system reaches "justified false beliefs" or not, is therefore irrelevant. It is fitness that is of essence, and in biology it may sometimes pay to act on "untrue" representations of the world if this acting promotes fitness.

MALFUNCTIONING

A distinctive part of the concept of function is that where there is function there may be malfunction – a failure of a system to perform the explanatory activity. A variety of causes, including genetic, developmental and environmental, can be amassed to explain proximate malfunctions. The immune system can, for example, be proximally caused to malfunction by malignancy, lack of resources, or infectious agents. Historical malfunctions are, on the other hand, more intricate to explain. Since traits retained in a lineage are the results of the trait's previous successes,

[1] The actual sentence reads "true belief with the addition of an account was knowledge, while belief without an account was outside its range."

Paul Sheldon Davies (2000) has argued that selected malfunctions cannot possibly occur. This was also the view of Charles Darwin:

> No organ will be formed, as Paley has remarked, for the purpose of causing pain or for doing injury to its possessor. If a fair balance be struck between the good and the evil caused by each part, each will be found on the whole advantageous. After the lapse of time, under changing conditions of life, if any part comes to be injurious, it will be modified; or if it be not so, the being will become extinct, as myriads have become extinct (Darwin 1859, p. 229).

But even though malfunctions cannot be selected for, they can still increase in frequency owing to the process of natural selection. This may occur if the malfunctional trait is coupled to a beneficial trait under strong selective pressure. Since different traits are coupled at the organismal level, the outcome of selection for any individual trait depends upon the other traits that the organism comes equipped with. In addition, the selection for a trait depends upon the level at which the trait is selected, and whether there are counter-selective forces at work on other levels. Traits can be selected for or against at several levels simultaneously and there is a real possibility that fitter traits may decrease in frequency while less fit traits increase. The result depends on the relative strengths of the selective pressures that are at work at each level.

Co-selection of traits may occur because of linkage, which occurs when the genes are close together on a chromosome, or because of pleiotropy which occurs when a single gene has two phenotypic effects. The effect of linkage in the evolution of malfunctional traits is well exemplified by the mammalian MHC complex. Since the genes that encode the MHC molecules are located close together on the same chromosome, they are usually inherited *en bloc*. One MHC gene in the cluster may thus confer protection against infectious disease, while a closely linked gene may confer susceptibility to disease.

The MHC genes also exemplify pleiotropy, but the classic example of this phenomenon is sickle-cell disease, which results from a point mutation in the gene that produces haemoglobin. The bearers of the mutation, who develop malformed blood cells, are simultaneously protected against malaria. Individuals that are homozygous for the mutated gene develop a severe haematological disease whereas heterozygous individuals are better off. Since both hetero- and homozygous individuals are protected against malaria, the best combination is to be heterozygous in malaria infested regions. In this case the optimal phenotype, normal blood cells and resistance to malaria, is not attainable owing to antagonistic pleiotropy; the same cause leads to two opposing effects – sickling and resistance.

Besides co-selection, malfunctioning can also increase owing to inevitable outcomes of the evolutionary process itself. During natural selection average fitness should increase with regards to the environment that existed previously, *i.e.* the selective environment. But the environment, which includes other members of the population as well as biotic and abiotic factors, changes all the time. And since fitness is always fitness relative to the environment, a trait that is at a local optimum in one environment may be rendered malfunctional if environmental change induces the trait's movement away from the local optimum. If environmental change leads to changes in developmental outcome, environmental change may even lead to novel phenotypes. Hence, organisms or lineages can grow unfit simply by virtue of standing still in a changing world.

THE DISPOSABLE ORGANISM

In stark contrast to the assertion that malfunctions are never evolutionary selected stands the common observation that organisms increasingly experience deterioration and malfunctional diseases as they age. For example, the relative increase in death rates for people over 65 compared to the rates for people aged 25–44 are 43-fold for cancer and 89-fold for pneumonia and influenza (Troen 2003). The unfavourable and near-universal phenomenon of senescence is perplexing for at least two reasons. First, cumulative natural selection should ordinarily proceed towards lengthening life, not shortening it. And second, it is truly remarkable that organisms which produce and maintain themselves during development should be unable to perform the much simpler task of merely maintaining what is already formed.

Still, the observation that the maximum lifespan of organisms is species specific suggests that some form of selection should be involved in senescence. To understand how this may come about, consider a trade-off that occurred when unicellular life gave rise to multicellularity. While single-celled individuals had to embody capacities for both reproduction and survival, cells in multicellular ensembles were offered the opportunity to specialise in one component or the other. This opportunity was seized by the ancestors of extant metazoans as they evolved a germ line that specialised in reproduction and a soma that specialised in vegetative functions. It is in this sense that the germ line, which has been reproducing for more than three billion years, owes its immortality to a vehicle, the ephemeral soma. Since reproductive costs can be lowered to the degree that the soma acquires mates, evolutionary theory predicts that the soma should evolve characteristics that would further enhance the reproductive component of fitness. This also implies that the soma should be maintained at least so far that it contributes to the germ line's reproduction.

But this line of reasoning does not account for why the soma deteriorates beyond the reproductive ages. To explain this, we have to take development of the soma into consideration. Selection for ageing can then be explained as an antagonistically pleiotropic phenomenon. This process may come about if a particular constellation of genes affects individual fitness in opposite ways at different ages, and if the beneficial effects become manifest at younger ages while the detrimental effects manifest at older ages (Williams 1957). The genes responsible for ageing would thus be kept in the gene pool by selection on their beneficial effects to the young that possess them and not owing to their detrimental effects in senescence. Natural selection can thus give rise to mechanisms that both create and destroy the organism.

Because organisms have a large capacity to maintain themselves in response to external perturbations, it is evident that ageing cannot be likened to a passive "wear and tear" process. Metchnikoff was well aware of this when in 1904 he envisioned old age as an

> infectious chronic disease which is manifested by a degeneration, or an enfeebling of the noble elements, and by the excessive activity of the macrophages. These modifications cause a disturbance of the equilibrium of the cells composing our body and set up a struggle within our organism which ends in a precocious ageing and in premature death, contrary to nature. (Citation from Bibel 1988).

While Metchnikoff speculated that old age is caused by cells of the innate immune system, Roy Walford (1969) further claimed that the adaptive immune system has a causal role in the ageing of vertebrates. He believed that the immune system incited the pathogenesis of ageing by bringing about "the unleasing of self-destroying processes of the nature of auto-immunity or transplantation disease." (Walford 1969, pp. 203–4).

In his theoretical outlook Walford distinguished between ultimate and proximate explanations of ageing. He held that proximate causes would differ amongst different species, but that the ultimate causes should be the same. These causes were postulated to reside within the dividing cell populations, but he did not elaborate further upon the mechanisms involved.

Building upon the works of Buss (1987) and Maynard-Smith and Szathmáry (1995), Bruce Charlton (1996) took Walford's idea one step further when he suggested that ageing was ultimately caused by "endogenous parasites". Charlton claimed that replicating lineages within an organism should inevitably evolve to elude surveillance, resist suppression and subvert organismal integration. Such lineages, which include malignant cells and clonally proliferating lymphocytes, should gradually adapt to the internal environment. The cells that elude the control mechanisms may then evolve in a, for the organism, maladaptive direction. In this perspective, ageing is the progressive deterioration of the soma owing to the exploitative behaviour of its component parts.

Little is still known about the cellular and molecular bases for longevity differences between species, and the immunologic hypotheses of Metchnikoff, Walford and others have so far not been vindicated. Still, several observations support the statement that the risk of suffering chronic diseases during senescence depends on influences acting early in life, and thus that ageing has a developmental origin. One mechanism that may promote ageing, and which fits nicely into the theoretical framework of Williams, is inflammation following organismal insults. Several observations support the view that while a strong proinflammatory immune response is necessary to resist otherwise fatal infections in early life, the overproduction of inflammatory molecules may cause inflammatory diseases and even death later in life (Licastro et al. 2005).

And in accordance with the views of Metchnikoff, Walford and Charlton, the factors that contribute to inflammation are mostly infectious agents, especially the chronic ones. For example, several viruses have been connected to an altered response pattern of the adaptive immune system. This is probably owing to their induction of a reconfiguration of T cell immunity, manifesting as the accumulation of senescent and dysfunctional cells (Pawelec et al. 2004), a shift in subpopulation frequency and expressed repertoire of antibodies and T cell receptors (Miller 1996), and a concomitant increase in regulatory T cells (Gregg et al. 2005).

These observations, combined with some general characteristics of senescence, including the decreased resistance to infectious disease and a decreased protection against cancer, are compatible with a view stating that higher demands are imposed on the regulatory elements of multicellular organisms during ageing. Deterioration occurs when these demands are exhausted in one way or other. Since the human organism was set to live 40 to 50 years during the course of evolution, it is perhaps little wonder that life beyond that point is marked by chronic diseases that slowly but inexorably damage all the organs, including Alzheimer's disease, atherosclerosis, and type II diabetes.

5.2. The re-enacting of ancient conflicts

THE EVOLUTION AND DISRUPTION OF INDIVIDUALITY

The concept of individuality, which is derived from the Latin word for non-divisible, *individuus*, has fared a tortuous road. Several attempts at determining necessary and sufficient conditions for individuality have been proposed (Santelices 1999; Wilson 1999), but the concept is still ambiguous. The properties most typically associated with biological individuality include genetic homogeneity, physiological autonomy, spatio-temporal continuity, rejection of grafts from non-self, and demarcation from other members of the same or different species. In the context of the current discussion, individuality is to be understood as an evolutionary concept that denotes entities that possess heritable variation in fitness and who have a complex structure with an internal organisation that develops through time.

Evolutionary individuals have appeared several times during evolutionary history. When for example prokaryotic cells joined to form the eukaryotic cell and when eukaryotic cells joined to form multicellular organisms, new evolutionary individuals emerged as a result of evolutionary transitions (Figure 5.1). Once two individuals joined to form a higher level individual, the two had to relinquish their individuality in favour of the higher level group of cells. The fitness of the previously independent entities were thus transferred to the higher level entity, and survival and reproduction came to depend upon their ability to cooperate for the good of the new individual (Michod 1999).

The multilevel selection approach to evolutionary transitions seeks to understand how a group of pre-existing individuals becomes a new evolutionary individual – how independent entities unite to form a multicellular adapted unit. For the entity

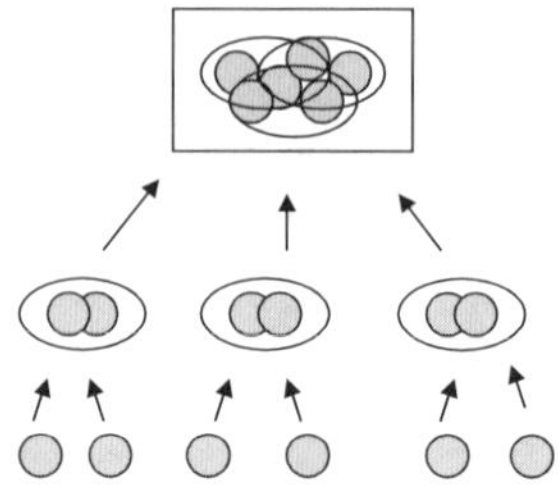

Figure 5.1. The evolutionary transitions. The figure depicts three levels of selection. If the higher level entity "allows" unrestrained reproductive ability of its constituting entities, there will be opportunity for selection at several levels and thus a risk for disruption of the higher level entity

to emerge as an individual, the new evolutionary unit must have found ways to modify the selfish tendency of competing entities while promoting their cooperative interactions for common benefit. In addition, the properties of the new entity must be made heritable to continue the evolutionary race. Given enough time this would lead to well-integrated and well-functioning organisms. What has been termed the paradox of the organism; that it is not torn apart by its conflicting constituents (Dawkins 1990), has thus been solved by evolution through the orchestration of appropriate developmental sequences and the hierarchical imposition of conflict-modifying mechanisms.

The sequestration of germ cells in early development, which in vertebrates takes place at the three or four cell stage (Soriano and Jaenisch 1986), has been proposed as one example of conflict-mediation between cells at the new organismal level (Buss 1987). Sequestration of germ cells implies that the genome of the somatic cells cannot be propagated unless they cooperate to enhance the reproductive potential of the germ. Since cooperation between the germ and the soma delimits the opportunity for cheating and thus disintegration of the new unit, a group consisting of specialised germ and somatic cells is thus no longer divisible – it has become an individual.

While evolutionary individuals are identified by the processes that have generated them, each evolutionary individual has to maintain its organismal integrity throughout life. The word integrity, which is derived from the Latin word *integer*, means the quality of being complete or undivided, and depicts the state of having an intact and uncorrupted self despite continuous conflicting motivations. Integrity thus entails individuality. If integrity is disrupted, evolutionary individuals may be let loose. Zlatko Dembic (2000, p. 563) has naturalised the concept of tissue integrity to be "a measure of all-possible adhesive and signalling contributions that a single cell accepts and sends in its normal, resting state." According to this outlook, the disruption of tissue integrity – not the experience of non-self or danger, is seen as the prime stimulus for immune system activation.

The maintenance of the integrity of evolutionary individuals is a precarious undertaking, and multiple defence mechanisms against disrupting challenges have evolved. The fact that vertebrate organisms develop from single cells means that, barring mutation, cells in these multicellular organisms are clones. This high level of relatedness is invoked to explain the high degrees of developmental coordination and extreme altruism observed in extant organisms, for example the relinquishing of reproductive capability by somatic cells. But even though kinship among members of the group tilts selection in favour of the group of cells and away from the entities that make up the group, germ-line sequestration is apparently not sufficient to oblige cooperation.

Several other means of directly suppressing the selfish tendencies of cheating entities have evolved. Such adaptations, which restrict the opportunity for conflict between the levels, are termed conflict modifiers (Michod and Nedelcu 2003). The uniparental inheritance of mitochondria, which ensures that there is no conflict between the mitochondria from the two parents, is one type of evolved conflict modifier. Others include the programmed cell death that occurs during development of the lymphocytes in the thymus, and the production of modifier cells, which are termed regulatory cells when they serve to modify immune system reactivity.

The most important disrupters of integrity are entities that are experienced as foreign to the individual. In the context of the present discussion there are three types of foreignness to consider; infection, chimerism and mosaicism. The parasitic entities that give rise to these phenotypic states display different degrees of disparity as compared to the host individual and thus elicit different degrees of host defence. Some characteristics of these inducers of proximate malfunctions are depicted in Figure 5.2 and summarised in the next section.

INTRAORGANISMAL PARASITISM

Infection

Infectious agents invade the tissues and cells of the host organism and sometimes even integrate their nucleic acids into the host's DNA. This generates intercellular heterogeneity and disruption of tissue integrity. While the immune system often manages to rid the exogenous agents before too much damage develops, several agents have evolved mechanisms to evade the immune system and thus to survive

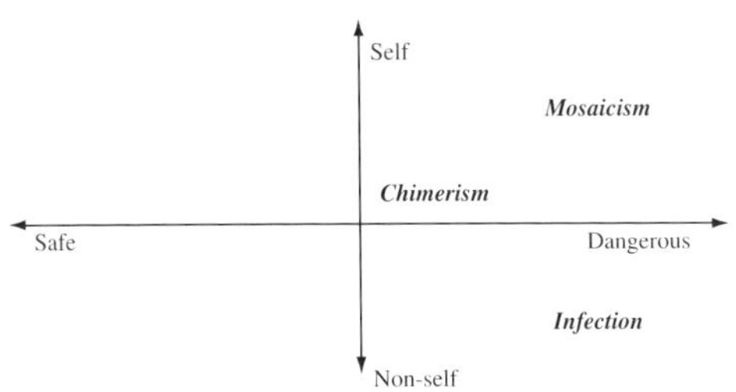

Figure 5.2. The three types of parasitism as delineated by gradations of danger and self. Infectious agents are more non-self than chimeric cells, which in turn are less self than mosaic cells. All three agents may signal danger to the immune system

and sometimes even reproduce during the entire life of the host. The agent that causes tuberculosis, *Mycobacterium tuberculosis*, has for example evolved mechanisms to maintain persistence, as have several viruses, including Epstein-Barr virus, HIV and hepatitis C virus.

It has recently become clear that infecting agents that proliferate within a host often display features typical of genetic heterogeneous populations, and that these populations may be in conflict with each other and with other infecting agents as well. The complicated population dynamics that emanate from such interactions has been well described in HIV infection, during which the massive overproduction of mutant viral particles becomes adjusted by the selective pressure exerted by the immune system. The result is a complex distribution of non-identical but closely related genomes that are sometimes termed quasispecies (Eigen 1996; Domingo 1998). The evolution of quasispecies is frequently observed during antiretroviral therapy when strains containing drug-resistance mutations emerge as a result of the selective pressure set off by the drugs.

Another characteristic of viral quasispecies is their frequent compartmentalisation in different cells and tissues of the same patient. This phenomenon, which has been observed during infections with both HIV (Itescu et al. 1994) and hepatitis C virus (Ducoulombier et al. 2004), probably occurs as a consequence of competition and differential survival of the different strains. The multi-strain phenomenon and differential predilection for specific tissues has also been observed in patients infected with Epstein-Barr virus, but in this case it appears that the patients become infected with multiple strains of the virus (Sitki-Green et al. 2004).

It has repeatedly been observed that infections with the Epstein-Barr virus lead to a dysregulated immune response. This is in part owing to the virus' peculiar capacity to induce transcriptional activation of genes from endogenous retroviruses that are able to stimulate or delete T cells with specific T cell receptors (Sutkowski et al. 2001). There is evidence that also other substances may activate endogenous retroviruses and thus induce poorly regulated immune responses. Since the human genome contains at least 22 retroviral families, some with virtually intact genomes (Bromham 2002), such transcriptional activation may lead to intraorganismal competition and probably also to inflammatory, autoimmune and malignant diseases (Goldberg et al. 2000).

Chimerism

Owing to the potent defence mechanisms elicited when attempting to transplant cells from one individual to another, the second type of parasitism, chimerism,[2] was, with

[2] In Greek mythology, the chimera was a monstrous creature with the head of a lion, the body of a goat and the tail of a serpent. Today, chimeras are routinely created in laboratories – mice and other animals with humanised immune systems are, for example, created for the purpose of better understanding health and disease.

one major exception, long believed to be a rare event in vertebrates. The exception to the rule was pregnancy, during which the mammalian foetus manages to survive despite being located within the uterus. But observations made in the late 1990s demonstrated that vertebrate chimerism may come naturally also in other settings.

The cells of the foetus have invasive potential and may establish themselves within the mother for life (Bianchi et al. 1996), and conversely, maternal cells can be detected in adult offspring up to 28 years of age (Maloney et al. 1999). In addition, as many as 8% of non-identical human twin pairs have chimerical blood (van Dijk et al. 1996). Establishment of microchimerism, which refers to the stable presence of a small number of non-host cells, usually stem cells or their progeny, is apparently only possible during the maturation of the immune system and during pregnancy-related immunosuppression (Anderson and Matzinger 2001). Hence, in contrast to infectious agents which establish themselves despite active host defences, chimerism is a result of host tolerance to foreignness.

Since haematopoietic stem cells may cross the placenta to establish in the mother and child, there is a possibility that the offspring's lymphocytes may attack the mother's tissue, and vice versa. Another possibility is that cells that cross the placenta may transdifferentiate into other types of cells, including hepatocytes, myocytes, keratinocytes and neural cells (Forbes et al. 2002). They may then become targets for attack by host lymphocytes, or alternatively, the chimeric lymphocytes may attack the host cells. Both scenarios will result in phenotypic autoimmune disease.

Several autoimmune diseases, including systemic sclerosis, primary biliary cirrhosis and Sjögren's syndrome, have been associated with microchimerism. These diseases have much in common with chronic graft-versus-host disease which occurs following transplantation. Since graft-versus-host disease is caused by mismatches between the MHC genes of the two individuals, and since autoimmunity associated with microchimerism appears to be linked to MHC molecules, it has been hypothesised that foetal microchimerism may induce autoimmunity in the mother (Nelson 2002). Since there is cellular migration from mother to child as well, and since maternal cells can be detected in adult offspring up to 28 years of age (Maloney et al. 1999), a similar argument may be used to explain autoimmunity in offspring.

Selection pressure exerted by parasites is believed to be the major cause of the polymorphism observed in the HLA, but direct evidence for this supposition is limited (Hughes and Yeager 1998). The diversity may also be owing to mating preferences in some species (Klein 2000). By extending these speculations, one may wonder whether interference with mate selection can have any influence on the prevalence of autoimmune disease. Several investigations have confirmed that MHC molecules are associated with women's choice of male odour, and that women prefer MHC-disparate males (Penn and Potts 1999; Jacob et al. 2002). When women are taking oral contraceptives, they seem to prefer the odour of MHC-similar men (Wedekind et al. 1995). Since contraceptives mimic the effects of pregnancy, these women may be attracted to MHC-similar men. Accordingly, females that choose mates while using contraceptives and later conceive a child with the same man, may be at increased risk of harbouring MHC-similar microchimeric cells. Whether or not there is an association between having MHC-similar microchimeric cells and autoimmune disease has, however, never been ascertained.

The chimerism discussed so far originates during the development of the individual organism. But chimerism can also be understood as being the result of evolutionary processes. In the evolutionary sense, all multicellular organisms are chimeric. The nucleus consists of molecules that are phylogenetically related to endogenous retroviruses (Li et al. 2001) as well as to archaebacterial and eubacterial ancestors (Staub et al. 2004). Quite analogously, the ribosomes (Staub et al. 2004) and mitochondria (Andersson and Kurland 1998) have an endosymbiotic origin. The endogenous retroviruses are able to encode and express proteins that may be targets of autoimmune attack (Krieg et al. 1992), and so are mitochondria (Baum 1995) and ribosomes (Stafford et al. 1998).

Mosaicism

Somatic mosaicism, which designates the presence of distinct populations of somatic cells in a given organism, results as a consequence of developmental processes. Somatic mosaicism is often caused by mutations of DNA, but may also occur owing to changes of the genome that do not affect the DNA sequence. Such changes, termed epigenetic, refer to modifications in gene expression that are controlled by heritable but potentially reversible changes in DNA transcription. The best known epigenetic signal is DNA methylation, in which cytosine, one of the four chemical bases that make up the genetic code, is tagged by a methyl group. DNA methylation is generally associated with silencing of gene expression, whereas active genes are usually unmethylated. But cytosine methylation is also important for genome stabilization, as made evident in people who suffer from immune system deficiencies owing to mutant genes for the DNA-methylation enzyme (Xu et al. 1999).

By metaphorical analogy to mosaics created by artists, Shoenfeld and Isenberg (1989) pictured the different phenotypic expressions that autoimmune diseases may take in different individuals as being the outcome of differential rearrangements of identical or nearly identical biological starting material. For example, the phenotypic expression of systemic lupus erythematosus and multiple sclerosis in monozygotic twins that are discordant for the disease may be owing to epigenetic alterations (Petronis 2001). Such ideas have led several researchers to emphasise the potential role of dysregulated epigenetics as being a susceptibility factor for the development of complex diseases. Epigenetic alterations have also been invoked to explain the age of onset, the phenotypic fluctuations and the differential susceptibility of males and females for disease manifestations. Interestingly, these ideas on development are concordant with Kirschner and Gerhart's (1998) view on evolvability, according to which evolution may take place without there being alterations in the genetic system.

Mosaicism is a characteristic feature of malignant diseases. Genetic alterations in malignant cells lead to profound changes, including immortalisation, blocking of terminal differentiation, invasiveness and potential for metastasising. Most antigens expressed by cancers are non-mutated self-antigens, and are thus ineffective at triggering immune responses. Hence, malignant cells also have the ability to evade the immune defences; the cells with the highest evasive potential are the ones that are selected for further proliferation. But mosaicism can also take clinical importance if it occurs at the level of haematopoietic progenitor cells in which case it may lead to increased susceptibility for disease, generation of immune diversity and phenotypic variability among monozygotic twins (Youssoufian and Pyeritz 2002).

A special type of mosaicism occurs in women. Early in female development cells randomly inactivate either the maternal or the paternal X chromosome. Each of these cells gives rise to a patch of cells in the adult female that maintains the same X chromosome in the inactive state. Thus, some regions of her body uses the X she inherited from her father, while the rest utilise the X inherited from her mother. A woman is thus a mosaic of two cell populations. Not all the genes in the inactivated X chromosome are inactive, though. About 15% of the genes in the inactivated X chromosome escape inactivation and are thus expressed at higher levels in females as compared to males (Carrel and Willard 2005).

Although not experimentally verified, the phenotypic differences that follow from random inactivation of X chromosomes could theoretically lead to internal conflicts between the two lineages during embryonic development. In line with this reasoning, Jeffrey Stewart has argued that the higher prevalence of autoimmune diseases in females as compared to males could be owing to their pattern of X inactivation mosaicism (Stewart 1998). He argued that if the dendritic cells of the thymus, which are derived from a small number of progenitor cells present at the time of X inactivation, all have the same X inactivation profile, there is a risk that maturing T cells that are reactive against self-molecules encoded by the inactivated X chromosome will not be negatively selected. These T cells will then emerge from the biased thymus as mature T cells. If they encounter peripheral cells using the X chromosome that was inactivated in the thymus, there is a chance that they will interpret the products of that other X chromosome as foreign and thus attack the cells. Thus, autoimmune disease in females is an internal conflict between the two fractions that make up a mosaic woman, much like the chronic graft-versus-host response that follows transplantation and chimerism.

Although the three types of parasitism – infection, chimerism and mosaicism – have traditionally been held to be distinct, some recent observations of transmissible tumours challenge this view. The first suggestive example comes from individuals that belong to the Australian marsupial, The Tasmanian devil. They develop facial-tumour disease following bites by conspecifics in the mouth regions. The disease, which is of high pathogenicity, eventually leads to the death of the animal. Evidence now indicates that the cancerous cells, which are karyoptypically similar in the afflicted animals, are derived from a clone capable of being transmitted amongst different members of the species (Pearse and Swift 2006).

The second example comes from a tumour that grows in dogs of all breads, and which can be transplanted into immunocompetent animals of other canine species, including foxes, coyotes and jackals. While the spread of cancerous cells amongst Tasmanian devils may be owing to a high degree of kinship amongst the members of the species, which reduces their immune responses to the implanted cancers, this is definitely not the case with the venereal tumour disease caused by transmission of cancerous cells between dogs. After coitus, the tumour cells firstly grow rapidly, after which they eventually regress spontaneously three to nine months later. The regression is apparently caused by the immune response as the afflicted animals are immune to rechallenge of the cancer, and since passive transfer of immunoglobulins from recovered dogs also confers immunity. The tumour, which probably arose from a common ancestral neoplastic cell originating in a wolf or a dog between 200 and 2500 years ago (Murgia et al. 2006), has reached a worldwide distribution. It is thus a successful parasite that harbours characteristics of infectious agents, chimeric transplanted cells and malignant mosaic cells.

AUTOIMMUNITY – A CONFLICT AMONGST EVOLUTIONARY INDIVIDUALS

Associations to be explained

The term autoimmune disease comprises a fairly heterogeneous group of diseases that are characterised by the occurrence of autoreactive T cells and autoantibodies in the sera and tissues of afflicted patients. The rich variety of autoreactive lymphocytes and antibodies are expressed to a limited degree in healthy individuals, while they are found at high concentrations and in distinct patterns in patients with specific diseases. Since there are few qualitative differences between naturally occurring autoreactive receptors and receptors that are involved in autoimmune disease, and since not all patients with autoimmune disease express autoreactive receptors, autoreactivity is neither necessary nor sufficient for the diagnosis autoimmune disease.

The autoimmune diseases are traditionally recognised as being systemic if the autoantigens are widely distributed, and organ-specific if they are restricted to a single type of tissue. The former group includes systemic lupus erythematosus, rheumatoid arthritis and many more, while the latter group comprises autoimmune thyroiditis and multiple sclerosis amongst others. The relationship between the autoreactive cells and disease etiopathogenesis is circumstantial in most cases. Hence, the term autoimmune disease depicts a nosological, not an etiological entity. There are exceptions to this, though. The autoimmune haemolytic anaemias, which are characterised by autoantibodies that react with membranous antigens, are etiologically defined.

Each autoimmune disease is a heterogeneous entity, and patients diagnosed with the same disease may exhibit a wide variety of symptoms and signs. These features are well displayed in systemic lupus erythematosus and multiple sclerosis. Systemic lupus erythematosus, which afflicts nearly every organ in the body, is associated with a characteristic pattern of autoantibodies reactive with nuclear, ribosomal, mitochondrial and membranous antigens. Like many other autoimmune diseases, including multiple sclerosis, it has a disproportionate preponderance in that it affects females several fold more often than males. The disease is also typical in that it seldom manifests before puberty but typically shows itself during early adult life (Beeson 1994).

Multiple sclerosis, an autoimmune disease that afflicts the central nervous system, displays similar age and sex-associations to lupus. Like lupus, patients with multiple sclerosis also express a variety of antibodies against intracellular autoantigens (Lu and Kalman 1999), and exhibit strong associations with genetic components, especially genes that encode certain MHC antigens. The increased disease concordance rate in monozygotic as compared to dizygotic twins, 30% versus 5%, is also similar to lupus. The prevalence of the two diseases, which is different at different locations of the Earth, is connected not only with the genetics of the populations who inhabit different regions, but also with environmental stimuli in the different regions (Keegan and Noseworthy 2002).

The unfolding of some infectious diseases is regularly accompanied by expansion of distinct autoantibodies and T cells (Ray et al. 1996). Quite analogously, the immune responses towards malignant cells are accompanied by autoreactive lymphocytes that are generated against antigens that need not even be expressed on the malignant cells (Lang et al. 2003). Also chimerism, for example following liver transplantation, is sometimes accompanied by autoreactive lymphocytes (Vergani and Mieli-Vergani 2002). A most peculiar observation, which contests the view that autoimmunity is a direct consequence of autoreactivity, is the observation that a variety of immunodeficiencies, both of the inborn and the acquired types, are accompanied by increased susceptibility to autoimmune manifestations (Etzioni 2003).

While several successful therapies that specifically target the disease-inducing mechanisms have been described for autoimmune disorders in animal studies, these have so far not been approved for the treatment of human disease. The fact that non-specific treatment, like corticosteroids, interferons and blockade of tumour necrosis factor, is successful while specific treatment is not, suggests that we do not understand the etiopathogenesis of the autoimmune diseases well. Despite claims to the contrary, we remain ignorant of what drives the chronicity of these diseases, and because "we do not understand the differences between the chronic and acute response, we cannot be sure which, if any, animal models of disease provide good reflections of the key processes that occur in human disease." (Feldmann and Steinmann 2005, p. 612).

When viewed from a biosemiotic angle, according to which sustained survival requires continuous monitoring of environmental signs followed by precise adaptive responses to the encountered stimuli, the relevance of animal models for human autoimmune disease becomes especially problematic. Whether the immune system of an animal detects a sign or not depends upon the evolutionary history of the species to which it belongs. Hence, there is a possibility that organisms of different species interpret the world differently, and that they in a sense inhabit different antigenic Umwelts. Differing Umwelts combined with multiply realised immune systems and conventional classification of immune system diseases suggest that proximate explanations of malfunctions may not generalise well.

Proximate explanations

In *The clonal selection theory* and in *Self and not-self* Burnet (1959, 1969) explained autoimmunity as an occurrence that followed failures of inactivation mechanisms for self-reactive T and B cells in the central lymphoid organs. The mechanisms that lead to central tolerance have now been worked out in detail, and several experimental investigations in animals testify to their importance. It is clear that self-reactive B and T cells normally escape the central inactivating mechanisms, but that they are rendered non-pathogenic by several mechanisms that lead to peripheral tolerance. These include failures at co-stimulation by accessory molecules, B cell receptor editing and several others. Hence, autoimmunity should occur when the suppressive mechanisms, be they central or peripheral, fail to perform.

Much recent work has been put into investigations of the regulatory aspects of the immune response, and many autoimmune diseases are now considered to be the result of a dysregulated immune system. Skewed functionality of regulatory T cells and the cytokines they produce has been amongst the strongest candidates to explain diseases like lupus and multiple sclerosis. Failures at regulation are connected to genetic factors in the afflicted individuals, but environmental factors are also important.

The association between autoimmune diseases and certain MHC molecules is for example believed to be owing to the ability of these MHC molecules to present disease-inducing peptides to autoreactive T cells. The peptides may be derived from self or from non-self that mimic self. Among the best characterised environmental factors that lead to autoimmune disease are infectious agents. For example, both lupus (McClain et al. 2005) and multiple sclerosis (Lang et al. 2002) express autoreactive receptors that cross-react with molecules of the Epstein-Barr virus. While some autoreactive responses may be explained by molecular mimicry, in which the pathogen exhibits molecules that are very similar to the molecules of the host, this is not always the case (Rose 1998). There are also convincing observations which suggest that the activation of endogenous retroviruses may be the proximate cause of multiple sclerosis (Fujinami and Libbey 1999) and lupus (Adelman and Marchalonis 2002).

While lupus is characterised by major disturbances in the B cell system, the major disturbances in multiple sclerosis are within the T cell system (Williams et al. 1994). And whereas treatment with interferon-α alleviates the symptoms of multiple sclerosis, interferon-α appears to have a pathogenic role in lupus. This phenomenon may be owing to the differential effect that interferon has on accessory cells and on the differentiation of T cells into T helper 1 or T helper 2 cells (Lohoff and Mak 2005).

Several mechanisms may account for the associations between autoreactivity and infection. These include release of sequestered auto-antigens following tissue damage, induction of inflammatory cytokines and costimulatory molecules, as well as trade-offs between the beneficial and deleterious effects of the regulatory T cells. The mechanisms can all lead to the expansion of autoreactive cells. Nevertheless, direct links between any cause and specific autoimmune diseases has rarely if ever been established. While infectious diseases have been associated with a variety of chronic autoimmune diseases, a true causal pattern between infectious agents and a specific disease has rarely been revealed in humans. The interactions are complex, and the causal chain goes in both directions; in some cases infections trigger autoimmunity, while infections may protect against autoimmunity in other cases.

In their discussion of self tolerance and autoimmunity John Rioux and Abul Abbas (2005, p. 584) summarised the contemporary status of proximate explanations in the three postulates

i) autoimmune disease develops when self-reactive lymphocytes escape from tolerance and are activated.
ii) the failure of self tolerance is the fundamental cause of autoimmunity
iii) autoimmunity is thought to result from a combination of genetic variants, acquired environmental triggers such as infections and stochastic events.

While the three postulates cogently summarise the received view on autoimmunity, a deeper analysis of the postulates discloses that they are better at revealing lack of knowledge than of explaining. As will become clear in the following sections, postulate i) has low explanatory value because activated self-reactive lymphocytes are prevalent also in healthy individuals; postulate ii) is tautological in that failure of self-tolerance is but another expression for autoimmunity and so cannot be its cause; and postulate iii) is too general to be of any explanatory value. Still, postulate iii), or so I will argue, parses the investigation in the right direction as it hints to the evolutionary genesis of the predisposition for autoimmunity.

An ultimate explanation

Contesting the explanandum

The paradoxical notion that the immune system serves as a vehicle for self-destroying capacities is currently regarded as the main explanandum of autoimmunity. It dates back to the 1950s when Ernest Witebsky, in an attempt to circumscribe the inconsistent concept of autoimmunity, designated etiological criteria for human autoimmune disease. The criteria, which were later revised to accommodate new knowledge derived from molecular and cellular biology, were graded into three evidential categories; direct, indirect and circumstantial (Rose and Bona 1993).

Direct evidence for autoimmune disease comes from transfer studies in which a disease is reproduced in a healthy individual following direct transfer of antibody, for example from mother to child during pregnancy. *Indirect evidence* is obtained if the disease can be reproduced in experimental animals or if autoantibodies or self-reactive T cells can be isolated from the afflicted organs, while *circumstantial evidence* includes statistical associations with autoimmune phenomena or susceptibility genes as well as favourable response to immunosuppression.

Fundamental to the revised criteria is the belief that "the disease is caused by an autoimmune response of the host without regard to the origin of that response." (Rose and Afanasyeva 2003, p. 136). This way of framing the problem has encouraged a wide range of investigations, but although much information has been collected concerning the detailed immunopathology of autoimmune diseases, most evidence for autoimmune disease remains indirect and circumstantial. It is therefore relevant to ask whether diseases subsumed by the two latter criteria really are on par with disease as identified by the first criterion, and whether exclusion of the origin of the autoimmune response from the explanandum is wise or not.

To give an example, the rheumatic diseases systemic lupus erythematosus, Sjögren's syndrome and rheumatoid arthritis are believed to be autoimmune, but the evidence for this is largely indirect and circumstantial. As is made evident from investigations demonstrating that the probability of disease given observed antibody is low, the autoantibodies that characterise the diseases are neither necessary nor sufficient constituents of the disease (Ulvestad et al. 2000). On the other hand, when combinations of several autoantibodies are tested by likelihood ratio testing, it is clear that patients with rheumatic disease have distinct immune system aberrations that are not manifest in control patients (Ulvestad 2003). The results of such investigations point out a major explanatory problem in autoimmunity research; the fact that autoreactivity is common and occurs naturally (Avrameas 1991), and that pathogenic autoantibodies are routinely generated during the response to foreign antigen (Ray et al. 1996). It even appears that autoreactivity may be beneficial for organ restitution and elimination of waste products (Maslloréns 2000; Schwartz 2002), and it has been postulated that autoreactivity is a special case of immunity – while immunity removes threats from without, autoreactivity removes threats from within (Nevo et al. 2003).

Observations that autoreactivity comes naturally, both following infections, malignancies and chimerism, and that autoimmune disease may occur in immunodeficient individuals, are not easily reconciled with the view that all diseases currently categorised as autoimmune diseases are caused by the immune system. On the other hand, the observations would be likely under a hypothesis stating that autoreactivity in some diseases is merely an epiphenomenon that occurs as a result of some other cause. As demonstrated in the next section, this idea allows us to discriminate between inflammatory epiphenomenal autoreactivity and true autoimmunity. The difference is dependent upon whether the disease emanates in the innate or the adaptive immune system.

Organismal integrity – a precarious undertaking

Since the concept of immunity signifies the idea that independent organisms stand against each other, the addition of the prefix auto- should designate the paradoxical conception that the competing organisms were integrated within one and the same organism. This view, which makes sense if the cells of the host fight an invading infectious agent, is not as straightforward if there are no foreigners to contend. However, if organism is understood as laid out in this book, as a composite entity made up from lower level individuals that once coalesced to form higher level individuals, autoimmunity can be put on par with immunity, the difference being that immunity designates contemporary conflicts whereas autoimmunity is the re-enacting of ancient conflicts between lower level evolutionary individuals.

This theoretical outlook entails that anything that has or gains heritable variation in fitness, including infectious agents, mitochondria, transopsons, malignant cells, chimeric cells and lymphocytes, should be able to precipitate autoimmune disease. Furthermore, it serves to explain a wide variety of perplexing phenomena in autoimmunity without having to invoke one type of reactivity for immunity and another for autoimmunity. And since maintenance of both integrity and identity are explained through a unified hypothesis, explanatory coherence is conserved. The view does, nevertheless, necessitate that we reconceptualise and disunify our ideas of what autoimmune diseases are.

The re-enacting of ancient conflicts is owing to the embodied drive inherent in any organism that is or once was an individual. Organisms are designed by natural selection to transmit their genes to future generations. Each composite organism consists of a variety of evolutionary individuals that are harmonised to enhance fitness. It is central to the hypothesis that when any of the evolutionary individuals undergo reproduction, there is a chance of conflict because not all genes, be they in mitochondria, somatic cells or lymphocytes, have equal opportunities to be transmitted to the offspring. Reproduction thus provides a chance for evolutionary individuals to exploit the organismal resources at some cost to the rest.

To ensure fairness in reproduction, countermeasures against exploitation have evolved. For example, most genomes contain suppressors that limit the activity of endogenous retroviruses; most cells have mechanisms that suppress the mitochondria; and most organisms have suppressors that regulate the activity of their immune system. Malfunction of any of these conflict modifying systems should lead to a reactivation of the selfish tendencies of the evolutionary individuals involved. If for example ancient retroviruses are left unchecked, there is a real chance that the internal workings of the cells involved would malfunction and that disease would ensue. And if tumour suppressor genes are malfunctional, malignancy may ensue.

In an analogous manner, the regulatory context becomes disarrayed when a conflict between hosts and pathogens occurs.

It is an interesting but largely unexplored possibility that activation of the immune system may be caused not only by non-self or danger, but also by the relaxation of control mechanisms. It is a common observation that cells grown *in vitro* take on an activated profile. This phenomenon is usually assumed to be owing to the fact that the cells become activated by the novel surroundings. While this is probably true in many cases, it is conceivable that some cells take on the activated profile because suppressive mechanisms that kept them silenced when *in situ* have been relaxed. To the extent that this is correct, the result obtained from *in vitro* studies need to be given a cautious interpretation as to their relevance for the *in vivo* situation.

There are several mechanisms that could deregulate the conflict modifying systems, but infectious disease is probably the most prevalent. This is made evident by the high proportion of autoimmune disease that are associated with infectious agents and by the observation that 15–20% of all cancers are related to infection. Infections may impinge on systemic, intercellular, cytoplasmic and nuclear suppressor systems, and the same infectious agent may thus impinge on many suppressor systems simultaneously. The outcome of the various competitive interactions will probably depend on the number of cells involved, and chronic disease is probably a chance occurrence that occurs if the dose is high, the struggle is intense, the different agents are mismatched, and the duration of the conflict is beyond a point of no return.

In accordance with this perspective, disease is the result of malrestricted competition between evolutionary individuals. The conflicts are seen as being the result of failures at conflict-modification. Deregulation of conflict modifiers would allow the suppressed individuals to regain their embodied drive and thus to partake in conflicts within. The three kinds of parasites previously discussed may themselves deregulate the regulators and thus induce conflicts. Continued activation of the immune response, either through chronic infections, chimerism or malignancy, should be able to modulate the embodied drive of evolutionary individuals. If these are the lymphocytes, it should lead to autoimmune disease. Autoimmune disease is, accordingly, a proximate manifestation of ultimate and unresolved conflicts.

While dysregulated conflict modification is the common feature of autoimmune disease, the variegated symptoms and immunophenotypic changes that accompany the various diseases are probably consequential to the level at which conflict modification is disrupted. Since the levels are entwined, it is conceivable that a disruption of conflict modifiers at one level should afflict the conflict modifiers at other levels. The stratified defence system, in which the adaptive immune system has been built upon the innate immune system, would imply that malfunctions of the innate system should have a readout in the adaptive system. But since this readout occurs as a consequence of innate dysregulation, it should not be designated autoimmunity; inflammatory activity would be a better term. Hence, in this case the expanded autoreactive T and B cells are only epiphenomena to the underlying unresolved conflict.

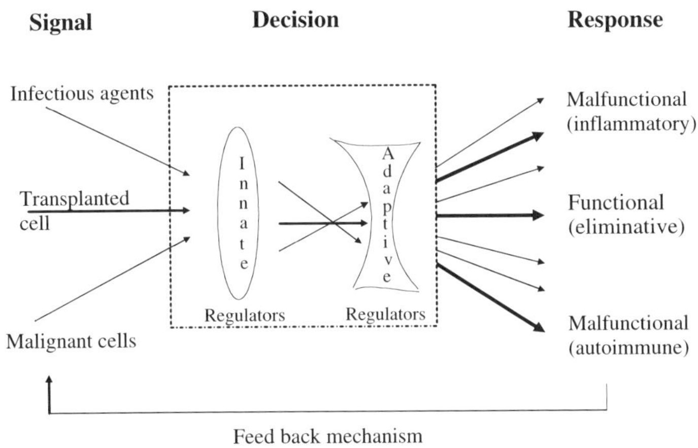

Figure 5.3. The common explanation for associations between immune system derangements and disease. The immune system has been evolutionary selected to function optimally. Malfunctions are not selected but occur owing to deregulation of regulators in either the innate or the adaptive immune systems. Any deregulation may lead to diseases and immune system derangements. The deranged response may thereafter influence the signal and thus establish a vicious circle

The associations between parasitism and disease as well as the difference between inflammatory and autoimmune diseases are depicted in Figure 5.3. If the conflicts emanate at the level of the innate immune system, the adaptive immune system derangements are but epiphenomena to the true conflict at the lower level. If, on the other hand, the conflict modifiers are severed at the level of the adaptive immune system, as occurs when central or peripheral tolerance-inducing mechanisms are disrupted, the ensuing disease is truly autoimmune as it is caused by the autoreactive receptors. Thus, of three evidential categories summarised by Rose and Bona (1993), direct evidence for autoimmunity emanates in the second lens, while the indirect and circumstantial evidence may as well be located to the first as to the second lens.

Stocktaking

The hypothesis stating that chronic diseases are caused by the re-enacting of ancient conflicts between evolutionary individuals emphasises the role of conflict modification and deregulation of conflict modifiers throughout development, and is built on three major assumptions. The first assumption is that conflict modifying mechanisms emerged and evolved as a consequence of evolutionary transitions, especially the transitions from prokaryotes to eukaryotes and from eukaryotes to multicellular animals, and that harmonising of the layered and entwined conflict modifying mechanisms was necessary at the higher transitory level. The second assumption is that some conflict modifiers, owing to their central function for cellular life, should have been retained throughout phylogeny. A third major assumption is that deregulation of the conflict modifiers should lead to malfunctioning.

There is much evidence to support the hypothesis, but since it was abducted from observations already at hand, its fit with observations can not by itself be used to evaluate its credentials. After all, evolutionary explanations are in many ways akin to "just-so" stories in that they may provide justified accounts that fit any observation whatsoever (Gould and Lewontin 1979). Hence, as any other evolutionary hypotheses, the conflict hypothesis needs to be tested by investigating the accuracy of its predicted observations and by comparing these predictions to predictions made by alternative hypotheses. In the following, several mechanisms and molecules involved in some sort of conflict modification that provide differential support for the hypothesis are enlisted. The record is not intended to give a comprehensive treatment of the subject, but is merely intended to hint at directions for further investigation.

The first group of defence-related conflict-modifying molecules to be considered, the *cyclophilins*, have been demonstrated in a wide variety of organisms, from prokaryotes to metazoans and plants (Romano et al. 2004). The cyclophilins are found in all tissue types and in all subcellular compartments of multicellular organisms, where they take part in the folding of proteins as well as in other cellular processes. Although much remains to be learned about the regulation of cyclophilins and their role in protective responses, they have been associated with the protection of cells from apoptosis (Lin and Lechleiter 2002), from retroviral infections (Towers et al. 2003), and from overreaction to bacterial endotoxins (Trahey and Weissman 1999).

When the immunosuppressive drug cyclosporine A, which binds to and inhibits the activity of cyclophilins, is administered to patients with transplanted tissues, the pronounced immunosuppressive effect induces acceptance of the graft. The drug can also induce remission of autoimmune diseases like lupus (Dammacco et al. 2000) and multiple sclerosis (Pette et al. 1997), while it enhances the development of cancer (Andre et al. 2004). Thus, cyclophilins seem to be involved in the inflammation or autoimmunity induced by infectious agents, chimeric cells and mosaic cells, although their exact role in these processes is not known.

Another group of widely distributed conflict modifying molecules, the *heat shock proteins*, is present in prokaryotes and in different subcellular compartments of eukaryotes, including the nucleus, mitochondria, endoplasmic reticulum and cytosol. Their synthesis is upregulated during infection, transplantation and cancer, and it has been suggested that they have become specialised to modify stressful events through their impact on cytokine production, triggering of Toll like receptors and deliverance of maturation signals to antigen presenting cells (Pockley 2003).

The heat shock proteins also facilitate protein folding of many important signal transduction elements; they influence morphogenetic responses to environmental cues, and buffer normal development from the destabilizing effects of perturbation processes. They are thus important for the ability of organisms to mount plastic responses to environmental signals. They may also buffer the impact of mutations, as proteins with amino acid changes become unstable unless heat shock proteins help their folding. In *Drosophila melanogaster* the importance of heat shock proteins is made evident by observations that flies with mutations in the heat shock proteins exhibit various phenotypic variations that are not observed in flies without these mutations (Rutherford and Lindquist 1998). While the effects of various amino acid changes are masked in flies with normal heat shock proteins, flies that are heterozygous for mutant heat shock proteins show unusual morphological abnormalities because their heat shock protein activity is insufficient. Hence, mutations that are masked in normal flies reveal themselves as morphological anomalies in heterozygous flies. An even more pronounced buffering effect of heat shock proteins has been observed in the plant *Arabidopsis thaliana* (Queitsch et al. 2002).

Heat shock proteins, which are upregulated at sites of inflammation, are strong inducers of the innate immune system (Prohaszka and Fust 2004), but are also involved in regulation of the adaptive immune system. The proteins are immunodominant, and a substantial amount of the T cell response to infectious agents is directed towards peptides derived from these proteins. It has been speculated that this response towards heat shock proteins from infectious agents leads to an anti-self response owing to molecular mimicry, and that this response precipitates autoimmune disease. Recent evidence does, nevertheless, indicate that T cells reactive with own heat shock proteins are regulatory T cells that serve to suppress autoimmunity (Van Eden et al. 2002; Pockley 2003). Upon this scenario, the heat shock proteins function as conflict modifiers of both the innate and the adaptive immune system.

Conflict mediation occurs also at the level of the nucleic acids. For example, the stability of RNA is regulated by small *interfering RNAs* whose nucleotide sequence can pair with messenger RNA, thus targeting this mRNA for destruction before it can be translated into protein. The small interfering RNA mechanism appears to be an evolutionary ancient mode of genome defence that has important regulatory roles in both prokaryote and eukaryote genomes (Gottesman 2005). It serves to protect the germline of *C. elegans* against retroviral elements (Vastenhouw and Plasterk 2004), and there are also indications that the system protects against viral infections (Carmichael 2002). The degree to which it has any role in malignancy is still not clarified.

While cyclophilins, heat shock proteins and interfering RNAs are present in all organisms, some conflict modifying molecules apparently made their phylogenetic appearance with the multicellular organisms. This includes the molecule *2-5A synthetase*, which is expressed in sponges as well as in vertebrates, but not in bacteria and yeast (Cayley et al. 1982; Grebenjuk et al. 2002). In human cells, the enzyme is localised to the nucleus, the nucleolus, the ribosomes and the mitochondria (Besse et al. 1998). The enzyme, which induces the cleavage of single-stranded RNA, has a functional role in intracellular homeostasis and defence against infecting agents. The function of the enzyme is best characterised in antiviral host defences, but the enzyme also functions in the regulation of the stability of RNAs that control host cell division, differentiation and apoptosis. It thus regulates the survival and reproductive ability of both self and non-self. The dual role of the enzyme requires that it should be tightly regulated; otherwise it might lead to disenchanting consequences. Too strong expression would lead to elimination of infectious agents but also to self-harm, while too low expression would lead to increased reproduction of the infectious agent at a cost to the host's reproduction and survival.

While the deleterious consequences of 2-5A synthetase are tolerable in organisms like sponges, in which cells are disposable to a higher degree than cells of vertebrates, higher metazoans with specialised organ systems would require either the disappearance of the enzyme, as has happened in insects, or it could become regulated by a higher level system, as took place in vertebrates where the genes for 2-5A synthetase came to be strongly regulated by interferon-α (Der et al. 1998).

Interferon-α has a multitude of effects, but only some of these are mediated through 2-5A synthetase. Interferon-α regulates the activity of a wide variety of genes, and it provides an important linkage between the innate and the adaptive immune systems, for example by acting as an adjuvant through the stimulation of dendritic cells (Le Bon and Tough 2002). Owing to potent anti-viral and antiproliferative effects, interferon-α has been used with success for the treatment of chronic hepatitis C viral infections, for leukaemia, and for autoimmune diseases such as multiple sclerosis. The exact mechanisms by which interferon-α treatment reduces disease activity in multiple sclerosis are largely unknown, but they may be mediated through receptors for interferon-α on microglial cells in the brain (Yamada and Yamanaka 1995). Microglia are macrophage-like cells that are able to stimulate and be stimulated by T cells, and who effectuate the degradation of myelin in multiple sclerosis brains (Ulvestad et al. 1994). But there may also be other mechanisms involved as interferon-α treatment of patients with multiple sclerosis increases the concentration of IgG and complement components and decreases the concentration of lymphocytes (Ulvestad et al. 2004).

Patients with lupus have serum levels of interferon-α that are sometimes as high as in patients with acute viral infections and in patients on treatment with interferon-α. Interestingly, a prominent side-effect of interferon-α treatment in patients with cancer and infectious disease is the development of autoantibodies against nuclear antigens and the development of autoimmune-like disease. Since the diseases are characterised by dysregulation of apoptosis, it has been hypothesised that the autoreactive antibodies are generated against material from apoptotic cells, and that these autoantibodies are rendered pathogenic in genetically predisposed individuals because of the adjuvant effects of interferon-α. This then leads to the formation of immune complexes that act as endogenous inducers of interferon-α, which again sustains the autoimmune process, thus establishing a vicious circle (Ronnblom and Alm 2001).

In a similar manner to the molecules involved in the innate immune response, the *receptors of the adaptive immune system* have a role in conflict modification. The evolutionary precursors of the receptors were probably non-rearranging molecules engaged in signalling and adhesion that became co-opted for defence-related functions in the vertebrate lineage (van den Berg et al. 2004). Their descendants are expressed on the so-called innate B and T lymphocytes, a subset of phylogenetically old lymphocytes that express germ-line encoded autoreactive antigen rcceptors. The cells, which recognise inducible self-ligands in conditions of tissue damage, cellular injury and stress, have probably been evolutionary selected to carry out a set of regulatory functions that involves cell-to-cell communication (Bendelac et al. 2001).

The precursors of the adaptive immune receptors, who very likely were mediators of conflicts amongst cells within the organism, evolved to become mediators of conflicts between the host and the parasitic world. This imposed a novel selection pressure on the receptors, and the adaptive immune system thereafter evolved largely as a result of the selective pressure imposed by infectious agents. But since the genes encoding the variable parts of the immunoglobulin and T cell receptors were anciently involved in mediating conflicts within, there were structural constraints on the molecules that restricted their future evolution. It is likely that the ancient function of the co-opted immune receptors could not be eliminated in vertebrates, and that this is why they have a relatively high affinity for endogenous evolutionary individuals, many of which needed to be controlled by a policing mechanism during the evolutionary transitions.

The autoreactive receptors that expand during infectious disease in current vertebrates may therefore be triggered, not by the disease-inducing agent, but by the danger experienced by the cells. The receptors are, in a sense, re-enacting their role as mediators of conflict between evolutionary individuals. This readily explains why the most common autoantigens are phylogenetically conserved molecules localised to distinct compartments within the cell, amongst others the nucleus, the nucleolus, the ribosomes and the mitochondria. The autoantigens in lupus are for example evolutionary conserved intracellular molecules that are involved in important functions such as DNA replication, chromosome segregation, DNA transcription and protein translation (Tan 1996). When viewed from an evolutionary angle it further emerges that the targeted antigens possess a common characteristic in addition to that of being evolutionary conserved; they are associated with ancient endosymbionts.

But if immunoglobulins were re-enacting ancient conflicts one would expect that the various antibodies against each autoantigen, say against double stranded DNA, should display some conserved characteristics in their antigen-binding sequences; that their usage of variable fragments should be restricted. This is not the case, and even though there are abnormalities in the B cell functioning of the systemic autoimmune diseases, there is little common variable-region gene usage (Dorner and Lipsky 2001). The heterogeneous repertoire of T cell and B cell receptors even suggests that there is an antigen-driven immune response in many of these diseases (Capra and Natvig 1993). This may, however, be owing to the fact that the receptors analysed were derived from diseased patients. Whether the pre-disease autoreactive receptors were similarly mutated is not known.

The heterogeneity of receptors observed may also be owing to the fact that each germline encoded antibody can bind to a large variety of unrelated antigenic determinants. While this ensures that the primary antibody response is composed of antibodies with a high degree of evolvability (Manivel et al. 2002), in some cases these antibodies can recognise epitopes both on infectious agents and self-molecules (Quinn et al. 1995). If the antibodies are triggered by the infectious agent, this may further explain the absence of observed gene-restriction of autoimmune receptors.

5.3. Environmental challenges

RESOURCES COUNT, DECISIONS AMOUNT

The resources of innate and adaptive immunity, which include the molecules and cells that constitute the immune system, are memory traces that have been shaped during previous immune responses, both in evolutionary and developmental time. While the memory traces of the innate immune system are engraved entirely in the genes, the memory traces of the adaptive immune system are in addition engraved in the variable antigen receptors and in the relative concentrations and distribution of cells and molecules that make up the system.

Deficiency of resources need not lead to malfunction since immunocompetence is multiply realisable. And although deficiency and malfunctioning are tightly associated within the innate immune system, the two do not have the same ontological status within the adaptive immune system. Malfunctions or deletions of the resources that make up the innate immune system are seldom encountered. This is partly owing to their deleterious consequences when absent and, hence, their fitness reducing effects. But it is also owing to the way that deleterious effects are buffered during development. For example, if specific resources of the innate immune system are absent from birth, as in gene knock-outs of certain cytokines, the consequences are often less serious than if the same deficiencies develop in adults. This is probably because the immune system self-organises to perform its function without them.

While total deficiencies of the innate immune system are rare, several variant molecules that are differentially associated with resistance to infectious agents exist. These include amongst others CD14, which is a receptor for bacterial lipopolysaccharide and probably also has a role in immune deviation towards allergic disorders (Baldini et al. 2002); a transcription factor for IL4, which is of importance for allergy deviation (Rockman et al. 2003); CCR5, which is a co-receptor for HIV and thus protects against HIV infection when absent (Paxton and Kang 1998); and caspase-12 which is associated with differential susceptibility to serious infectious disease (Saleh et al. 2004). Certain variants of the polymorphic defence molecules have also been associated with malfunctions of the adaptive immune system. The polymorphisms of these genes have a decisive effect on the response patterns of the adaptive immune system and are therefore closely associated with its functioning.

Within the adaptive immune system malfunctioning signifies a relational property that needs not become manifest until the organism encounters an unfavourable environment. Deletions of components of the adaptive immune system are quite frequent. They often go unnoticed, and their fitness reducing effects are not of the same magnitude as failures of the innate immune system. For example, deficiency of secretory IgA, which is the most common of the primary immunodeficiencies as it afflicts one in seven hundred individuals, is a heterogeneous condition with symptoms ranging from none to recurrent respiratory or gastrointestinal diseases (Edwards et al. 2004). While secretory IgA has a defined biological role, most individuals deficient for IgA are healthy owing to compensating defence mechanisms. There are alternative pathways to reach the same functional result, and IgA deficient individuals therefore seldom experience serious symptoms. While these alternative pathways may not be as efficient for protecting the mucosal membranes as IgA, they may still perform the feat as required.

Allergic, autoimmune and other chronic inflammatory diseases are sometimes said to be caused by an imbalance between the various types of T helper cells. This claim is unfortunate because the balance metaphor, which is derived from the ancient Hippocratic teachings, signifies a passive system in which the basic cure of any malfunction is to restore the balance between the elements. A better metaphor would be to envision the adaptive immune system as a decision-maker. Decisions involve recognition, regulation and thresholds, and signify activity as opposed to passivity. In addition, a decision-making system is compatible with the bearings of the embodied drive hypothesis. Deficiency of resources is seldom encountered in patients with allergic and autoimmune diseases, but biased decisions amount. Hence, the adaptive immune system may be responding in a functional or malfunctional manner according to how the decisions are effectuated.

ALLERGY – THE SPATIO-TEMPORAL MISMATCH HYPOTHESIS

Immunologists have characterised the proximate causal interactions of allergic disease in detail. But while they have been successful at explaining the mechanisms that lead to specific symptoms, including mast cell release of histamines upon triggering with allergens and IgE, they are less able to explain why the prevalence and severity of allergic disease have increased during the 20th century (Jackson 2001). It is clear that neither the human population nor the allergens that are the proximate causes of allergic symptoms have undergone any radical evolutionary process during this short time, and the role played by natural selection should therefore be negligible.

Nevertheless, if one acknowledges that the functionality of selected traits are spatio-temporally restricted, the increasing prevalence of allergic disease can still be understood and explained within an evolutionary framework. Since natural selection favours different genes in different places, organisms that are adapted to one environment and then move to a different environment may produce less adapted offspring in the new environment. The classical case of this is sickle cell anaemia, but the argument goes as well for temporally separated environments; if the current environment differs from previous selective environments that gave rise to the trait, there is a real chance that adapted traits may be rendered malfunctional in contemporaneous surroundings.

In contrast to genetic resistance to malaria, which is unrelated to the developmental age of the individual, a growing body of evidence suggests that resistance to allergic, autoimmune and infectious disease is laid down *in utero* and during early childhood. Resistance is governed via a series of control mechanisms associated with the functional transition of the relatively quiescent foetal immune system to the more competent state required to survive in the extrauterine environment. For example, the capacity of the immune system to produce Th1 cytokines is selectively constrained during foetal life, probably owing to the cytokines' toxicity towards the placenta (Chaouat et al. 2002). This cytokine deviation is achieved in part via the secretion of Th1-inhibitory mediators by the placenta.

After birth, recognition of microbial signals from the extrauterine environment, particularly via stimulation from the intestinal microflora, progressively induces maturation of the immune system towards the adult-equivalent range. The kinetics of this developmental process is highly variable, and slow or deviated maturation appears to be associated with reduced microbial stimulation. The belief that absence of microbial stimulation during early life events leads to immune system malfunctions has been epitomised as the hygiene hypothesis (Strachan 1989). One of it's current versions holds that the Western world's preoccupation with cleanliness has led to an altered intestinal microbial colonisation (Kim and Drake-Lee 2003). The microbes in the gut, which have coevolved with humans throughout evolutionary time, provide important signs to the developing immune system. And when the evolutionary selected interactions are altered, it alters the development of the immune system as well. This induces a lack of tolerance to otherwise harmless food proteins and inhaled antigens, as well as towards self-antigens. Immune system diseases like allergy are thus induced because the adjuvancy of our society has changed.

Traditionally, infectious agents are thought to precipitate disease because they compete with the host for resources. But those who take the Heideggerian view of the world seriously would argue that it is equally evident that microbes may do as much harm when absent. According to this view, only agents that are already being-with-others in the world can experience a loss; or stated differently, one cannot loose something that does not belong to oneself. And when individuals loose their evolutionary selected microflora, they simultaneously loose a part of themselves. This loss is proximally owing to the stimulating activity of microbial factors on the developing immune system. Because the microbes set up the network properties and response dispositions of the immune system, they perform a formative rather than, or in addition to, an efficient causal role.

The original hygiene hypothesis centred on the role that Th1-inducing microbial infections had on inhibiting Th2 mediated allergies. The balancing metaphor was often utilised to explain the effects of the different Th subsets, and disease was thought to be an imbalance between Th1 and Th2 responses. However, it has recently become clear that regulatory T cells are able to control the effector mechanisms of both Th1 and Th2 cells, and so the hygiene hypothesis has taken the form of a regulatory, not a balancing hypothesis (Maizels 2005). For example, multicellular worm helminth parasites have exerted a major influence on the evolution of the vertebrates. Worm-infection, which is highly prevalent in non-Western societies, is often asymptomatic, suggesting that the helminths have evolved sophisticated methods of immune evasion. Infection with helminths leads to a predominant Th2 response with production of high levels of IgE, but nevertheless, the infected individuals are protected from allergic manifestations. This, according to the regulatory hygiene hypothesis, is owing to the activity of T regulatory cells, which suppress effector mechanisms of both Th1 and Th2 cells (Wilson and Maizels 2006).

But individuals also differ in their genetic predisposition towards allergy. A recent observation in humans has intriguing bearings on the possibility of natural selection working on the Th1/Th2 network (Rockman et al. 2003). A single nucleotide polymorphism in the promoter of the cytokine IL4 was found to affect the binding of a transcriptional activator of IL4 in T cells. The allele, which leads to a threefold increase in IL4 production compared with the other allele, is common in some populations. Although the allele is currently associated with increased risk of allergic disease as it shifts T cells towards the allergic Th2 type, it might have conferred resistance against disease caused by infectious agents at some point in the evolution of the human species. This interpretation is supported by results that indicate that natural selection, and not genetic drift, drove the differentiation of the alleles. Selection on the IL4 promoter, which alters the cytokine pattern within the immune system, illustrates the importance of regulatory variation as well as the ease with which new beneficial regulatory interactions can occur by point mutation (Rockman et al. 2003).

The increasing prevalence of allergy has been characterised both as "the modern plague" and "epidemic". Thus, in a sense the epidemic has become more virulent. That Westernisation of the world leads to more hygienic environments is well supported by data, but epidemiological data further indicate that cleanliness is not a sufficient cause for allergy to increase. The critical factor appears to be the interaction between relevant patterns of the environment and critical developmental stages of the individual organism. If the intersection between the two is uncoordinated in some way or other, if there is a spatio-temporal mismatch, allergic disease may develop (Figure 5.4).

The spatiotemporal mismatch hypothesis is able to explain both autoimmunity and allergy as being the results of evolutionary contingencies. There is much evidence that the increased incidence of autoimmunity and allergy that emerged during the twentieth century is directly associated with altered relations between hosts and parasites. The proximate causes involved in the two types of disease are thus similar, as both are caused by dysregulated conflict modifiers and because infectious agents are the main offenders of the conflict modifying mechanisms. While the roles of chimerism and mosaicism have been much explored in autoimmunity, their role in allergy is still largely uncharted, and the studies that have been performed are permeated with inconsistent findings (Turner et al. 2006).

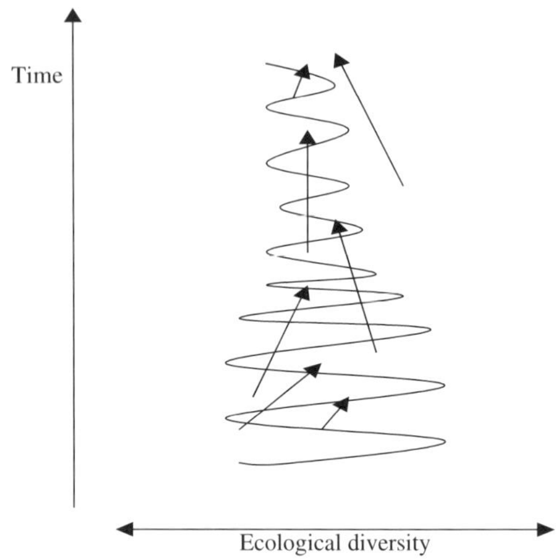

Figure 5.4. According to the time-space model of organismal interactions with the environment, environmental change can change developmental outcome, thus leading to novel phenotypes. The life course of individual organisms are depicted as arrows, the length of each arrow stipulates relative length of individual life. The meandering line stipulates important environmental signals that impinge on the developing organism. Each individual encounters environmental stimuli at different periods of their life cycle, as marked by the intersection between arrow and environmental meandering. The figure depicts the narrowing diversity of environmental stimuli, the Westernisation of society, and indicates at least one individual (upper right) that does not receive the appropriate developmental triggers

CROSSING BARRIERS

The most dramatic effect of host-pathogen interactions occurs when the spread of pathogenic microbes reaches epidemic proportions, in case of which an entire group of individuals or even a whole species may become extinct. This may happen if host organisms encounter infectious agents not previously present in their environment. The defence systems will then not be properly adapted to the new agent and catastrophic consequences may ensue. With the arrival of Christopher Columbus to America in 1492 the Indians, who had remained largely isolated from people on other continents for more than 30.000 years, experienced a mortality from infectious disease of more than 90% in many places (Naranjo 1992). The Indians had developed free from the epidemics that had been attacking Europe, Asia and Africa, and were therefore defenceless against infections caused by influenza, smallpox, measles, yellow fever, malaria, diphtheria and typhus.

Extensive human travelling all over the globe, increased contacts between previously separated species, increases in population sizes and changes in ecology are believed to be responsible for the emergence of new infectious disease and their rapid spread throughout the world, leaving few populations naive for long periods of time. An illustrious case was the outbreak of severe acute respiratory syndrome (SARS) in Southeast Asia in the spring of 2003 caused by a new type of coronavirus.

The virus, which passed over from an animal host to people, infected about 8000 individuals of which 10% died. The economic costs of SARS were estimated to nearly US$100 billion, mostly as a result of cancelled travel and decreased investment in the afflicted areas (Pearson et al. 2003).

The SARS epidemic was an astounding example of how quickly international collaborative efforts can identify causal infectious agents and provide diagnostic tests to identify infected but not yet diseased patients. The World Health Organisation (WHO) issued its first global warning on March 12th, and about a month later a detailed paper about SARS and the coronavirus responsible for the syndrome was published (Peiris et al. 2003). Solid background information helped the researchers to select workable hypotheses and methods to identify the infectious agent, and within three months methods for diagnosing of SARS were established around the world. Knowledge of the causative agent and its mode of spread established that isolation of infected patients would limit the dissemination of SARS. This measure was a success and by early July 2003 the WHO declared that the human chains of SARS virus transmission had been broken everywhere in the world.

Epidemic diseases of humans and their agricultural products may come to threaten the entire human enterprise if not proper measures are instantiated to limit their spread. About 40 new infectious agents responsible for human disease have been identified during the last 30 years (Desselberger 2000). The recurrence of old infectious agents and the emergence of novel diseases are cause for concern. When vaccines or specific treatment are unavailable, one has to rely on preventive measures or on the capability of the immune system to clear the infectious agent. That this is not always feasible has been shown repeatedly. A better understanding of the tactics used by novel pathogens as well as of the immune system's defensive strategies is thus needed to handle the threats from without.

Ignorance of evolutionary and ecological aspects of host-parasite relations is a hindrance to the invention of alternative explanations for immune phenomena. For example, the explaining of host-pathogen interactions as a combat between the immune system and the microbe has severely biased the explanatory focal point. This is because the war-metaphor, which emphasises organismal survival, fails to take notice of the reproductive aspects of fitness. Since trade-offs involved in optimising one or the other fitness component may lead functional systems to evolve in quite opposite directions to the ones predicted, both kinds of fitness should be investigated.

A science that is comfortable with studying nature's entities, the *natura naturata*, without aiming to understand the processes that generate them, the *natura naturans*, is in-comprehensive and runs the risk of seeing but aspects of the phenomena it aims to understand. In biology essentially new events continuously take place, events that defy prediction by virtue of inductive methodology. Based on previous observations and inductive methodology it is, for example, relatively straightforward to predict the likely occurrence of a new influenza epidemic next year. But the same methodology is not suitable to predict the novelties which are of practical concern – the antigenic characteristics of next years' influenza virus. Such prediction would require a more fine-grained understanding of evolutionary and ecological premises as well as better surveillance strategies.

The entities we observe have already happened and can therefore happen again – they are determinate possible. In biological evolution, however, novel phenomena may take place that were not immanent in those phenomena that preceded them. These are emergent phenomena – they are thus indeterminate possible (Østerberg 1988). The distinction between the determinate and indeterminate possible is important both epistemologically and ontologically. For the researcher that wants to predict novelty it is of essence to transcend the determinate possible that belongs to the *natura natrurata.* Knowledge of the indeterminate possible can hardly be realised by studying *natura naturata* in isolation, but by investigating mechanisms, *natura naturans*, he may hope to be able to predict the indeterminate possible. The researcher that tries to outsmart the influenza virus by making a vaccine is in the business of predicting the virus' next move up the evolutionary ladder. By getting a grip of *natura naturans*, his hopes of predicting the evolution of the virus and thus to produce a vaccine before the virus has finished its evolution, may come within reach.

To get an idea of the complexities involved, a comparison with a game of chess between a computer and a human opponent is appropriate. Because of the strict and unchanging rules of chess, computer scientists have managed to create successful computer programs that outperform their human competitors by their ability to predict the biological opponent's next move. But while chess is performed according to fixed rules that generate determinate possible outcomes, the rules of host-pathogen interactions are constantly changing, thus making the outcomes indeterminate possible. These outcomes are still far too complex to simulate in a computer model.

The virus that tries to outsmart the host's immune system by evolving novel antigenic structures is acting out the game of life, in which the only reward is paid in the currency of fitness. Like the researcher who needs a theory with predictive power, the host organism needs a defence system that is capable of acting in accordance with the same principles. While it is clear that the immune system is empowered with this capability when it encounters infectious agents with evolutionary conserved structures, several observations indicate that the predictive powers of the adaptive immune system are restrained. The fact that infecting agents manage to seclude themselves from the host's defence systems and thus induce chronic inflammatory diseases attests to this (Lorber 1996; Zimmer 2001).

The post-SARS warning by the director general of WHO, that the global population could expect to face a new infectious disease every year, underscores the need to be alert to and continuously survey for emerging diseases. The burdens of health and the economic costs involved when infectious disease reaches epidemic proportions automatically launch questions like; What type of infectious agent causes the disease? By which mechanisms does the agent inflict disease? Why is the agent pathogenic? Are there any behavioural alterations that can limit the spread of the disease? Is specific therapy available? When will a vaccine be provided?

Clarification of the queries requires extensive interdisciplinary and international collaboration between scientists trained in a variety of biological and medical specialities, including specialists trained in immunology, microbiology, evolutionary biology, ecology, molecular biology, epidemiology, pharmacology and vaccinology. The specialists should investigate the co-evolution of man and specific microbes as well as ecological interactions between man and a variety of other species, and all these observations should be related to the developmental stage of the individual. Investigation of the genetics of the interacting species will not suffice; the spatio-temporal appearance of reciprocal signals should also be evaluated. These signals may for example be developmentally shaped, and two infectious agents with the same genetic architecture may thus differ in pathogenicity even in identical twins. Semiosis and phenotypic plasticity are thus of essence.

Evolutionary theory in its contemporary form is not the kind of theory that can reveal the necessity of what happens or what will happen. A fundamental question that needs to be asked is whether it can become such a theory. Based on results from empirical and conceptual investigations outlined several places in this book, I think not. The basic arguments upon which I base this belief were outlined already in the 19th centrury, with Darwin's theory of natural selection and Hegel's conceptualisation of history. Kierkegaard (1849, p. 59), who adapted many of Hegel's insights, gave a precise formulation of these when claiming that "freedom is the dialectical element in the categories of possibility and necessity". In our discussions of the immune self it became clear that the self is a kind of being that eludes attempts at objectivation; it cannot be caught in the act, so to say. The self is thus elusive and transcendental. And so is life, both of parasites and hosts. Since both the self and life itself are essentially free, they are transcendental in nature, always ready to cross barriers.

Freedom is mediated between semiosis (possibility) and law (necessity). The dialectic taking place between the two is extremely difficult to investigate and conceptualise. Whether or not science, even if it manages to grip the *natura naturans* through the *natura naturata*, will be capable of crossing the epistemological barrier between the determinate and the indeterminate possible is therefore an open question. Perhaps the most we can hope for is to be able to cope adaptively with the challenges of infectious agents and the malfunctional immune system on a short term basis. But even this achievement, which will require outstanding research efforts of both empirical and conceptual sorts, will keep scientists occupied for decades.

As long as life continuously transcends its boundary conditions, the philosophical void that Orosz (2000) would like to see filled, between the satisfying accumulation of immunologic facts and their unsatisfactory understanding, can in principle not be fulfilled until the history of life has reached its completion. Hence, contrary to Jerne's 1969-statement, the scientific era does not draw to an end (unless the same goes for life); the most we can say as of today is that the end of immunology might have begun.

These ideas on the course of history, which I have adopted to answer Orosz' query for new perspectives on reality as well as to explain the ongoing history of host-parasite relations, were eloquently formulated by Hegel in his preface to the *Philosophy of right*:

> Philosophy, as the thought of the world, does not appear until reality has completed its formative process, and made itself ready. History thus corroborates the teaching of the conception that only in the maturity of reality does the ideal appear as counterpart to the real, apprehends the real world in its substance, and shapes it into an intellectual kingdom. When philosophy paints its grey in grey, one form of life has become old, and by means of grey it cannot be rejuvenated, but only known. The owl of Minerva takes its flight only when the shades of night are gathering. (Hegel 1821, p. xxx).

Appendix

As laid out in the previous chapters, the innate and the adaptive immune systems consist of molecules and cells connected in a network (Figure A.1). Some of the major players are described in more detail in the appendix.

Innate immunity

Several molecules within the innate system react with specific patterns on the surface of infectious organisms, and are therefore termed pattern recognition receptors. These comprise the fluid phase molecules of the complement system, including the classical, alternative and lectin pathways, as well as the Toll-like receptors that are expressed on the surface of antigen presenting cells. While Toll-like receptors detect extracellular microbial patterns, another class of innate pattern-recognition receptors have recently been described to detect intracellular pathogens (Martinon and Tschopp 2005).

The **Toll-like receptors**, of which thirteen variants have been identified in mammals, are able to sense molecular patterns on organisms as diverse as protozoa, bacteria, fungi and viruses, and make up the proximate sensing apparatus of mammals and insects (Beutler 2004). Vertebrate Toll-like receptors are slowly evolving molecules, indicating a strong selective pressure for maintenance of function (Roach et al. 2005).

Toll-like receptors are exposed primarily on the surface of **dendritic cells**, a class of cells which present antigenic peptides in a highly immunogenic form to naive T cells. Because of their important role in immune system activation, much research has recently been performed to understand the biology of dendritic cells. The cells, which are found in both lymphoid and non-lymphoid sites and comprise several subtypes, are derived from precursors in the bone marrow. Upon maturation they emigrate to the tissues of the organism where they constitute the major population of antigen-presenting cells.

Dendritic cells express many types of receptors that enable them to sense life, in the form of infectious agents, and death, in the form of necrotic cells, and they rapidly become activated in response to such signals. They then migrate to the nearest lymph node, carrying molecular pieces of the activation signals with them. Within the lymph node they orchestrate the developing immune response by integrating and relaying various signals through interactions with T cells, B cells and NK cells (Creusot and Mitchison 2004).

Other important cells of the vertebrate innate immune system, besides the dendritic cells, are the phagocytes and natural killer cells. The **phagocytic**

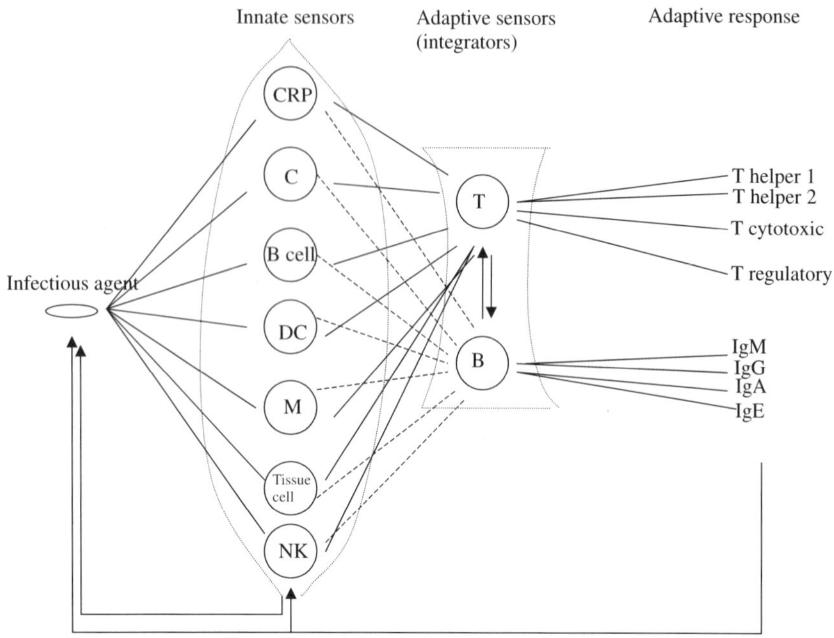

Figure A.1. The immune network superimposed on the lens metaphor. The infectious agent triggers receptors of the innate immune system. The responding of the components of the innate immune system generates a context for the triggering of antigen-specific lymphocytes, which differentiate into context-adapted specific effector cells. T and B lymphocytes integrate the information derived from the innate sensors. Potent feed-back mechanisms serve to regulate the response. (CRP, C reactive protein; C, complement; DC, dendritic cell; M, macrophage; NK, natural killer cell)

cells, which are made up of the polymorphonuclear granulocytes and the monocytes/macrophages, are professional phagocytes whose predominant role is to remove particulate antigens. This function is mediated through several receptors, such as IgG Fc receptors and complement receptors. Monocytes/macrophages may be activated by several T cell derived molecules, and may upon stimulation release several inflammatory mediators, such as cytokines and enzymes. Monocytes, which comprise 2–10% of blood leukocytes, migrate through the blood-vessel walls into the various tissues to become macrophages. These cells are efficient antigen-presenting cells for T cells.

Natural killer cells (NK cells) account for up to 15% of blood lymphocytes. They have several surface antigens in common with T cells, but do not express T cell antigen receptors. NK cells have a "natural" capability to kill tumour cells and are also cytotoxic to virus-infected cells. In contrast to T cells, which recognise antigen in the context of MHC, NK cells recognise reduced expression of MHC, missing self, on the surface of the altered cell. NK cells also have a receptor for the Fc part of IgG and are able to lyse IgG coated cells by a mechanism referred to as antibody-dependent cellular cytotoxicity.

Adaptive immunity

The adaptive immune system is made up of the MHC molecules and the T and B lymphocytes. The polymorphic **major histocompatibility complex (MHC)** genes encode the molecules that present peptides to T cells. Certain of the genes have extraordinary high levels of polymorphism, among the highest known in any organism. The functions of MHC molecules is to present self-antigens to T cells in the thymus, and so induce immunocompetence or cell death in the maturing T cells, and to present foreign antigens in the periphery, thus inducing an immune response. The MHC genes in humans are called human leukocyte antigen (HLA) genes and are of two types, class I and II. HLA Class II molecules are expressed mainly on specialised antigen presenting cells, such as dendritic cells, macrophages, and B cells. This limited expression, in contrast to HLA class I molecules that are expressed on all nucleated cells, ensures a strict control with the initiation of an immune response.

There are several sorts of lymphocytes that make up the cellular elements of the immune response. These include the **T lymphocytes**, which mediate cellular immunity. T cells mature in the thymus, and comprise 70–80% of lymphocytes in peripheral blood. They express two types of antigen receptors. Approximately 85–95% of T cells in blood express the $\alpha\beta$ receptor ($\alpha\beta$ T cells), and up to 15% express the $\gamma\delta$ receptor ($\gamma\delta$ T cells). T cells come in two distinct functional subtypes. The **CD8** T cells, which induce death in infected cells and hence are called cytotoxic T cells, recognise peptide presented by HLA class I. The **CD4** T helper cells, which are the main conductors and facilitators of the immune response, recognise antigen bound by HLA class II molecules.

The CD4 T helper cells have been further divided into T helper 1, T helper 2 and T regulatory cells on the basis of the cytokines they produce. Th1 cells produce interleukin (IL) 2 and interferon γ, and activate macrophages and cytotoxic CD8 T cells. Th2 cells produce IL-4, IL-5, IL-10 and IL-13, the secretion of which leads to the activation and differentiation of B cells into antibody producing plasma cells. Regulatory T cells, some of which are CD4/CD25 positive, regulate the immune response either through cell contact or through the secretion of cytokines.

The **B lymphocytes**, which comprise the other major group of lymphocytes, mature in the bone marrow and account for 5–15% of blood lymphocytes. The receptor for antigen on B cells is an immunoglobulin molecule. Upon activation, B cells mature into plasma cells that secrete soluble immunoglobulins with the same specificity as the membrane bound immunoglobulins. There are five immunoglobulin classes (isotypes). IgG, which is the major immunoglobulin in serum, consists of four subclasses. IgA, which is predominant in seromucous secretions, consists of two subclasses. IgM, which is expressed on the B cell as its antigen-binding receptor, is the first immunoglobulin to appear in serum during a primary immune response. IgD is mainly a B cell antigen receptor. IgE, usually present in small quantities, is of major importance in allergic diseases.

While **immunoglobulins** may perform their effects far away from the B cell producer, such as in plasma and secretions, **T cell receptors** are bound to the T cell – they are never secreted. The effects of T cell receptor stimulation are therefore restricted to the immediate vicinity of the T cell, and is performed either through secretion of regulatory molecules or through cytotoxicity. When the antigen receptors of B cells react with antigen, several effects ensue. B cells may differentiate into plasma cells and start to secrete immunoglobulins. IgG and IgM bound to antigen may activate the complement cascade, a system of soluble plasma proteins, leading to cytolysis of intact cells. Alternatively, immunoglobulins and complement may opsonise (make edible) antigens for specific phagocytosis via receptors for the Fc part of IgG and complement expressed on the surface of macrophages and granulocytes.

In contrast to the **genes that encode antigen-receptors** of the innate immune system, the T cell receptors and immunoglobulins come in non-functional bits and pieces that need to be adaptively joined within the developing lymphocyte before they become functional. It is remarkable that the two types of adaptive receptors, which probably evolved from the same ancestral molecule, utilise very different modes of recognising their antigen. Whereas the immunoglobulin of a B cell reacts with three-dimensional conformation-dependent shapes, such as whole bacteria, the $\alpha\beta$ T cell receptor reacts with linear peptides of eight to 25 amino acids, derived from degraded antigen. These fragments must be exposed within the peptide-binding groove of the MHC-molecules exposed on the surface of antigen presenting cells.

References

Abkowitz, J. L., S. N. Catlin and P. Guttorp (1996). Evidence that hematopoiesis may be a stochastic process in vivo. Nat Med 2: 190–7

Ada, G. and D. Isaacs (2003). Carbohydrate-protein conjugate vaccines. Clin Microbiol Infect 9: 79–85

Adelman, M. K. and J. J. Marchalonis (2002). Endogenous retroviruses in systemic lupus erythematosus: candidate lupus viruses. Clin Immunol 102: 107–16

Ader, R., N. Cohen and D. Felten (1995). Psychoneuroimmunology: interactions between the nervous system and the immune system. Lancet 345: 99–103

Agrawal, A., Q. M. Eastman and D. G. Schatz (1998). Transposition mediated by RAG1 and RAG2 and its implications for the evolution of the immune system. Nature 394: 744–51

Ahmed, R. and D. Gray (1996). Immunological memory and protective immunity: understanding their relation. Science 272: 54–60

Ailus, K. and T. Palosuo (1995). IgM class autoantibodies in human cord serum. J Reprod Immunol 29: 61–7

Alexander, R. D. and G. Borgia (1978). Group selection, altruism, and the levels of organization of life. Ann Rev Ecol Syst 9: 449–74.

Allen, T. M., J. S. Lanchbury, A. L. Hughes and D. I. Watkins (1996). The T-cell receptor beta chain-encoding gene repertoire of a New World primate species, the cotton-top tamarin. Immunogenetics 45: 151–60

Amadori, A., R. Zamarchi, G. De Silvestro, G. Forza, G. Cavatton, et al. (1995). Genetic control of the CD4/CD8 T-cell ratio in humans. Nat Med 1: 1279–83

Amsen, D., J. M. Blander, G. R. Lee, K. Tanigaki, T. Honjo, et al. (2004). Instruction of distinct CD4 T helper cell fates by different notch ligands on antigen-presenting cells. Cell 117: 515–26

Amundson, R. and G. Lauder (1994). Function without purpose: the uses of causal role function in evolutionary biology. Biology and Philosophy 9: 443–69

Anderson, C. C. and P. Matzinger (2001). Immunity or tolerance: opposite outcomes of microchimerism from skin grafts. Nat Med 7: 80–7

Andersson, S. G. and C. G. Kurland (1998). Ancient and recent horizontal transfer events: the origins of mitochondria. APMIS Suppl 84: 5–14

Andre, N., B. Roquelaure and J. Conrath (2004). Molecular effects of cyclosporine and oncogenesis: a new model. Med Hypotheses 63: 647–52

Andrews, J. H. (1998). Bacteria as modular organisms. Annu Rev Microbiol 52: 105–26

Antia, R., S. S. Pilyugin and R. Ahmed (1998). Models of immune memory: on the role of cross-reactive stimulation, competition, and homeostasis in maintaining immune memory. Proc Natl Acad Sci U S A 95: 14926–31

Aravind, L. and G. Subramanian (1999). Origin of multicellular eukaryotes – insights from proteome comparisons. Curr Opin Genet Dev 9: 688–94

Aristotle (1984). History of animals. In: The complete works of Aristotle. Volume 1. J. Barnes (Ed.). Princeton University Press, Princeton, 774–993

Arthur, W. (2001). Developmental drive: an important determinant of the direction of phenotypic evolution. Evol Dev 3: 271–8

Athanassakis, I. and S. Vassiliadis (2002). T-regulatory cells: are we re-discovering T suppressors? Immunol Lett 84: 179–83

Avrameas, S. (1991). Natural autoantibodies: from 'horror autotoxicus' to 'gnothi seauton'. Immunol Today 12: 154–9

Azumi, K., R. De Santis, A. De Tomaso, I. Rigoutsos, F. Yoshizaki, et al. (2003). Genomic analysis of immunity in a Urochordate and the emergence of the vertebrate immune system: "waiting for Godot". Immunogenetics 55: 570–81

Baccala, R., T. V. Quang, M. Gilbert, T. Ternynck and S. Avrameas (1989). Two murine natural polyreactive autoantibodies are encoded by nonmutated germ-line genes. Proc Natl Acad Sci U S A 86: 4624–8

Baehrecke, E. H. (2002). How death shapes life during development. Nat Rev Mol Cell Biol 3: 779–87

Baldini, M., D. Vercelli and F. D. Martinez (2002). CD14: an example of gene by environment interaction in allergic disease. Allergy 57: 188–92

Baltimore, D. (2001). Our genome unveiled. Nature 409: 814–6

Barkow, J. H., L. Cosmides and J. Tooby, Eds. (1992). The adapted mind: Evolutionary psychology and the generation of culture. Oxford University Press New York.

Bassler, B. L. (1999). How bacteria talk to each other: regulation of gene expression by quorum sensing. Curr Opin Microbiol 2: 582–7

Baum, D. and M. Donoghue (2001). A likelihood framework for the phylogenetic analysis of adaptation. In: Adaptationism and optimality. S. Orzack and E. Sober (Eds.). Cambridge University Press, Cambridge, 24–44

Baum, H. (1995). Mitochondrial antigens, molecular mimicry and autoimmune disease. Biochim Biophys Acta 1271: 111–21

Beeson, P. B. (1994). Age and sex associations of 40 autoimmune diseases. Am J Med 96: 457–62

Behe, M. (1996). Darwin's black box. The biochemical challenge to evolution. Simon & Schuster, New York

Belkaid, Y., C. A. Piccirillo, S. Mendez, E. M. Shevach and D. L. Sacks (2002). CD4 + CD25+ regulatory T cells control Leishmania major persistence and immunity. Nature 420: 502–7

Ben Jacob, E., I. Becker, Y. Shapira and H. Levine (2004). Bacterial linguistic communication and social intelligence. Trends Microbiol 12: 366–72

Bendelac, A., M. Bonneville and J. F. Kearney (2001). Autoreactivity by design: innate B and T lymphocytes. Nat Rev Immunol 1: 177–86

Benner, R., A. van Oudenaren, J. J. Haaijman, J. Slingerland-Teunissen, B. S. Wostmann, et al. (1981). Regulation of the "spontaneous' (background) immunoglobulin synthesis. Int Arch Allergy Appl Immunol 66: 404–15

Berentsen, S., K. Bo, F. V. Shammas, A. O. Myking and E. Ulvestad (1997). Chronic cold agglutinin disease of the "idiopathic" type is a premalignant or low-grade malignant lymphoproliferative disease. APMIS 105: 354–62

Bergman, M. (2003). Peirce's derivations of the interpretant. Semiotica 144: 1–17

Bernasconi, N. L., E. Traggiai and A. Lanzavecchia (2002). Maintenance of serological memory by polyclonal activation of human memory B cells. Science 298: 2199–202

Besse, S., D. Rebouillat, I. Marie, F. Puvion-Dutilleul and A. G. Hovanessian (1998). Ultrastructural localization of interferon-inducible double-stranded RNA-activated enzymes in human cells. Exp Cell Res 239: 379–92

Best, S. M., S. V. Collins and P. J. Kerr (2000). Coevolution of host and virus: cellular localization of virus in myxoma virus infection of resistant and susceptible European rabbits. Virology 277: 76–91

Beutler, B. (2004). Inferences, questions and possibilities in Toll-like receptor signalling. Nature 430: 257–63

Bianchi, D. W., G. K. Zickwolf, G. J. Weil, S. Sylvester and M. A. DeMaria (1996). Male fetal progenitor cells persist in maternal blood for as long as 27 years postpartum. Proc Natl Acad Sci U S A 93: 705–8

Bibel, D. (1988). Elie Metchnikoff's bacillus of long life. ASM News 54: 661–65

Billingham, R. E., L. Brent and P. B. Medawar (1953). Actively acquired tolerance of foreign cells. Nature 172: 603–6

Blackstone, N. W. and D. R. Green (1999). The evolution of a mechanism of cell suicide. Bioessays 21: 84–8

Blanke, S. R. (2005). Micro-managing the executioner: pathogen targeting of mitochondria. Trends Microbiol 13: 64–71.

Boes, M., J. Cerny, R. Massol, M. Op den Brouw, T. Kirchhausen, et al. (2002). T-cell engagement of dendritic cells rapidly rearranges MHC class II transport. Nature 418: 983–8

Boissinot, S., J. Davis, A. Entezam, D. Petrov and A. V. Furano (2006). Fitness cost of LINE-1 (L1) activity in humans. Proc Natl Acad Sci U S A 103: 9590–94

Bonneaud, C., J. Mazuc, G. Gonzalez, C. Haussy, O. Chastel, et al. (2003). Assessing the cost of mounting an immune response. American Naturalist 161: 367–379

Braga, L. (1999). Peirce and biology. Semiotica 127: 5–21.

Bray, M. (2004). Henry Kempe and the birth of vaccinia immune globulin. Clin Infect Dis 39: 767–9

Brent, L. (1997). A history of transplantation immunology. Academic Press, San Diego

Bretscher, P. and M. Cohn (1970). A theory of self-nonself discrimination. Science 169: 1042–9

Bromham, L. (2002). The human zoo: endogenous retroviruses in the human genome. Trends Ecol Evol 17: 91–97.

Brookfield, J. (1998). Quorum sensing and group selection. Evolution 52: 1263–69

Brown, J. L. and A. Eklund (1994). Kin recognition and the major histocompatibility complex: An integrative review. American Naturalist 143: 435–61

Brown, S. P. and R. A. Johnstone (2001). Cooperation in the dark: signalling and collective action in quorum-sensing bacteria. Proc R Soc Lond B Biol Sci 268: 961–5

Bull, J. (1994). Virulence. Evolution 48: 1423–37

Burnet, M. (1959). The clonal selection theory of acquired immunity. Cambridge University Press, Cambridge

Burnet, M. (1969). Self and not-self. Cambridge University Press, London

Burnet, M. and D. White (1972). Natural history of infectious disease. Cambridge University Press, Cambridge

Burton, D. R. and P. W. Parren (2000). Vaccines and the induction of functional antibodies: time to look beyond the molecules of natural infection? Nat Med 6: 123–5

Buss, L. (1987). The evolution of individuality. Princeton University Press, Princeton

Callard, R., A. J. George and J. Stark (1999). Cytokines, chaos, and complexity. Immunity 11: 507–13

Cann, S. A. H., J. P. van Netten, C. van Netten and D. W. Glover (2002). Spontaneous regression: a hidden treasure buried in time. Med Hypotheses 58: 115–19

Cannon, J. P., R. N. Haire and G. W. Litman (2002). Identification of diversified genes that contain immunoglobulin-like variable regions in a protochordate. Nat Immunol 3: 1200–7

Cantor, A. B. and S. H. Orkin (2001). Hematopoietic development: a balancing act. Curr Opin Genet Dev 11: 513–9

Caporale, L. H. (2000). Mutation is modulated: implications for evolution. Bioessays 22: 388–95

Caporale, L. H. (2003). Natural selection and the emergence of a mutation phenotype: an update of the evolutionary synthesis considering mechanisms that affect genome variation. Annu Rev Microbiol 57: 467–85

Capra, J. D. and J. B. Natvig (1993). Is there V region restriction in autoimmune disease? The Immunologist. 1: 16–19.

Carmichael, G. G. (2002). Medicine: silencing viruses with RNA. Nature 418: 379–80

Carrel, L. and H. F. Willard (2005). X-inactivation profile reveals extensive variability in X-linked gene expression in females. Nature 434: 400–4

Carroll, L. (1965). Alice's adventures in Wonderland & Through the looking glass. Airmont Publishing Company, Inc., New York

Casadesus, J. and R. D'Ari (2002). Memory in bacteria and phage. Bioessays 24: 512–8

Casadevall, A. and L. A. Pirofski (1999). Host-pathogen interactions: redefining the basic concepts of virulence and pathogenicity. Infect Immun 67: 3703–13

Casadevall, A. and L. A. Pirofski (2003). Exploiting the redundancy in the immune system: vaccines can mediate protection by eliciting 'unnatural' immunity. J Exp Med 197: 1401–4

Casellas, R., T. A. Shih, M. Kleinewietfeld, J. Rakonjac, D. Nemazee, et al. (2001). Contribution of receptor editing to the antibody repertoire. Science 291: 1541–4

Cayley, P. J., R. F. White, J. F. Antoniw, N. J. Walesby and I. M. Kerr (1982). Distribution of the ppp(A2'p)nA-binding protein and interferon-related enzymes in animals, plants, and lower organisms. Biochem Biophys Res Commun 108: 1243–50

Chaouat, G., S. Zourbas, S. Ostojic, G. Lappree-Delage, S. Dubanchet, et al. (2002). A brief review of recent data on some cytokine expressions at the materno-foetal interface which might challenge the classical Th1/Th2 dichotomy. J Reprod Immunol 53: 241–56

Charlton, B. (1996). Endogenous parasitism: a biological process with implications for senescence. Evolutionary Theory 11: 119–24

Chmielowski, B., P. Muranski and L. Ignatowicz (1999). In the normal repertoire of CD4+ T cells, a single class II MHC/peptide complex positively selects TCRs with various antigen specificities. J Immunol 162: 95–105

Christ, E. and A. I. Tauber (1997). Debating humoral immunity and epistemology: The rivalry of the immunochemists Jules Bordet and Paul Ehrlich. J Hist Biol 30: 321–56

Chun, C. K., E. A. Ozer, M. J. Welsh, J. Zabner and E. P. Greenberg (2004). Inactivation of a Pseudomonas aeruginosa quorum-sensing signal by human airway epithelia. Proc Natl Acad Sci U S A 101: 3587–90

Cohen, I. R. (2000). Tending Adam's garden. Academic Press, San Diego

Cohen, M. L. (2000). Changing patterns of infectious disease. Nature 406: 762–7

Cohen, M. N. (1989). Health and the rise of civilization. Yale University press, New Haven

Cohen, S., W. J. Doyle, D. P. Skoner, B. S. Rabin and J. M. Gwaltney, Jr. (1997). Social ties and susceptibility to the common cold. JAMA 277: 1940–4

Cohn, M. (1972). Immunology: what are the rules of the game? Cell Immunol 5: 1–20

Cohn, M. (1998). The self-nonself discrimination in the context of function. Theor Med Bioeth 19: 475–84; discussion 485–504

Cohn, M. (2003). Does complexity belie a simple decision–on the Efroni and Cohen critique of the minimal model for a self-nonself discrimination. Cell Immunol 221: 138–42

Cohn, M. and R. E. Langman (1990). The protecton: the unit of humoral immunity selected by evolution. Immunol Rev 115: 11–147

Combes, C. (2001). Parasitism. The ecology and evolution of intimate interactions. The University of Chicago Press, Chicago

Conan Doyle, A. (1976). The illustrated Sherlock Holmes treasury. Avenel Books, New York

Copleston, F. (1962). A history of philosophy. Volume I. Image Books, New York

Crespi, B. J. (2001). The evolution of social behavior in microorganisms. Trends Ecol Evol 16: 178–183

Creusot, R. and N. A. Mitchison (2004). How DCs control cross-regulation between lymphocytes. Trends Immunol 25: 126–31.

Cummins, R. (1975). Functional analysis. Journal of Philosophy 72: 741–64

Damasio, A. (1994). Descartes' error. Vintage, London

Damasio, A. (2001). Fundamental feelings. Nature 413: 781

Dammacco, F., O. Della Casa Alberighi, G. Ferraccioli, V. Racanelli, L. Casatta, et al. (2000). Cyclosporine-A plus steroids versus steroids alone in the 12-month treatment of systemic lupus erythematosus. Int J Clin Lab Res 30: 67–73

Dao, D. N., R. H. Kessin and H. L. Ennis (2000). Developmental cheating and the evolutionary biology of Dictyostelium and Myxococcus. Microbiology 146 (Pt 7): 1505–12

Darwin, C. (1859). The origin of species. Penguin Books (1985), London

Davies, P. (2000). Malfunctions. Biology and Philosophy 15: 19–38

Davies, S. J., J. L. Grogan, R. B. Blank, K. C. Lim, R. M. Locksley, et al. (2001). Modulation of blood fluke development in the liver by hepatic CD4+ lymphocytes. Science 294: 1358–61

Dawkins, R. (1982). The extended phenotype. The long reach of the gene. Oxford University Press, Oxford

Dawkins, R. (1989). The selfish gene. 2nd edition. Oxford University Press, Oxford

Dawkins, R. (1990). Parasites, desiderata lists and the paradox of the organism. Parasitology 100 Suppl: S63–73

Day, T. (2002). On the evolution of virulence and the relationship between various measures of mortality. Proc R Soc Lond B Biol Sci 269: 1317–23 de Groot, N. G., N. Otting, G. G. Doxiadis, S. S. Balla-Jhagjhoorsingh, J. L. Heeney, et al. (2002). Evidence for an ancient selective sweep in the MHC class I gene repertoire of chimpanzees. Proc Natl Acad Sci U S A 99: 11748–53

de Kievit, T. R. and B. H. Iglewski (2000). Bacterial quorum sensing in pathogenic relationships. Infect Immun 68: 4839–49

Deacon, T. (1997). The symbolic mind: The co-evolution of language and the brain. Penguin Books, London

Deininger, P. L., J. V. Moran, M. A. Batzer and H. H. Kazazian, Jr. (2003). Mobile elements and mammalian genome evolution. Curr Opin Genet Dev 13: 651–8

Delsuc, F., H. Brinkmann and H. Philippe (2005). Phylogenomics and the reconstruction of the tree of life. Nature Rev Genet 6: 361–75

Demas, G. E., D. L. Drazen and R. J. Nelson (2003). Reductions in total body fat decrease humoral immunity. Proc R Soc Lond B Biol Sci 270: 905–11

Dembic, Z. (2000). Immune system protects integrity of tissues. Mol Immunol 37: 563–9

Dembski, W. (1999). Intelligent design. The bridge between science & theology. InterVarsity Press, Downers Grove

Der, S. D., A. Zhou, B. R. Williams and R. H. Silverman (1998). Identification of genes differentially regulated by interferon alpha, beta, or gamma using oligonucleotide arrays. Proc Natl Acad Sci U S A 95: 15623–8

Deragon, J. M. and P. Capy (2000). Impact of transposable elements on the human genome. Ann Med 32: 264–73

Descartes, R. (1637). Discourse on method and the mediations. Penguin Books (1968), London

Desselberger, U. (2000). Emerging and re-emerging infectious diseases. J Infect 40: 3–15

Dickinson, W. J. and J. Seger (1999). Cause and effect in evolution. Nature 399: 30

Diez Roux, A. V. (2004). The study of group-level factors in epidemiology: rethinking variables, study designs, and analytical approaches. Epidemiol Rev 26: 104–11

Dixon, B. (2003). Harmonious coexistence. Lancet Infect Dis 3: 178

Dixon, F. J. and D. W. Fisher, Eds. (1983). The biology of immunologic disease., Sinauer Associates, Inc. Sunderland.

Domingo, E. (1998). Quasispecies and the implications for virus persistence and escape. Clin Diagn Virol 10: 97–101

Dong, Y. H., L. H. Wang, J. L. Xu, H. B. Zhang, X. F. Zhang, et al. (2001). Quenching quorum-sensing-dependent bacterial infection by an N-acyl homoserine lactonase. Nature 411: 813–7

Doolittle, W. F. and C. Sapienza (1980). Selfish genes, the phenotype paradigm and genome evolution. Nature 284: 601–3

Dorner, T. and P. E. Lipsky (2001). Immunoglobulin variable-region gene usage in systemic autoimmune diseases. Arthritis Rheum 44: 2715–27

Doulatov, S., A. Hodes, L. Dai, N. Mandhana, M. Liu, et al. (2004). Tropism switching in Bordetella bacteriophage defines a family of diversity-generating retroelements. Nature 431: 476–81

Dover, G. A., T. Strachan, E. S. Coen and S. D. Brown (1982). Molecular drive. Science 218: 1069

Downward, J. (2001). The ins and outs of signalling. Nature 411: 759–62

Dreyer, W., W. Gray and L. Hood (1967). The genetic, molecular, and cellular basis of antibody formation: some facts and a unifying hypothesis. Cold Spring Harbor Symposium on Quantitative Biology. 32: 353–67

Dreyer, W. J. and J. C. Bennett (1965). The molecular basis of antibody formation: a paradox. Proc Natl Acad Sci U S A 54: 864–9

Dreyfus, H. (1991). Being-in-the-world. A commentary on Heidegger's Being and Time, Division I. MIT Press, Cambridge

Du Pasquier, L. and G. W. Litman, Eds. (2000). Origin and evolution of the vertebrate immune system. Current Topics in Microbiology and Immunology, Springer-Verlag, Berlin.

Ducoulombier, D., A. M. Roque-Afonso, G. Di Liberto, F. Penin, R. Kara, et al. (2004). Frequent compartmentalization of hepatitis C virus variants in circulating B cells and monocytes. Hepatology 39: 817–25

Dupré, J. (2001). Human nature and the limits of science. Oxford University Press, Oxford

Dustin, M. L. and A. C. Chan (2000). Signaling takes shape in the immune system. Cell 103: 283–94

Dwyer, J. (1988). The body at war. Unwin Hyman Limited, Australia

Ebert, D. (1999). The evolution and expression of parasite virulence. In: Evolution in health and disease. S. C. Stearns (Ed.). Oxford University Press, Oxford, 161–172

Eco, U. (1988). On semiotics and immunology. In: The semiotics of cellular communication in the immune system. E. Sercarz, F. Celada, N. A. Mitchison and T. Tada (Eds.). Springer-Verlag, Berlin, 3–15

Edelman, G. M. and J. A. Gally (2001). Degeneracy and complexity in biological systems. Proc Natl Acad Sci U S A 98: 13763–8

Edwards, E., S. Razvi and C. Cunningham-Rundles (2004). IgA deficiency: clinical correlates and responses to pneumococcal vaccine. Clin Immunol 111: 93–7

Efroni, S. and I. R. Cohen (2002). Simplicity belies a complex system: a response to the minimal model of immunity of Langman and Cohn. Cell Immunol 216: 23–30

Efroni, S. and I. R. Cohen (2003). The heuristics of biologic theory: the case of self-nonself discrimination. Cell Immunol 223: 87–9

Eigen, M. (1996). On the nature of virus quasispecies. Trends Microbiol 4: 216–8

Eisen, H. N. (2001). Specificity and degeneracy in antigen recognition: yin and yang in the immune system. Annu Rev Immunol 19: 1–21

Elitzur, A. C. (1994). Let there be life. Thermodynamic reflections on biogenesis and evolution. J Theor Biol 168: 429–59

Elster, J. (1979). Ulysses and the sirens. Studies in rationality and irrationality. Cambridge University Press, Cambridge

Emmeche, C. (1991). A semiotical reflection on biology, living signs and artificial life. Biology & Philosophy 6: 325–40

Endler, J. (1992). Signals, signal conditions, and the direction of evolution. American Naturalist 139: S125–S153

Etzioni, A. (2003). Immune deficiency and autoimmunity. Autoimmun Rev 2: 364–9

Ewald, P. (1994). Evolution of infectious disease. Oxford University Press, Oxford

Ewald, P. W. (1995). The evolution of virulence: a unifying link between parasitology and ecology. J Parasitol 81: 659–69

Feldmann, M. and L. Steinman (2005). Design of effective immunotherapy for human autoimmunity. Nature 435: 612–9

Fenner, F. and D. O. White (1976). Medical virology. Academic Press., New York

Ferguson, N. M., A. P. Galvani and R. M. Bush (2003). Ecological and immunological determinants of influenza evolution. Nature 422: 428–33

Flajnik, M. F. (2002). Comparative analyses of immunoglobulin genes: surprises and portents. Nat Rev Immunol 2: 688–98

Fleck, L. (1935). Genesis and development of a scientific fact. The University of Chicago Press (1979), Chicago

Fleischmann, R. D., M. D. Adams, O. White, R. A. Clayton, E. F. Kirkness, et al. (1995). Whole-genome random sequencing and assembly of Haemophilus influenzae Rd. Science 269: 496–512

Flynn, J. L. and J. Chan (2001). Immunology of tuberculosis. Annu Rev Immunol 19: 93–129

Folkerts, G., G. Walzl and P. J. Openshaw (2000). Do common childhood infections 'teach' the immune system not to be allergic? Immunol Today 21: 118–20

Folstad, I. and A. J. Karter (1992). Parasites, Bright Males, and the Immunocompetence Handicap. American Naturalist 139: 603–622

Foote, J. and C. Milstein (1994). Conformational isomerism and the diversity of antibodies. Proc Natl Acad Sci U S A 91: 10370–4

Forbes, S. J., P. Vig, R. Poulsom, N. A. Wright and M. R. Alison (2002). Adult stem cell plasticity: new pathways of tissue regeneration become visible. Clin Sci (Lond) 103: 355–69

Forsdyke, D. R., C. A. Madill and S. D. Smith (2002). Immunity as a function of the unicellular state: implications of emerging genomic data. Trends Immunol 23: 575–9

Forster, M. and E. Sober (1994). How to Tell When Simpler, More Unified, or Less Ad Hoc Theories will Provide More Accurate Predictions. The British Journal for the Philosophy of Science 45: 1–35
Foster, K. R., G. Shaulsky, J. E. Strassmann, D. C. Queller and C. R. Thompson (2004). Pleiotropy as a mechanism to stabilize cooperation. Nature 431: 693–6
Frank, S. A. (1995). Mutual policing and repression of competition in the evolution of cooperative groups. Nature 377: 520–2
Frank, S. A. (1996). Models of parasite virulence. Q Rev Biol 71: 37–78
Frank, S. A. (1997). Models of symbiosis. American Naturalist 150: S80–S99
Frank, S. A. (2002). Immunology and evolution of infectious disease. Princeton University Press, Princeton
Frauwirth, K. A. and C. B. Thompson (2002). Activation and inhibition of lymphocytes by costimulation. J Clin Invest 109: 295–9
Fray, R. (2002). Altering plant-microbe interaction through artificially manipulating bacterial quorum sensing. Annals of Botany 89: 245–253
Freitas, A. A. and B. Rocha (2000). Population biology of lymphocytes: the flight for survival. Annu Rev Immunol 18: 83–111
Frew, A. J. (2003). Immunotherapy of allergic disease. J Allergy Clin Immunol 111: S712–9
Fujinami, R. S. and J. E. Libbey (1999). Endogenous retroviruses: are they the cause of multiple sclerosis? Trends Microbiol 7: 263–4
Fuqua, W. C., S. C. Winans and E. P. Greenberg (1994). Quorum sensing in bacteria: the LuxR-LuxI family of cell density-responsive transcriptional regulators. J Bacteriol 176: 269–75
Fåhræus, R. (1924). Blodet i lækekonstens historia. Bokförlaget Natur och Kultur, Stockholm
Gaillard, R. C. and E. Spinedi (1998). Sex- and stress-steroids interactions and the immune system: evidence for a neuroendocrine-immunological sexual dimorphism. Domest Anim Endocrinol 15: 345–52
Galdston, I. (1981). Social and historical foundations of modern medicine. Brunner/Mazel, New York
Gandon, S., M. J. Mackinnon, S. Nee and A. F. Read (2001). Imperfect vaccines and the evolution of pathogen virulence. Nature 414: 751–6
Gapin, L., Y. Bravo de Alba, A. Casrouge, J. P. Cabaniols, P. Kourilsky, et al. (1998). Antigen presentation by dendritic cells focuses T cell responses against immunodominant peptides: studies in the hen egg-white lysozyme (HEL) model. J Immunol 160: 1555–64
Garcia, K. C., L. Teyton and I. A. Wilson (1999). Structural basis of T cell recognition. Annu Rev Immunol 17: 369–97
Gardner, A., S. West and A. Buckling (2004). Bacteriocins, spite and virulence. Proc R Soc Lond B Biol Sci 271: 1529–35.
Genereux, D. P. and J. M. Logsdon, Jr. (2003). Much ado about bacteria-to-vertebrate lateral gene transfer. Trends Genet 19: 191–5
Germain, R. N. (2004). An innately interesting decade of research in immunology. Nat Med 10: 1307–20
Gershon, R. K. (1975). A disquisition on suppressor T cells. Transplant Rev 26: 170–85
Gevers, D., F. M. Cohan, J. G. Lawrence, B. G. Spratt, T. Coenye, et al. (2005). Opinion: Re-evaluating prokaryotic species. Nat Rev Microbiol 3: 733–9
Gibbs, R. A., G. M. Weinstock, M. L. Metzker, D. M. Muzny, E. J. Sodergren, et al. (2004). Genome sequence of the Brown Norway rat yields insights into mammalian evolution. Nature 428: 493–521
Gibbs, W. W. (2001). Cybernetic cells. Sci Am 285: 43–7
Gilbert, S. F. (2001). Ecological developmental biology: developmental biology meets the real world. Dev Biol 233: 1–12
Giske, J., M. Mangel, P. Jakobsen, G. Huse, C. Wilcox, et al. (2003). Explicit trade-off rules in proximate adaptive agents. Evolutionary Ecology Research 5: 835–865
Glass, W. G., D. H. McDermott, J. K. Lim, S. Lekhong, S. F. Yu, et al. (2006). CCR5 deficiency increases risk of symptomatic West Nile virus infection. J Exp Med 203: 35–40
Gleeson, M., D. C. Nieman and B. K. Pedersen (2004). Exercise, nutrition and immune function. J Sports Sci 22: 115–25

Godfrey-Smith, P. (1993). Functions: consensus without unity. Pacific Philosophical Quarterly 74: 196–208.

Goksøyr, J. (1967). Evolution of eucaryotic cells. Nature 214: 1161

Goldberg, B., H. B. Urnovitz and R. B. Stricker (2000). Beyond danger: unmethylated CpG dinucleotides and the immunopathogenesis of disease. Immunol Lett 73: 13–8

Golub, E. (1988). Semiosis for the immune system but not the immune response or what can be learned about language by studying the immune system? In: The semiotics of cellular communication in the immune system. E. Sercarz, F. Celada, N. A. Mitchison and T. Tada (Eds.). Springer-Verlag, Berlin, 65–69

Golub, E. (1997). The limits of medicine. How science shapes our hope for the cure. The University of Chicago Press, Chicago

Gottesman, S. (2005). Micros for microbes: non-coding regulatory RNAs in bacteria. Trends Genet 21: 399–404

Gould, S. J. (2000). Wonderful life. Vintage, London

Gould, S. J. and R. C. Lewontin (1979). The spandrels of San Marco and the Panglossian paradigm: a critique of the adaptationist programme. Proc R Soc Lond B Biol Sci 205: 581–98

Gould, S. J. and E. A. Lloyd (1999). Individuality and adaptation across levels of selection: how shall we name and generalize the unit of Darwinism? Proc Natl Acad Sci U S A 96: 11904–9

Gowans, J. L. (1996). The lymphocyte–a disgraceful gap in medical knowledge. Immunol Today 17: 288–91

Gray, K. M. and J. R. Garey (2001). The evolution of bacterial LuxI and LuxR quorum sensing regulators. Microbiology 147: 2379–87

Gray, M. W., G. Burger and B. F. Lang (1999). Mitochondrial evolution. Science 283: 1476–81

Grebenjuk, V. A., A. Kuusksalu, M. Kelve, J. Schutze, H. C. Schroder, et al. (2002). Induction of (2′–5′)oligoadenylate synthetase in the marine sponges Suberites domuncula and Geodia cydonium by the bacterial endotoxin lipopolysaccharide. Eur J Biochem 269: 1382–92

Gregg, R., C. M. Smith, F. J. Clark, D. Dunnion, N. Khan, et al. (2005). The number of human peripheral blood CD4 CD25 regulatory T cells increases with age. Clin Exp Immunol 140: 540–6

Grindstaff, J. L., E. D. Brodie, 3rd and E. D. Ketterson (2003). Immune function across generations: integrating mechanism and evolutionary process in maternal antibody transmission. Proc R Soc Lond B Biol Sci 270: 2309–19

Grossman, Z. and W. E. Paul (2001). Autoreactivity, dynamic tuning and selectivity. Curr Opin Immunol 13: 687–98

Grünbaum, A. (1903). Theories of immunity and their clinical application. Lancet 161: 775–779.

Guiney, D. G. (1997). Regulation of bacterial virulence gene expression by the host environment. J Clin Invest 99: 565–9

Gupta, R. S. and G. B. Golding (1996). The origin of the eukaryotic cell. Trends Biochem Sci 21: 166–71

Habermas, J. (1968). Knowledge and human interests. Beacon Press, Boston

Hakelien, A. M., H. B. Landsverk, J. M. Robl, B. S. Skalhegg and P. Collas (2002). Reprogramming fibroblasts to express T-cell functions using cell extracts. Nat Biotechnol 20: 460–6

Hall, M. A., K. R. Ahmadi, P. Norman, H. Snieder, A. J. MacGregor, et al. (2000). Genetic influence on peripheral blood T lymphocyte levels. Genes Immun 1: 423–7

Hamburger, V. (1980). Embryology and the modern synthesis in evolutionary theory. In: The evolutionary synthesis. E. Mayr and W. B. Provine (Eds.). Harvard University Press, Cambridge, 96–112

Hamilton, W. D. (1964a). The genetical evolution of social behaviour. I. J Theor Biol 7: 1–16

Hamilton, W. D. (1964b). The genetical evolution of social behaviour. II. J Theor Biol 7: 17–52

Hamilton, W. D. (1970). Selfish and spiteful behaviour in an evolutionary model. Nature 228: 1218–20

Hamilton, W. D. and M. Zuk (1982). Heritable true fitness and bright birds: a role for parasites? Science 218: 384–7

Hanssen, S., D. Hasselquist, I. Folstad and K. Erikstad (2004). Costs of immunity: immune responsiveness reduces survival in a vertebrate. Proc R Soc Lond B Biol Sci 271: 925–30

Harada, Y., M. Muramatsu, T. Shibata, T. Honjo and K. Kuroda (2003). Unmutated immunoglobulin M can protect mice from death by influenza virus infection. J Exp Med 197: 1779–85

Harboe, M., R. van Furth, H. Schubothe, K. Lind and R. S. Evans (1965). Exclusive occurrence of K chains in isolated cold haemagglutinins. Scand J Haematol 2: 259 66

Harris, H. (2002). Things come to life. Spontaneous generation revisited. Oxford University Press, Oxford

Hashimoto, K., H. Handa, K. Umehara and S. Sasaki (1978). Germfree mice reared on an "antigen-free" diet. Lab Anim Sci 28: 38–45

Hatfield, G. (1992). Descartes' physiology and its relation to his psychology. In: The Cambridge companion to Descartes. J. Cottingham (Ed.). Cambridge University Press, Cambridge, 335–370

Heath, W. R. and F. R. Carbone (2003). Immunology: dangerous liaisons. Nature 425: 460–1

Hegel, G. W. F. (1807). Phenomenology of spirit. Oxford University Press (1977), Oxford

Hegel, G. W. F. (1821). Philosophy of right. Prometheus Books (1996), Amherst

Heidegger, M. (1927). Being and time. Blackwell Publishers Ltd. (1962), Oxford

Heidegger, M. (1995). The fundamental concepts of metaphysics. World, finitude, solitude. Indiana University Press, Bloomington

Hendrix, R. W., J. G. Lawrence, G. F. Hatfull and S. Casjens (2000). The origins and ongoing evolution of viruses. Trends Microbiol 8: 504–8

Hill, A. V. (1999). Genetics and genomics of infectious disease susceptibility. Br Med Bull 55: 401–13

Hillson, J. L., N. S. Karr, I. R. Oppliger, M. Mannik and E. H. Sasso (1993). The structural basis of germline-encoded VH3 immunoglobulin binding to staphylococcal protein A. J Exp Med 178: 331–6

Hoffmann, M. and J. W. Kappler (1972). The antigen specificity of thymus-derived helper cells. J Immunol 108: 261–3

Hoffmeyer, J. (1996). Signs of meaning in the universe. Indiana University Press, Bloomington.

Hoffmeyer, J. (2005). Biosemiotik. Forlaget Ries, Charlottenlund

Hood, L. (2003). Systems biology: integrating technology, biology, and computation. Mech Ageing Dev 124: 9–16

Hughes, A. (1999). Adaptive evolution of genes and genomes. Oxford University Press, Oxford

Hughes, A. L. (2002). Natural selection and the diversification of vertebrate immune effectors. Immunol Rev 190: 161–8

Hughes, A. L. and M. Yeager (1998). Natural selection at major histocompatibility complex loci of vertebrates. Annu Rev Genet 32: 415–35

Hull, D. L., R. E. Langman and S. S. Glenn (2001). A general account of selection: biology, immunology, and behavior. Behav Brain Sci 24: 511–28; discussion 528–73

Hurst, L. D. (2002). The Ka/Ks ratio: diagnosing the form of sequence evolution. Trends Genet 18: 486

Itano, A. and E. Robey (2000). Highly efficient selection of CD4 and CD8 lineage thymocytes supports an instructive model of lineage commitment. Immunity 12: 383–9

Itescu, S., P. F. Simonelli, R. J. Winchester and H. S. Ginsberg (1994). Human immunodeficiency virus type 1 strains in the lungs of infected individuals evolve independently from those in peripheral blood and are highly conserved in the C-terminal region of the envelope V3 loop. Proc Natl Acad Sci U S A 91: 11378–82

Jackson, M. (2001). Allergy: the making of a modern plague. Clin Exp Allergy 31: 1665–71

Jackson, M. (2003). Allergy and history. Stud. Hist. Phil. Biol. & Biomed. Sci. 34: 383–98

Jacob, S., M. K. McClintock, B. Zelano and C. Ober (2002). Paternally inherited HLA alleles are associated with women's choice of male odor. Nat Genet 30: 175–9

James, W. (1907). Pragmatism. Dover Publications, Inc. (1995), New York

Janeway, C. A., Jr. (1989). Approaching the asymptote? Evolution and revolution in immunology. Cold Spring Harb Symp Quant Biol 54 Pt 1: 1–13

Janeway, C. A., Jr. (2001). How the immune system works to protect the host from infection: a personal view. Proc Natl Acad Sci U S A 98: 7461–8

Janeway, C. A., Jr. and P. Travers (1994). Immunobiology. The immune system in health and disease. Current Biology Ltd., London

Jensen, K. H., T. Little, A. Skorping and D. Ebert (2006). Empirical Support for Optimal Virulence in a Castrating Parasite. PLoS Biol 4: e197

Jerne, N. K. (1955). The natural selection theory of antibody formation. Proc Natl Acad Sci U S A 41: 849–57

Jerne, N. K. (1969). The complete solution of immunology. Australas Ann Med 18: 345–8

Jerne, N. K. (1974). Towards a network theory of the immune system. Ann Immunol (Paris) 125C: 373–89

Jerne, N. K. (1984). Idiotypic networks and other preconceived ideas. Immunol Rev 79: 5–24

Jerne, N. K. (1993). The Nobel Lectures in Immunology. The Nobel Prize for Physiology or Medicine, 1984. The generative grammar of the immune system. Scand J Immunol 38: 1–9

Johnston, T. (2001). Toward a systems view of development: An appraisal of Lehrman's critique of Lorenz. In: Cycles of contingency. Developmental systems and evolution. S. Oyama, P. Griffiths and R. Gray (Eds.). The MIT Press, Cambridge, 15–23

Joint, I., K. Tait, M. E. Callow, J. A. Callow, D. Milton, et al. (2002). Cell-to-cell communication across the prokaryote-eukaryote boundary. Science 298: 1207

Jokela, J. and E. Haukioja (2000). Evolution of strategies to stay in the game. Biology & Philosophy 15: 177–196

Jordan, J. D., E. M. Landau and R. Iyengar (2000). Signaling networks: the origins of cellular multitasking. Cell 103: 193–200

Kaech, S. M., S. Hemby, E. Kersh and R. Ahmed (2002). Molecular and functional profiling of memory CD8 T cell differentiation. Cell 111: 837–51

Kant, I. (1781). Critique of pure reason. Everyman (1993), London

Katz, L. (1998). Changing perspectives on the origin of eukaryotes. Trends Ecol Evol 13: 493–497

Kauffman, S. (1995). At home in the universe. Penguin Books, London

Kawai, T. and S. Akira (2005). Toll-like receptor downstream signalling. Arthritis Res Ther 7: 12–9

Keating, P. and A. Cambrosio (1997). Helpers and suppressors: on fictional characters in immunology. J Hist Biol 30: 381–96

Keegan, B. M. and J. H. Noseworthy (2002). Multiple sclerosis. Ann Rev Med 53: 285–302.

Keller, L., Ed. (1999). Levels of selection in evolution. Princeton University Press, Princeton.

Kerr, B., M. A. Riley, M. W. Feldman and B. J. Bohannan (2002). Local dispersal promotes biodiversity in a real-life game of rock-paper-scissors. Nature 418: 171–4

Khakoo, S. I., R. Rajalingam, B. P. Shum, K. Weidenbach, L. Flodin, et al. (2000). Rapid evolution of NK cell receptor systems demonstrated by comparison of chimpanzees and humans. Immunity 12: 687–98

Kidwell, M. G. and D. R. Lisch (2000). Transposable elements and host genome evolution. Trends in Ecology and Evolution 15: 95–99

Kierkegaard, S. (1849). The sickness unto death. Penguin Books (1989), London

Kim, D. S. and A. B. Drake-Lee (2003). Infection, allergy and the hygiene hypothesis: historical perspective. J Laryngol Otol 117: 946–50

Kimbrell, D. A. and B. Beutler (2001). The evolution and genetics of innate immunity. Nat Rev Genet 2: 256–67

Kimura, M. (1991). Recent development of the neutral theory viewed from the Wrightian tradition of theoretical population genetics. Proc Natl Acad Sci U S A 88: 5969–73

Kinoshita, K. and T. Honjo (2001). Linking class-switch recombination with somatic hypermutation. Nat Rev Mol Cell Biol 2: 493–503

Kirk, G., J. Raven and M. Schofield (1983). The presocratic philosophers. Cambridge University Press, Cambridge

Kirschner, M. and J. Gerhart (1998). Evolvability. Proc Natl Acad Sci U S A 95: 8420–7

Kitano, H. (2002). Systems biology: a brief overview. Science 295: 1662–4

Klein, J. (1982). Immunology. The science of self-nonself discrimination. John Wiley & Sons, New York

Klein, J. and C. O'Huigin (1994). MHC polymorphism and parasites. Philos Trans R Soc Lond B Biol Sci 346: 351–7; discussion 357–8

Klein, J. and A. Sato (1998). Birth of the major histocompatibility complex. Scand J Immunol 47: 199–209

Klein, J. and N. Takahate (2002). Where do we come from? The molecular evidence for human descent. Springer, Berlin

Klein, S. L. (2000). Hormones and mating system affect sex and species differences in immune function among vertebrates. Behav Processes 51: 149–166

Korber, B., M. Muldoon, J. Theiler, F. Gao, R. Gupta, et al. (2000). Timing the ancestor of the HIV-1 pandemic strains. Science 288: 1789–96

Kraj, P., D. F. Friedman, F. Stevenson and L. E. Silberstein (1995). Evidence for the overexpression of the VH4-34 (VH4.21) Ig gene segment in the normal adult human peripheral blood B cell repertoire. J Immunol 154: 6406–20

Krakauer, D. C. and M. Pagel (1996). Selection by somatic signals: the advertisement of phenotypic state through costly intercellular signals. Philos Trans R Soc Lond B Biol Sci 351: 647–58

Krieg, A. M., M. F. Gourley and A. Perl (1992). Endogenous retroviruses: potential etiologic agents in autoimmunity. Faseb J 6: 2537–44

Kroos, L. and J. R. Maddock (2003). Prokaryotic development: emerging insights. J Bacteriol 185: 1128–46

Kuhn, T. (1962). The structure of scientific revolutions. The University of Chicago Press (1996), Chicago

Kull, K. (1999). Biosemiotics in the twentieth century: A view from biology. Semiotica 127: 385–414

Kull, K. (2001). Jakob von Uexküll: An introduction. Semiotica 134: 1–59

Kurtz, J. and K. Franz (2003). Innate defence: evidence for memory in invertebrate immunity. Nature 425: 37–8

Lafferty, K. J. and A. J. Cunningham (1975). A new analysis of allogeneic interactions. Aust J Exp Biol Med Sci 53: 27–42

Laird, D. J., A. W. De Tomaso, M. D. Cooper and I. L. Weissman (2000). 50 million years of chordate evolution: seeking the origins of adaptive immunity. Proc Natl Acad Sci U S A 97: 6924–6

Lander, E. S., L. M. Linton, B. Birren, C. Nusbaum, M. C. Zody, et al. (2001). Initial sequencing and analysis of the human genome. Nature 409: 860–921

Landsteiner, K. (1945). The specificity of serological reactions. Dover Publications, Inc., New York

Lang, B., R. C. Dale and A. Vincent (2003). New autoantibody mediated disorders of the central nervous system. Curr Opin Neurol 16: 351–7

Lang, H. L., H. Jacobsen, S. Ikemizu, C. Andersson, K. Harlos, et al. (2002). A functional and structural basis for TCR cross-reactivity in multiple sclerosis. Nat Immunol 3: 940–3

Langman, R. (1992). Comment. Immuology Today 13: 399–400

Langman, R. E. and M. Cohn (1986). The "complete" idiotype network is an absurd immune system. Immunology Today 7: 100–1

Langman, R. E. and M. Cohn (1996). Cutting edge: terra firma: a retreat from "danger". J Immunol 157: 4273–6

Langman, R. E. and M. Cohn (2002). If the immune repertoire evolved to be large, random, and somatically generated, then. Cell Immunol 216: 15–22

LaPorte, J. (2002). Must signals handicap? The Monist 85: 86–104

Le Bon, A. and D. F. Tough (2002). Links between innate and adaptive immunity via type I interferon. Curr Opin Immunol 14: 432–6

Lederberg, J. (1959). Genes and antibodies. Science 129: 1649–53

Lederberg, J. (2000). Infectious history. Science 288: 287–93

LeGrand, E. K. (1997). An adaptationist view of apoptosis. The Quarterly Review of BIology 72: 135–147

Lehrman, D. (1953). A critique of Konrad Lorenz's theory of instinctive behavior. The Quarterly Review of Biology 28: 337–63

Leibniz, G. W. (1714). Monadology. In: Leibniz and the Monadology. A. Savile (Ed.). Routledge (2000), London

Leigh, E. G. (1999). Levels of selection, potential conflicts and their resolution: The role of the "common good". In: Levels of selection in evolution. L. Keller (Ed.). Princeton University Press, Princeton, 15–30

Lemke, H., A. Coutinho and H. Lange (2004). Lamarckian inheritance by somatically acquired maternal IgG phenotypes. Trends Immunol 25: 180–6

Lemke, H. and H. Lange (1999). Is there a maternally induced immunological imprinting phase a la Konrad Lorenz? Scand J Immunol 50: 348–54

Lennox, J. (2001). Aristotle's philosophy of biology. Cambridge University Press, Cambridge

Levine, B. and D. J. Klionsky (2004). Development by self-digestion: molecular mechanisms and biological functions of autophagy. Dev Cell 6: 463–77

Levine, M. and R. Tjian (2003). Transcription regulation and animal diversity. Nature 424: 147–51

Levins, R. and R. Lewontin (1985). The dialectical biologist. Harvard University Press, Cambridge

Lewis, K. (2000). Programmed death in bacteria. Microbiol Mol Biol Rev 64: 503–14

Lewis, S. M. (1999). Evolution of immunoglobulin and T-cell receptor gene assembly. Ann N Y Acad Sci 870: 58–67

Lewontin, R. (1970). The units of selection. Ann Rev Ecol Syst 1: 1–18

Lewontin, R. C. (1974). Annotation: the analysis of variance and the analysis of causes. Am J Hum Genet 26: 400–11

Lewontin, R. C. (1978). Adaptation. Sci Am 239: 212–8, 220, 222 passim

Lewontin, R. (2000). The triple helix. Gene, organism, and environment. Harvard University Press, Cambridge

Li, H., A. Llera, E. L. Malchiodi and R. A. Mariuzza (1999). The structural basis of T cell activation by superantigens. Annu Rev Immunol 17: 435–66

Li, W. H., Z. Gu, H. Wang and A. Nekrutenko (2001). Evolutionary analyses of the human genome. Nature 409: 847–9

Li, Y., M. B. Spellerberg, F. K. Stevenson, J. D. Capra and K. N. Potter (1996). The I binding specificity of human VH 4–34 (VH 4–21) encoded antibodies is determined by both VH framework region 1 and complementarity determining region 3. J Mol Biol 256: 577–89

Licastro, F., G. Candore, D. Lio, E. Porcellini, G. Colonna-Romano, et al. (2005). Innate immunity and inflammation in ageing: a key for understanding age-related diseases. Immun Ageing 2: 8

Lin, D. T. and J. D. Lechleiter (2002). Mitochondrial targeted cyclophilin D protects cells from cell death by peptidyl prolyl isomerization. J Biol Chem 277: 31134–41

Litman, G. W., M. K. Anderson and J. P. Rast (1999). Evolution of antigen binding receptors. Annu Rev Immunol 17: 109–47

Livak, F. and H. T. Petrie (2002). Access roads for RAG-ged terrains: control of T cell receptor gene rearrangement at multiple levels. Semin Immunol 14: 297–309

Lloyd, E. A. and S. J. Gould (1993). Species selection on variability. Proc Natl Acad Sci U S A 90: 595–9

Lloyd, G. (Ed). (1978). Hippocratic writings. Penguin Books, London

Lobkowicz, N. (1967). Theory and Practice: History of a concept from Aristotle to Marx. University of Notre Dame Press, Notre Dame

Lochmiller, R. and C. Deerenberg (2000). Trade-offs in evolutionary immunology: just what is the cost of immunity? Oikos 88: 87–98

Lockyer, C. R. and D. A. Gillatt (2001). BCG immunotherapy for superficial bladder cancer. J R Soc Med 94: 119–23

Lohm, J., M. Grahn, A. Langefors, O. Andersen, A. Storset, et al. (2002). Experimental evidence for major histocompatibility complex-allele-specific resistance to a bacterial infection. Proc R Soc Lond B Biol Sci 269: 2029–33

Lohoff, M. and T. W. Mak (2005). Roles of interferon-regulatory factors in T-helper-cell differentiation. Nat Rev Immunol 5: 125–35

Longrigg, J. (1993). Greek rational medicine. Routledge, London

Lorber, B. (1996). Are all diseases infectious? Ann Intern Med 125: 844–51

Lorenz, K. (1965). Evolution and modification of behavior. The University of Chicago Press, Chicago

Lorenz, K. (1982). Kant's doctrine of the a priori in the light of contemporary biology (Original 1941). In: Learning, development and culture. H. C. Plotkin (Ed.). John Wiley & Sons, Chichester, 121–143

Lu, F. and B. Kalman (1999). Autoreactive IgG to intracellular proteins in sera of MS patients. J Neuroimmunol 99: 72–81

Luisi, P. L. (2003). Autopoiesis: a review and a reappraisal. Naturwissenschaften 90: 49–59

Lundin, B. S., A. Dahlman-Hoglund, I. Pettersson, U. I. Dahlgren, L. A. Hanson, et al. (1999). Antibodies given orally in the neonatal period can affect the immune response for two generations: evidence for active maternal influence on the newborn's immune system. Scand J Immunol 50: 651–6

Lutz, M. B. and G. Schuler (2002). Immature, semi-mature and fully mature dendritic cells: which signals induce tolerance or immunity? Trends Immunol 23: 445–9

Lübcke, P. (1982). Husserl: Filosofi som streng videnskab. In: Vor tids filosofi: Engagement og forståelse. P. Lübcke (Ed.). Politikens Forlag A/S, Copenhagen, 35–68

Lynch, M. and J. L. Blanchard (1998). Deleterious mutation accumulation in organelle genomes. Genetica 102–103: 29–39

Mackie, J. L. (1980). The cement of the universe. A study of causation. Oxford at the Clarendon Press, Oxford

Madeo, F., S. Engelhardt, E. Herker, N. Lehmann, C. Maldener, et al. (2002). Apoptosis in yeast: a new model system with applications in cell biology and medicine. Curr Genet 41: 208–16

Magor, B. G., D. A. Ross, L. Pilstrom and G. W. Warr (1999). Transcriptional enhancers and the evolution of the IgH locus. Immunol Today 20: 13–7

Mahner, M. and M. Bunge (1997). Foundations of biophilosophy. Springer, Berlin

Maizels, R. M. (2005). Infections and allergy – helminths, hygiene and host immune regulation. Curr Opin Immunol 17: 656–61

Malanchere, E., F. Huetz and A. Coutinho (1997). Maternal IgG stimulates B lineage cell development in the progeny. Eur J Immunol 27: 788–93

Maloney, S., A. Smith, D. E. Furst, D. Myerson, K. Rupert, et al. (1999). Microchimerism of maternal origin persists into adult life. J Clin Invest 104: 41–7

Manivel, V., F. Bayiroglu, Z. Siddiqui, D. M. Salunke and K. V. Rao (2002). The primary antibody repertoire represents a linked network of degenerate antigen specificities. J Immunol 169: 888–97

Marcos, A., E. Nova and A. Montero (2003). Changes in the immune system are conditioned by nutrition. Eur J Clin Nutr 57 Suppl 1: S66–9

Margulis, L., M. F. Dolan and R. Guerrero (2000). The chimeric eukaryote: origin of the nucleus from the karyomastigont in amitochondriate protists. Proc Natl Acad Sci U S A 97: 6954–9

Martin, L. B., 2nd, A. Scheuerlein and M. Wikelski (2003). Immune activity elevates energy expenditure of house sparrows: a link between direct and indirect costs? Proc R Soc Lond B Biol Sci 270: 153–8

Martinon, F. and J. Tschopp (2005). NLRs join TLRs as innate sensors of pathogens. Trends Immunol 26: 447–54

Maruyama, M., K. P. Lam and K. Rajewsky (2000). Memory B-cell persistence is independent of persisting immunizing antigen. Nature 407: 636–42

Masllorêns, F. O. (2000). Autoimmune disease and physiological autoimmunity: recognition of self. Alergol Immunol Clin 15: 5–12

Mason, D. (1998). A very high level of crossreactivity is an essential feature of the T-cell receptor. Immunol Today 19: 395–404

Matricardi, P. M. and S. Bonini (2000). Mimicking microbial 'education' of the immune system: a strategy to revert the epidemic trend of atopy and allergic asthma? Respir Res 1: 129–32

Matsunaga, T. and A. Rahman (1998). What brought the adaptive immune system to vertebrates?–The jaw hypothesis and the seahorse. Immunol Rev 166: 177–86

Matthen, M. and E. Levy (1984). Teleology, error, and the human immune system. Journal of Philosophy 81: 351–72

Maturana, H. and F. Varela (1987). The tree of knowledge. The biological roots of human understanding. Shambhala Publications, Inc., Boston

Matzinger, P. (1994). Tolerance, danger, and the extended family. Annu Rev Immunol 12: 991–1045

Matzinger, P. (2002). The danger model: a renewed sense of self. Science 296: 301–5

Mayer, W. E., C. O'Huigin, H. Tichy, J. Terzic and M. Saraga-Babic (2002). Identification of two Ikaros-like transcription factors in lamprey. Scand J Immunol 55: 162–70
Maynard-Smith, J. and E. Szathmáry (1995). The major transitions in evolution. Oxford University Press, Oxford
Mayr, E. (1961). Cause and effect in biology. Science 134: 1501–6
Mayr, E. (1988). Toward a new philosophy of biology. Observations of an evolutionist. The Belknap Press of Harvard University Press, Cambridge
Mazumdar, P. M. (1995). Species and specificity. An interpretation of the history of immunology. Cambridge University Press, Cambridge
McClain, M. T., L. D. Heinlen, G. J. Dennis, J. Roebuck, J. B. Harley, et al. (2005). Early events in lupus humoral autoimmunity suggest initiation through molecular mimicry. Nat Med 11: 85–9
McGuirk, P. and K. H. Mills (2002). Pathogen-specific regulatory T cells provoke a shift in the Th1/Th2 paradigm in immunity to infectious diseases. Trends Immunol 23: 450–5
McHugh, R. S. and E. M. Shevach (2002). The role of suppressor T cells in regulation of immune responses. J Allergy Clin Immunol 110: 693–702
McKean, K. A. and M. Zuk (1995). An evolutionary perspective on signaling in behavior and immunology. Naturwissenschaften 82: 509–16
Medawar, P. B. (1969). Induction and intuition in scientific thought. The American Philosophical Society, Philadelphia
Medzhitov, R. and C. A. Janeway, Jr. (2002). Decoding the patterns of self and nonself by the innate immune system. Science 296: 298–300
Melino, G. (2001). The Sirens' song. Nature 412: 23
Mendoza, H. L. and I. Faye (1999). Physiological aspects of the immunoglobulin superfamily in invertebrates. Dev Comp Immunol 23: 359–74
Messaoudi, I., J. A. Guevara Patino, R. Dyall, J. LeMaoult and J. Nikolich-Zugich (2002). Direct link between mhc polymorphism, T cell avidity, and diversity in immune defense. Science 298: 1797–800
Metchnikoff, E. (1892). The struggle for existence between parts of the animal organism. In: The evolutionary biology papers of Elie Metchnikoff. H. Gourko, D. I. Williamson and A. I. Tauber (Eds.). Kluwer Academic Publishers (2000), Dordrecht, 207–16
Michelson, A. and R. Kopan (2002). Differentiation and gene regulation: toward a holistic understanding of animal development: intercellular communication and transcriptional regulation are two sides of the same coin. Curr Opin Genet Dev 12: 499–502
Michod, R. (1999). Darwinian dynamics. Evolutionary transitions in fitness and individuality. Princeton University Press, Princeton
Michod, R. E. and A. M. Nedelcu (2003). On the reorganization of fitness during evolutionary transitions in individuality. Integr Comp Biol 43: 64–73.
Miller, D. J. and M. Rodriguez (1995). A monoclonal autoantibody that promotes central nervous system remyelination in a model of multiple sclerosis is a natural autoantibody encoded by germline immunoglobulin genes. J Immunol 154: 2460–9
Miller, M. J., S. H. Wei, M. D. Cahalan and I. Parker (2003). Autonomous T cell trafficking examined in vivo with intravital two photon microscopy. Proc Natl Acad Sci U S A 100: 2604–9
Miller, R. A. (1996). The aging immune system: primer and prospectus. Science 273: 70–4
Milstein, C. and M. S. Neuberger (1996). Maturation of the immune response. Adv Protein Chem 49: 451–85
Mims, C. (2000). The war within us: Everyman's guide to infection and immunity. Academic Press, San Diego
Mishler, B. D. and M. J. Donoghue (1982). Species concepts: A case for pluralism. Systematic Zoology 31: 491–503
Mitchell, G. F. and J. F. Miller (1968). Cell to cell interaction in the immune response. II. The source of hemolysin-forming cells in irradiated mice given bone marrow and thymus or thoracic duct lymphocytes. J Exp Med 128: 821–37
Mitchison, N. A. (1971). The carrier effect in the secondary response to hapten-protein conjugates. II. Cellular cooperation. Eur J Immunol 1: 18–27

Moller, A. P., P. Christe and E. Lux (1999). Parasitism, host immune function, and sexual selection. Q Rev Biol 74: 3–20

Moore, C. B., M. John, I. R. James, F. T. Christiansen, C. S. Witt, et al. (2002). Evidence of HIV-1 adaptation to HLA-restricted immune responses at a population level. Science 296: 1439–43

Moran, N. (1992). The evolutionary maintenance of alternative phenotypes. American Naturalist 139: 971–89

Moret, Y. and P. Schmid-Hempel (2000). Survival for immunity: the price of immune system activation for bumblebee workers. Science 290: 1166–8

Moret, Y. and M. T. Siva-Jothy (2003). Adaptive innate immunity? Responsive-mode prophylaxis in the mealworm beetle, Tenebrio molitor. Proc R Soc Lond B Biol Sci 270: 2475–80

Morris, S. (2003). Life's solution. Inevitable humans in a lonely universe. Cambridge University Press, Cambridge

Mosmann, T. R. and R. L. Coffman (1989). TH1 and TH2 cells: different patterns of lymphokine secretion lead to different functional properties. Annu Rev Immunol 7: 145–73

Moulin, A. M. (1989). The immune system: a key concept for the history of immunology. Hist Philos Life Sci 11: 221–36

Murgia, C., J. K. Pritchard, S. Y. Kim, A. Fassati and R. A. Weiss (2006). Clonal origin and evolution of a transmissible cancer. Cell 126: 477–87

Nachtomy, O., A. Shavit and J. Smith (2002). Leibnizian organisms, nested individuals, and units of selection. Theory in Biosciences 121: 205–30

Naranjo, P. (1992). Epidemic hecatomb in the New World. Allergy Proc 13: 237–41

Nei, M., X. Gu and T. Sitnikova (1997). Evolution by the birth-and-death process in multigene families of the vertebrate immune system. Proc Natl Acad Sci U S A 94: 7799–806

Nelson, J. L. (2002). Microchimerism: incidental byproduct of pregnancy or active participant in human health? Trends Mol Med 8: 109–13

Neveu, P. J. (1977a). Stimulation and inhibition of macrophages prior to immunization with a hapten-carrier conjugate. Immunology 33: 269–73

Neveu, P. J. (1977b). Are cellular and humoral components of the immune response linked by a dispatcher cell? Int Arch Allergy Appl Immunol 53: 88–92

Nevo, U., J. Kipnis, I. Golding, I. Shaked, A. Neumann, et al. (2003). Autoimmunity as a special case of immunity: removing threats from within. Trends Mol Med 9: 88–93

Nonaka, M. (2001). Evolution of the complement system. Curr Opin Immunol 13: 69–73

Nordin, I. (1999). The role of science in medicine. Theor Med Bioeth 20: 229–44

Norris, K. and M. R. Evans (2000). Ecological immunology: life history trade-offs and immune defense in birds. Behavioral Ecology 11: 19–26

Nossal, G. J. (2001). A purgative mastery. Nature 412: 685–6

O'Brien, S. J., M. E. Roelke, L. Marker, A. Newman, C. A. Winkler, et al. (1985). Genetic basis for species vulnerability in the cheetah. Science 227: 1428–34

O'Brien, S. J. and N. Yuhki (1999). Comparative genome organization of the major histocompatibility complex: lessons from the Felidae. Immunol Rev 167: 133–44

Oberholzer, T., R. Wick, P. L. Luisi and C. K. Biebricher (1995). Enzymatic RNA replication in self-reproducing vesicles: an approach to a minimal cell. Biochem Biophys Res Commun 207: 250–7

Ockham, W. (1990). Philosophical writings. A selection. Hackett Publishing Company, Indianapolis

Ohlin, M. and C. A. Borrebaeck (1996). Characteristics of human antibody repertoires following active immune responses in vivo. Mol Immunol 33: 583–92

Olsen, G. J., C. R. Woese and R. Overbeek (1994). The winds of (evolutionary) change: breathing new life into microbiology. J Bacteriol 176: 1–6

Orgel, L. E. and F. H. Crick (1980). Selfish DNA: the ultimate parasite. Nature 284: 604–7

Orosz, C. (2000). Tending Adam's garden. Graft 3: 339–40.

Orosz, C. (2001). An introduction to immuno-ecology and immuno-informatics. In: Design Principles for the Immune System and Other Distributed Autonomous Systems. L. Segel and I. R. Cohen (Eds.). Oxford University Press, Oxford, 125–49

Østerberg, D. (1988). Metasociology. An inquiry into the origins and validity of social thought. Norwegian University Press, Oslo

Ostertag, E. and H. J. Kazazian (2001). Biology of mammalian L1 retrotransposons. Ann Rev Genet 35: 501–38

Osteryoung, K. W. and J. Nunnari (2003). The division of endosymbiotic organelles. Science 302: 1698–704

Owen, R. (1945). Immunogenetic consequences of vascular anastomoses between bovine twins. Science 102: 400–1

Owen, R., H. Davis and R. Morgan (1946). Quintuplet calves and erythrocyte mosaicism. J Heredity 37: 291–97

Owens, I. P. (2002). Ecology and evolution. Sex differences in mortality rate. Science 297: 2008–9

Oyama, S., P. Griffiths and R. Gray, Eds. (2001). Cycles of contingency. Developmental systems and evolution. The MIT press, Cambridge

Paabo, S. (2003). The mosaic that is our genome. Nature 421: 409–12

Padgett, D. A. and R. Glaser (2003). How stress influences the immune response. Trends Immunol 24: 444–8

Paley, W. (1802). Natural theology. Coachwhip Publications (2005), Landisville

Palis, J. and M. C. Yoder (2001). Yolk-sac hematopoiesis: the first blood cells of mouse and man. Exp Hematol 29: 927–36

Pancer, Z., C. T. Amemiya, G. R. Ehrhardt, J. Ceitlin, G. L. Gartland, et al. (2004). Somatic diversification of variable lymphocyte receptors in the agnathan sea lamprey. Nature 430: 174–80

Parker, D. C. (1993). T cell-dependent B cell activation. Annu Rev Immunol 11: 331–60

Parnes, O. (2003). 'Trouble from within': allergy, autoimmunity, and pathology in the first half of the twentieth century. Stud Hist Phil Biol & Biomed Sci 34: 425–54

Pascual, V., K. Victor, M. Spellerberg, T. J. Hamblin, F. K. Stevenson, et al. (1992). VH restriction among human cold agglutinins. The VH4-21 gene segment is required to encode anti-I and anti-i specificities. J Immunol 149: 2337–44

Pattee, H. (1995). Evolving self-reference: Matter, Symbols, and semantic closure. Communication and Cognition – Artificial Intelligence 12: 9–28

Pawelec, G., A. Akbar, C. Caruso, R. Effros, B. Grubeck-Loebenstein, et al. (2004). Is immunosenescence infectious? Trends Immunol 25: 406–10

Paxton, W. A. and S. Kang (1998). Chemokine receptor allelic polymorphisms: relationships to HIV resistance and disease progression. Semin Immunol 10: 187–94

Pearse, A. M. and K. Swift (2006). Allograft theory: transmission of devil facial-tumour disease. Nature 439: 549

Pearson, H., T. Clarke, A. Abbott, J. Knight and D. Cyranoski (2003). SARS: what have we learned? Nature 424: 121–6

Peirce, C. (1955). The general theory of probable inference. In: Philosophical writings of Peirce. J. Buchler (Ed.). Dover, New York, 190–217

Peirce, C. (1992). The essential Peirce. Volume 1. Indiana University Press, Bloomington

Peirce, C. (1998). The essential Peirce. Volume 2. Indiana University Press, Bloomington

Peiris, J. S., S. T. Lai, L. L. Poon, Y. Guan, L. Y. Yam, et al. (2003). Coronavirus as a possible cause of severe acute respiratory syndrome. Lancet 361: 1319–25

Penn, D. and W. Potts (1999). The evolution of mating preferences and major histocompatibility genes. American Naturalist 153: 145–64

Perlman, R. (2000). The concept of organism in physiology. Theory in Biosciences 119: 174–86

Petronis, A. (2001). Human morbid genetics revisited: relevance of epigenetics. Trends Genet 17: 142–6

Pette, M., D. F. Pette, P. A. Muraro, R. Martin and H. F. McFarland (1997). In vitro modulation of human, autoreactive MBP-specific CD4 + T-cell clones by cyclosporin A. J Neuroimmunol 76: 91–9

Philippe, H. and A. Germot (2000). Phylogeny of eukaryotes based on ribosomal RNA: long-branch attraction and models of sequence evolution. Mol Biol Evol 17: 830–4

Pineda-Krch, M. and K. Lehtila (2004). Costs and benefits of genetic heterogeneity within organisms. J Evol Biol 17: 1167–77

Plato (1989). Theaetetus. In: Plato: The collected dialogues. E. Hamilton and H. Cairns (Eds.). Princeton University Press, Princeton, 845–919

Plotkin, H. C. (1982). Evolutionary epistemology and evolutionary theory. In: Learning, development, and culture. H. C. Plotkin (Ed.). John Wiley & Sons, Chichester, 3–13

Pockley, A. G. (2003). Heat shock proteins as regulators of the immune response. Lancet 362: 469–76

Pogo, A. O. and A. Chaudhuri (2000). The Duffy protein: a malarial and chemokine receptor. Semin Hematol 37: 122–9

Pomiankowski, A. (1999). Intragenomic conflict. In: Levels of selection in evolution. L. Keller (Ed.). Princeton University Press, Princeton, 121–52

Popper, K. (1935). The logic of scientific discovery. Routledge (1959), London

Popper, K. (1979). Objective knowledge. An evolutionary approach. Oxford University Press, Oxford

Popper, K. (1989). Conjectures and refutations. The growth of scientific knowledge. Routledge, London

Poulin, R. and C. Combes (1999). The concept of virulence: Interpretations and implications – Comment. Parasitology Today 15: 474–475

Poulin, R. and D. Mouillot (2003). Parasite specialization from a phylogenetic perspective: a new index of host specificity. Parasitology 126: 473–80

Price, G. R. (1970). Selection and covariance. Nature 227: 520–1

Price, G. R. (1972). Extension of covariance selection mathematics. Ann Hum Genet 35: 485–90

Price, G. R. (1995). The nature of selection. (Written circa 1971, published posthumously). J Theor Biol 175: 389–96

Price, P. (1996). Biological evolution. Saunders College Publishing, Fort Worth

Prince, V. E. and F. B. Pickett (2002). Splitting pairs: the diverging fates of duplicated genes. Nat Rev Genet 3: 827–37

Prodi, G. (1988). Signs and codes in immunology. In: The semiotics of cellular communication in the immune system. E. Sercarz, F. Celada, N. A. Mitchison and T. Tada (Eds.). Springer-Verlag, Berlin, 53–64

Prohaszka, Z. and G. Fust (2004). Immunological aspects of heat-shock proteins-the optimum stress of life. Mol Immunol 41: 29–44

Pugh-Bernard, A. E., G. J. Silverman, A. J. Cappione, M. E. Villano, D. H. Ryan, et al. (2001). Regulation of inherently autoreactive VH4-34 B cells in the maintenance of human B cell tolerance. J Clin Invest 108: 1061–70

Purnell, M. A. (2002). Feeding in extinct jawless heterostracan fishes and testing scenarios of early vertebrate evolution. Proc R Soc Lond B Biol Sci 269: 83–8

Queitsch, C., T. A. Sangster and S. Lindquist (2002). Hsp90 as a capacitor of phenotypic variation. Nature 417: 618–24

Quine, W. (1951). Two Dogmas of Empiricism. The Philosophical Review 60: 20–43

Quinn, A., E. E. Adderson, P. G. Shackelford, W. L. Carroll and M. W. Cunningham (1995). Autoantibody germ-line gene segment encodes VH and VL regions of a human anti-streptococcal monoclonal antibody recognizing streptococcal M protein and human cardiac myosin epitopes. J Immunol 154: 4203–12

Raff, R. A. (2000). Evo-devo: the evolution of a new discipline. Nat Rev Genet 1: 74–9

Rand, D. (2001). The units of selection on mitochondrial DNA. Annu Rev Ecol Syst 32: 415–48

Rand, D., R. Haney and A. Fry (2004). Cytonuclear coevolution: the genomics of cooperation. Trends Ecol Evol 19: 645–53

Rao, K. V. (1999). Selection in a T-dependent primary humoral response: new insights from polypeptide models. APMIS 107: 807–18

Rao, S. P., J. M. Riggs, D. F. Friedman, M. S. Scully, T. W. LeBien, et al. (1999). Biased VH gene usage in early lineage human B cells: evidence for preferential Ig gene rearrangement in the absence of selection. J Immunol 163: 2732–40

Ray, S. K., C. Putterman and B. Diamond (1996). Pathogenic autoantibodies are routinely generated during the response to foreign antigen: a paradigm for autoimmune disease. Proc Natl Acad Sci U S A 93: 2019–24

Read, A. F. and J. E. Allen (2000). Evolution and immunology. The economics of immunity. Science 290: 1104–5

Redfield, R. J. (2002). Is quorum sensing a side effect of diffusion sensing? Trends Microbiol 10: 365–70

Reeve, H. (2000). Multi-level selection and human cooperation. Evol Human Behavior 21: 65–72

Richards, M. H. and J. L. Nelson (2000). The evolution of vertebrate antigen receptors: a phylogenetic approach. Mol Biol Evol 17: 146–55

Rigby, M. C., R. F. Hechinger and L. Stevens (2002). Why should parasite resistance be costly? Trends Parasitol 18: 116–20

Rigby, M. C. and J. Jokela (2000). Predator avoidance and immune defence: costs and trade-offs in snails. Proc R Soc Lond B Biol Sci 267: 171–6

Rioux, J. D. and A. K. Abbas (2005). Paths to understanding the genetic basis of autoimmune disease. Nature 435: 584–9

Ritchie, A. J., A. O. Yam, K. M. Tanabe, S. A. Rice and M. A. Cooley (2003). Modification of in vivo and in vitro T- and B-cell-mediated immune responses by the Pseudomonas aeruginosa quorum-sensing molecule N-(3-oxododecanoyl)-L-homoserine lactone. Infect Immun 71: 4421–31

Roach, J. C., G. Glusman, L. Rowen, A. Kaur, M. K. Purcell, et al. (2005). The evolution of vertebrate Toll-like receptors. Proc Natl Acad Sci U S A 102: 9577–82

Rockman, M. V., M. W. Hahn, N. Soranzo, D. B. Goldstein and G. A. Wray (2003). Positive selection on a human-specific transcription factor binding site regulating IL4 expression. Curr Biol 13: 2118–23

Roitt, I. M., M. F. Greaves, G. Torrigiani, J. Brostoff and J. H. Playfair (1969). The cellular basis of immunological responses. A synthesis of some current views. Lancet 2: 367–71

Romano, P., Z. He and S. Luan (2004). Introducing immunophilins. From organ transplantation to plant biology. Plant Physiol 134: 1241–3

Ronnblom, L. and G. V. Alm (2001). An etiopathogenic role for the type I IFN system in SLE. Trends Immunol 22: 427–31

Rook, G. A. and J. L. Stanford (1998). Give us this day our daily germs. Immunol Today 19: 113–6

Roost, H. P., M. F. Bachmann, A. Haag, U. Kalinke, V. Pliska, et al. (1995). Early high-affinity neutralizing anti-viral IgG responses without further overall improvements of affinity. Proc Natl Acad Sci U S A 92: 1257–61

Rose, N. R. (1998). The role of infection in the pathogenesis of autoimmune disease. Semin Immunol 10: 5–13

Rose, N. R. and M. Afanasyeva (2003). From infection to autoimmunity: the adjuvant effect. ASM News 69: 132–37

Rose, N. R. and C. Bona (1993). Defining criteria for autoimmune diseases (Witebsky's postulates revisited). Immunol Today 14: 426–30

Rose, S. (1997). Lifelines. Penguin Books, London

Rosebury, T. (1962). Microorganisms indigenous to man. McGraw-Hill Book Company, Inc., New York

Rosenberg, A. (1994). Instrumental biology or the disunity of science. The University of Chicago Press, Chicago

Rothenberg, E. V. (2000). Stepwise specification of lymphocyte developmental lineages. Curr Opin Genet Dev 10: 370–9

Rotrosen, D., J. B. Matthews and J. A. Bluestone (2002). The immune tolerance network: a new paradigm for developing tolerance-inducing therapies. J Allergy Clin Immunol 110: 17–23

Roy, S., M. T. Scherer, T. J. Briner, J. A. Smith and M. L. Gefter (1989). Murine MHC polymorphism and T cell specificities. Science 244: 572–5

Royall, R. (1997). Statistical evidence. A likelihood paradigm. Chapman & Hall/CRC, Boca Raton

Rutherford, S. L. and S. Lindquist (1998). Hsp90 as a capacitor for morphological evolution. Nature 396: 336–42

Råberg, L., M. Grahn, D. Hasselquist and E. Svensson (1998). On the adaptive significance of stress-induced immunosuppression. Proc R Soc Lond B Biol Sci 265: 1637–41

Råberg, L., J. C. de Roode, A. S. Bell, P. Stamou, D. Gray, et al. (2006). The role of immune-mediated apparent competition in genetically diverse malaria infections. American Naturalist 168: 41–53

Sakaguchi, S., N. Sakaguchi, M. Asano, M. Itoh and M. Toda (1995). Immunologic self-tolerance maintained by activated T cells expressing IL-2 receptor alpha-chains (CD25). Breakdown of a single mechanism of self-tolerance causes various autoimmune diseases. J Immunol 155: 1151–64

Saleh, M., J. P. Vaillancourt, R. K. Graham, M. Huyck, S. M. Srinivasula, et al. (2004). Differential modulation of endotoxin responsiveness by human caspase-12 polymorphisms. Nature 429: 75–9

Salvetti, M., G. Ristori, R. Bomprezzi, P. Pozzilli and R. D. Leslie (2000). Twins: mirrors of the immune system. Immunol Today 21: 342–7

Salzet, M. (2001). Vertebrate innate immunity resembles a mosaic of invertebrate immune responses. Trends Immunol 22: 285–8

Sambrano, G. R., G. Chandy, S. Choi, D. Decamp, R. Hsueh, et al. (2002). Unravelling the signal-transduction network in B lymphocytes. Nature 420: 708–10

Santelices, B. (1999). How many kinds of individual are there? Trends in Ecology and Evolution 14: 152–155

Sanz, I., P. Kelly, C. Williams, S. Scholl, P. Tucker, et al. (1989). The smaller human VH gene families display remarkably little polymorphism. Embo J 8: 3741–8

Sartre, J. P. (1943). Being and nothingness. An essay on phenomenological ontology. Routledge (1958), London

Sato, A. and A. Iwasaki (2004). Induction of antiviral immunity requires Toll-like receptor signaling in both stromal and dendritic cell compartments. Proc Natl Acad Sci U S A 101: 16274–9

Schlichting, C. and M. Pigliucci (1998). Phenotypic evolution. A reaction norm perspective. Sinauer Associates, Inc., Sunderland

Schlosser, G. (2002). Modularity and the units of evolution. Theory in Biosciences 121: 1–80

Schluter, S. F., R. M. Bernstein, H. Bernstein and J. J. Marchalonis (1999). 'Big Bang' emergence of the combinatorial immune system. Dev Comp Immunol 23: 107–11

Schluter, S. F. and J. J. Marchalonis (2003). Cloning of shark RAG2 and characterization of the RAG1/RAG2 gene locus. Faseb J 17: 470–2

Schwartz, M. (2002). Autoimmunity as the body's defense mechanism against the enemy within: Development of therapeutic vaccines for neurodegenerative disorders. J Neurovirol 8: 480–5

Secombes, C. J. and L. Pilstrom (2000). Evolution of adaptive immunity. Biologist (London) 47: 44–8

Segerstråle, U. (2000). Defenders of the truth. The sociobiology debate. Oxford University Press, Oxford

Semple, S., G. Cowlishaw and P. M. Bennett (2002). Immune system evolution among anthropoid primates: parasites, injuries and predators. Proc R Soc Lond B Biol Sci 269: 1031–7

Sen, J. (2001). Signal transduction in thymus development. Cell Mol Biol (Noisy-le-grand) 47: 197–215

Sercarz, E., F. Celada, N. A. Mitchison and T. Tada, Eds. (1988). The semiotics of cellular communication in the immune system. Springer-Verlag, Berlin

Sercarz, E. E. (1998). Immune focusing vs diversification and their connection to immune regulation. Immunol Rev 164: 5–10

Serwold, T., F. Gonzalez, J. Kim, R. Jacob and N. Shastri (2002). ERAAP customizes peptides for MHC class I molecules in the endoplasmic reticulum. Nature 419: 480–3

Shaheen, S. O., P. Aaby, A. J. Hall, D. J. Barker, C. B. Heyes, et al. (1996). Measles and atopy in Guinea-Bissau. Lancet 347: 1792–6

Shapiro, J. A. (1998). Thinking about bacterial populations as multicellular organisms. Annu Rev Microbiol 52: 81–104

Sharov, A. (1999). The origin and evolution of signs. Semiotica 127: 521–35.

Sheldon, B. and S. Verhulst (1996). Ecological immunology: costly parasite defences and trade-offs in evolutionary ecology. Trends Ecol Evol 11: 317–21

Shi, Y., J. E. Evans and K. L. Rock (2003). Molecular identification of a danger signal that alerts the immune system to dying cells. Nature 425: 516–21

Shimamura, M., Y. Y. Huang and H. Goji (2003). Antibody production in early life supported by maternal lymphocyte factors. Biochim Biophys Acta 1637: 55–8

Shintani, S., J. Terzic, A. Sato, M. Saraga-Babic, C. O'HUigin, et al. (2000). Do lampreys have lymphocytes? The Spi evidence. Proc Natl Acad Sci U S A 97: 7417–22

Shoenfeld, Y. and D. A. Isenberg (1989). The mosaic of autoimmunity. Immunol Today 10: 123–6

Shudo, E. and Y. Iwasa (2001). Inducible defense against pathogens and parasites: optimal choice among multiple options. J Theor Biol 209: 233–47

Shudo, E. and Y. Iwasa (2002). Optimal defense strategy: storage vs. new production. J Theor Biol 219: 309–23

Shudo, E. and Y. Iwasa (2004). Dynamic optimization of host defense, immune memory, and post-infection pathogen levels in mammals. J Theor Biol 228: 17–29

Silverstein, A. M. (1989). A history of immunology. Academic Press Inc., San Diego

Silverstein, A. M. (1991). The dynamics of conceptual change in twentieth century immunology. Cell Immunol 132: 515–31

Silverstein, A. M. (1999). Paul Ehrlich's passion: the origins of his receptor immunology. Cell Immunol 194: 213–21

Silverstein, A. M. (2002). The Clonal Selection Theory: what it really is and why modern challenges are misplaced. Nat Immunol 3: 793–6

Simpson, E. H. (1951). The interpretation of interaction in contingency tables. J Royal Stat Soc B 2: 238–41

Singh, S. M., B. Murphy and R. O'Reilly (2002). Epigenetic contributors to the discordance of monozygotic twins. Clin Genet 62: 97–103

Sitki-Green, D. L., R. H. Edwards, M. M. Covington and N. Raab-Traub (2004). Biology of Epstein-Barr virus during infectious mononucleosis. J Infect Dis 189: 483–92

Skjervheim, H. (1959). Objectivism and the study of man. Universitetsforlaget, Oslo.

Skjervheim, H. (1996). Selected essays. The Department of Philosophy, Bergen

Smith, D. C. (1979). From extracellular to intracellular: the establishment of a symbiosis. Proc R Soc Lond B Biol Sci 204: 115–30

Sober, E. (1984). The nature of selection. Evolutionary theory in philosophical focus. The University of Chicago Press, Chicago

Sober, E. (1993). Philosophy of biology. Oxford University Press, Oxford

Sober, E. (1999). Testability. Proceedings and Addresses of the American Philosophical Association. 73: 47–76

Sober, E. (2002). Instrumentalism, parsimony, and the Akaike framework. Philosophy of Science 69: S112–S123

Sober, E. and D. S. Wilson (1998). Unto others. The evolution and psychology of unselfish behaviour. Harvard University Press, Cambridge

Soriano, P. and R. Jaenisch (1986). Retroviruses as probes for mammalian development: allocation of cells to the somatic and germ cell lineages. Cell 46: 19–29

Spinoza, B. D. (1677). Ethics. Everyman (1993), London

Stafford, H. A., C. J. Anderson, D. B. Blalock and M. Reichlin (1998). Development of the anti-ribosomal P autoantibody response. Clin Exp Rheumatol 16: 119–24

Starzl, T. E. and R. M. Zinkernagel (1998). Antigen localization and migration in immunity and tolerance. N Engl J Med 339: 1905–13

Staub, E., P. Fiziev, A. Rosenthal and B. Hinzmann (2004). Insights into the evolution of the nucleolus by an analysis of its protein domain repertoire. Bioessays 26: 567–81

Stauffer, Y., S. Marguerat, F. Meylan, C. Ucla, N. Sutkowski, et al. (2001). Interferon-alpha-induced endogenous superantigen. a model linking environment and autoimmunity. Immunity 15: 591–601

Stebbins, C. E. and J. E. Galan (2001). Structural mimicry in bacterial virulence. Nature 412: 701–5

Sterelny, K. (2003). Thought in a hostile world. The evolution of human cognition. Blackwell Publishing, Oxford

Sterelny, K. and P. Griffiths (1999). Sex and death. An introduction to philosophy of biology. The University of Chicago Press, Chicago

Stewart, J. (1992). Immunoglobulins did not arise in evolution to fight infection. Immunol Today 13: 396–9; discussion 399–400

Stewart, J. J. (1998). The female X-inactivation mosaic in systemic lupus erythematosus. Immunol Today 19: 352–7

Stoner, D. S., B. Rinkevich and I. L. Weissman (1999). Heritable germ and somatic cell lineage competitions in chimeric colonial protochordates. Proc Natl Acad Sci U S A 96: 9148–53

Stoner, D. S. and I. L. Weissman (1996). Somatic and germ cell parasitism in a colonial ascidian: possible role for a highly polymorphic allorecognition system. Proc Natl Acad Sci U S A 93: 15254–9

Strachan, D. P. (1989). Hay fever, hygiene, and household size. BMJ 299: 1259–60

Strassmann, J. E., Y. Zhu and D. C. Queller (2000). Altruism and social cheating in the social amoeba Dictyostelium discoideum. Nature 408: 965–7

Sutkowski, N., B. Conrad, D. A. Thorley-Lawson and B. T. Huber (2001). Epstein-Barr virus transactivates the human endogenous retrovirus HERV-K18 that encodes a superantigen. Immunity 15: 579–89

Swets, J. (1996). Signal detection theory and ROC analysis in psychology and diagnostics: Collected papers. Lawrence Erlbaum Associates, Publishers, Mahwah

Tada, T. (1988). Contexts in the cellular communications: How the mobile cells recognize right partner cells to talk with? In: The semiotics of cellular communication in the immune system. E. Sercarz, F. Celada, N. A. Mitchison and T. Tada (Eds.). Springer-Verlag, Berlin, 83–104

Talmage, D. W. (1959). Immunological specificity, unique combinations of selected natural globulins provide an alternative to the classical concept. Science 129: 1643–8

Talmage, D. W. (1986). The acceptance and rejection of immunological concepts. Annu Rev Immunol 4: 1–11

Talmage, D. W. (1988). A century of progress: beyond molecular immunology. J Immunol 141: S5–16

Tamas, I., L. Klasson, B. Canback, A. K. Naslund, A. S. Eriksson, et al. (2002). 50 million years of genomic stasis in endosymbiotic bacteria. Science 296: 2376–9

Tan, E. M. (1996). Autoantibodies–what do they recognize? Verh Dtsch Ges Pathol 80: 1–11

Tanchot, C., M. M. Rosado, F. Agenes, A. A. Freitas and B. Rocha (1997). Lymphocyte homeostasis. Semin Immunol 9: 331–7

Tang, G., Y. Minemoto, B. Dibling, N. H. Purcell, Z. Li, et al. (2001). Inhibition of JNK activation through NF-kappaB target genes. Nature 414: 313–7

Tanigaki, K., M. Tsuji, N. Yamamoto, H. Han, J. Tsukada, et al. (2004). Regulation of alphabeta/gammadelta T cell lineage commitment and peripheral T cell responses by Notch/RBP-J signaling. Immunity 20: 611–22

Tauber, A. I. (1990). Metchnikoff, the modern immunologist. J Leukoc Biol 47: 561–7

Tauber, A. I. (1994). The immune self. Theory or metaphor? Cambridge University Press, Cambridge

Tauber, A. I. (2000). Moving beyond the immune self? Semin Immunol 12: 241–8; discussion 257–344

Tauber, A. I. (2003). Metchnikoff and the phagocytosis theory. Nat Rev Mol Cell Biol 4: 897–901

Tauber, A. and S. Podolsky (1997). The generation of diversity. Clonal selection theory and the rise of molecular immunology. Harvard University Press, Cambridge

Taylor, C. (1995). Overcoming epistemology. In: Philosophical arguments. C. Taylor (Ed.). Harvard University Press, Cambridge, 1–19

Thornton, C. A., S. Welle, R. C. Griggs and G. N. Abraham (1996). Human IgG production in vivo: determination of synthetic rate by nonradioactive tracer incorporation. J Immunol 157: 950–5

Tollrian, R. and C. D. Harvell, Eds. (1999). The ecology and evolution of inducible defences. Princeton University Press, Princeton

Tonegawa, S. (1983). Somatic generation of antibody diversity. Nature 302: 575–81

Towers, G. J., T. Hatziioannou, S. Cowan, S. P. Goff, J. Luban, et al. (2003). Cyclophilin A modulates the sensitivity of HIV-1 to host restriction factors. Nat Med 9: 1138–43

Trahey, M. and I. L. Weissman (1999). Cyclophilin C-associated protein: a normal secreted glycoprotein that down-modulates endotoxin and proinflammatory responses in vivo. Proc Natl Acad Sci U S A 96: 3006–11

Troen, B. R. (2003). The biology of aging. Mt Sinai J Med 70: 3–22

Turner, M. C., Y. Chen, D. Krewski and P. Ghadirian (2006). An overview of the association between allergy and cancer. Int J Cancer 118: 3124–32

Uexküll, J. (1982). The theory of meaning (Original 1940). Semiotica 42: 25–82

Uinuk-Ool, T., W. E. Mayer, A. Sato, R. Dongak, M. D. Cooper, et al. (2002). Lamprey lymphocyte-like cells express homologs of genes involved in immunologically relevant activities of mammalian lymphocytes. Proc Natl Acad Sci U S A 99: 14356–61

Ulvestad, E. (2003). Modelling autoimmune rheumatic disease: a likelihood rationale. Scand J Immunol 58: 106–11

Ulvestad, E., K. Williams, R. Bjerkvig, K. Tiekotter, J. Antel, et al. (1994). Human microglial cells have phenotypic and functional characteristics in common with both macrophages and dendritic antigen-presenting cells. J Leukoc Biol 56: 732–40

Ulvestad, E., S. Berentsen, K. Bo and F. V. Shammas (1999). Clinical immunology of chronic cold agglutinin disease. Eur J Haematol 63: 259–66

Ulvestad, E., A. Kanestrom, T. M. Madland, E. Thomassen, H. J. Haga, et al. (2000). Evaluation of diagnostic tests for antinuclear antibodies in rheumatological practice. Scand J Immunol 52: 309–15

Ulvestad, E., A. Kanestrom, T. M. Madland, E. Thomassen and H. J. Haga (2001). Clinical utility of diagnostic tests for rheumatoid factor. Scand J Rheumatol 30: 87–91

Ulvestad, E., J. H. Aarseth, C. Vedeler, H. Nyland and K. M. Myhr (2004). The effects of interferon-alpha2a on concentrations of immunoglobulins, complement and lymphocytes in patients with multiple sclerosis. Scand J Immunol 59: 103–8

van de Lagemaat, L. N., J. R. Landry, D. L. Mager and P. Medstrand (2003). Transposable elements in mammals promote regulatory variation and diversification of genes with specialized functions. Trends Genet 19: 530–6

van den Berg, T. K., J. A. Yoder and G. W. Litman (2004). On the origins of adaptive immunity: innate immune receptors join the tale. Trends Immunol 25: 11–6

van Dijk, B. A., D. I. Boomsma and A. J. de Man (1996). Blood group chimerism in human multiple births is not rare. Am J Med Genet 61: 264–8

Van Eden, W., R. Van Der Zee, P. Van Kooten, S. E. Berlo, P. M. Cobelens, et al. (2002). Balancing the immune system: Th1 and Th2. Ann Rheum Dis 61 Suppl 2: ii25–8

van Gent, D. C., K. Mizuuchi and M. Gellert (1996). Similarities between initiation of V(D)J recombination and retroviral integration. Science 271: 1592–4

van Inwagen, P. (1993). Metaphysics. Oxford University Press, Oxford

Van Regenmortel, M. H. (1998). From absolute to exquisite specificity. Reflections on the fuzzy nature of species, specificity and antigenic sites. J Immunol Methods 216: 37–48

van Valen, L. (1973). A new evolutionary law. Evol Theory 1: 1–30

Varela, F. (1979). Principles of biological autonomy. North Holland, New York

Varela, F. (1991). Organism: A meshwork of selfless selves. In: Organism and the origins of self. A. I. Tauber (Ed.). Kluwer Academic Publishers, Dordrecht, 79–107

Vastenhouw, N. L. and R. H. Plasterk (2004). RNAi protects the Caenorhabditis elegans germline against transposition. Trends Genet 20: 314–9

Velicer, G. J. (2003). Social strife in the microbial world. Trends Microbiol 11: 330–7

Velicer, G. J., L. Kroos and R. E. Lenski (2000). Developmental cheating in the social bacterium Myxococcus xanthus. Nature 404: 598–601

Ventura-Holman, T. and C. J. Lobb (2002). Structural organization of the immunoglobulin heavy chain locus in the channel catfish: the IgH locus represents a composite of two gene clusters. Mol Immunol 38: 557–64

Vergani, D. and G. Mieli-Vergani (2002). Autoimmunity after liver transplantation. Hepatology 36: 271–6

von Bertalanffy, L. (1962). General system theory – a critical review. General Systems 7: 1–20

Wagner, G. P. and L. Altenberg (1996). Perspective: Complex adaptations and the evolution of evolvability. Evolution 50: 967–976

Walford, R. (1969). The immunologic theory of aging. Munksgaard, Copenhagen

Watson, F. L., R. Puttmann-Holgado, F. Thomas, D. L. Lamar, M. Hughes, et al. (2005). Extensive diversity of Ig-superfamily proteins in the immune system of insects. Science 309: 1874–8

Wedekind, C., T. Seebeck, F. Bettens and A. J. Paepke (1995). MHC-dependent mate preferences in humans. Proc R Soc Lond B Biol Sci 260: 245–9

Wei, X., J. M. Decker, S. Wang, H. Hui, J. C. Kappes, et al. (2003). Antibody neutralization and escape by HIV-1. Nature 422: 307–12

Welsh, R. M. and L. K. Selin (2002). No one is naive: the significance of heterologous T-cell immunity. Nat Rev Immunol 2: 417–26

Wherry, E. J., V. Teichgraber, T. C. Becker, D. Masopust, S. M. Kaech, et al. (2003). Lineage relationship and protective immunity of memory CD8 T cell subsets. Nat Immunol 4: 225–34

Williams, G. C. (1957). Pleiotropy, natural selection, and the evolution of senescence. Evolution 11: 398–411

Williams, G. C. (1966). Adaptation and natural selection. Princeton University Press, Princeton

Williams, K. C., E. Ulvestad and W. F. Hickey (1994). Immunology of multiple sclerosis. Clin Neurosci 2: 229–45

Wilson, D. B., D. H. Wilson, K. Schroder, C. Pinilla, S. Blondelle, et al. (2004). Specificity and degeneracy of T cells. Mol Immunol 40: 1047–55

Wilson, D. (1979). Structured demes and trait-group variation. American Naturalist 113: 606–10

Wilson, D. S. and E. Sober (1989). Reviving the superorganism. J Theor Biol 136: 337–56

Wilson, E. O. (1975). Sociobiology: The new synthesis. Harvard University Press, Cambridge

Wilson, J. (1999). Biological individuality. Cambridge University Press, Cambridge

Wilson, M. S. and R. M. Maizels (2006). Regulatory T cells induced by parasites and the modulation of allergic responses. Chem Immunol Allergy 90: 176–95

Wilson, R. A. (2004). Test cases, resolvability, and group selection: a critical examination of the myxoma case. Philosophy of Science 71: 380–401

Winther, R. G. (2001). Varieties of modules: kinds, levels, origins, and behaviors. J Exp Zool 291: 116–29

Wittgenstein, L. (1960). The blue and brown books. Harper & Row, New York

Woese, C. (1998b). The universal ancestor. Proc Natl Acad Sci U S A 95: 6854–9

Woese, C. R. (1998a). Default taxonomy: Ernst Mayr's view of the microbial world. Proc Natl Acad Sci U S A 95: 11043–6

Wolfson, H. A. (1934). The philosophy of Spinoza. Vol II. Harvard University Press, Cambridge

Wright, L. (1973). Functions. Philosophical Review 82: 139–68

Xu, G. L., T. H. Bestor, D. Bourc'his, C. L. Hsieh, N. Tommerup, et al. (1999). Chromosome instability and immunodeficiency syndrome caused by mutations in a DNA methyltransferase gene. Nature 402: 187–91

Yamada, T. and I. Yamanaka (1995). Microglial localization of alpha-interferon receptor in human brain tissues. Neurosci Lett 189: 73–6

Yoo, J. Y. and S. Desiderio (2003). Innate and acquired immunity intersect in a global view of the acute-phase response. Proc Natl Acad Sci U S A 100: 1157–62

Youssoufian, H. and R. E. Pyeritz (2002). Mechanisms and consequences of somatic mosaicism in humans. Nat Rev Genet 3: 748–58

Yu, C. Y. and C. C. Whitacre (2004). Sex, MHC and complement C4 in autoimmune diseases. Trends Immunol 25: 694–9

Yu, Z., M. R. Theoret, C. E. Touloukian, D. R. Surman, S. C. Garman, et al. (2004). Poor immunogenicity of a self/tumor antigen derives from peptide-MHC-I instability and is independent of tolerance. J Clin Invest 114: 551–9

Zahavi, A. and A. Zahavi (1997). The handicap principle. A missing piece of Darwin's puzzle. Oxford University Press, Oxford

Zepik, H. H., E. Blochliger and P. L. Luisi (2001). A Chemical Model of Homeostasis. Angew Chem Int Ed Engl 40: 199–202

Zhang, J., C. Niu, L. Ye, H. Huang, X. He, et al. (2003). Identification of the haematopoietic stem cell niche and control of the niche size. Nature 425: 836–41

Zhang, S. M., C. M. Adema, T. B. Kepler and E. S. Loker (2004). Diversification of Ig superfamily genes in an invertebrate. Science 305: 251–4

Ziegler, A., H. Kentenich and B. Uchanska-Ziegler (2005). Female choice and the MHC. Trends Immunol 26: 496–502

Zimmer, C. (2001). Microbiology. Do chronic diseases have an infectious root? Science 293: 1974–7

Zinkernagel, R. M. (2002). On differences between immunity and immunological memory. Curr Opin Immunol 14: 523–36

Zinkernagel, R. M. and P. C. Doherty (1974). Restriction of in vitro T cell-mediated cytotoxicity in lymphocytic choriomeningitis within a syngeneic or semiallogeneic system. Nature 248: 701–2

Zuk, M. and A. M. Stoehr (2002). Immune defense and host life history. American Naturalist 160: S9–S22

Zuniga, M. C. (2002). A pox on thee! Manipulation of the host immune system by myxoma virus and implications for viral-host co-adaptation. Virus Res 88: 17–33

Index

Abduction, 26–28
Adapter molecules, 77
Adaptive, 11, 15, 16, 22, 34, 37, 45–48, 51, 53, 55, 59, 67, 68, 73, 76–79, 81, 82, 89, 90, 103, 104, 108, 110–112, 124, 127, 128, 139, 145, 146, 149–151, 156, 159, 160, 162–166, 168, 170, 173–175, 178, 179, 187, 191–193, 195–199, 204, 207–210
Adaptive plasticity, 81–115, 127, 175
Adjuvant, 13, 100, 196
Ageing, 16, 177–179
Akaike's theorem, 41, 42
Allergy, 17–25, 78, 79, 106, 198–201
Altruism, 14, 136, 138–140, 181
Anaphylaxis, 17
Anthropocentric bias, 65, 66
Antibodies, 8–10, 12, 13, 17, 18, 28, 38, 39, 49, 51, 53–55, 57, 59–64, 99, 113, 114, 142, 163, 165, 170, 171, 179, 186, 187, 190, 196–198, 208, 209
Antigen presenting cells, 12–14, 43, 55, 77, 107, 109, 194, 207, 209, 210
Antigens, 9–14, 10, 16, 26, 39, 42, 43, 49–64, 76–79, 90–92, 96, 99, 107, 109, 110, 113, 138, 158, 161, 162, 164–171, 174, 185–187, 189, 190, 194, 196–109, 200, 207–210
Antitoxin, 8, 9
Apoptosis, 52, 134–136, 140, 160, 194–196
Aristotle, 72, 119, 133
Astrobiology, 119
Atomistic fallacy, 75
Autoimmune disease, 8, 48, 63, 64, 68, 78, 90, 91, 109, 113, 169, 183–197, 199
Autoimmunity, 16, 17, 79, 96, 117, 183, 186–195, 201
Autophagy, 134, 135
Autopoiesis, 132–134
Autoreactive, 165, 169–172, 186–189, 192, 193, 196, 197
Autoreactivity, 90, 149, 171, 186, 187, 189–191
Auxiliary assumptions, 32

Bacteriophages, 155
Bayes' theorem, 32–34, 103, 104
BCG, 17
Behavioural ecologists, 45
Behe, M., 35, 36
Behring, E., 8, 9, 38
Being-in, 77, 86
Being-in-the-world, 26, 65, 80, 83, 87, 94, 103, 173
Benign coexistence, 159
Big bang, 162–164
Biosemiotics, 101, 102, 107, 120, 127, 187
Birth and death model, 165, 167, 169, 170
Bloodletting, 4
Bone marrow, 12, 51, 64, 65, 133, 135, 207, 209
Bordet, J., 18–20
Botryllus schlosseri, 143
Brent, L., 63
Burnet, M., 10, 11, 29, 63, 64, 88, 159, 160
Buss, L., 126, 142, 152, 178, 180

Carbapenem, 141, 142
Carrier, 11–13, 102
Caspase-12, 198
CCR5, 147, 198
CD14, 198
CD4 T helper cells, 209
CD8 T cells, 51, 62, 209
Central tolerance, 64, 188
Charlton, B., 178, 179
Cheating, 108, 136, 143, 145, 180, 181
Cheetah, 162
Chimera, 63, 73, 182
Chimerism, 63, 65, 143, 144, 181–185, 187, 190, 192, 201
Chimpanzee, 162, 169
Chronic infections, 2, 56, 192
Clonal selection theory, 10, 11, 13, 28, 29, 110
Co-operation, 14, 15, 17, 99, 107, 126, 136–138, 140, 143, 144, 152, 180, 181
Co-selection, 176
Coevolution, 37, 153, 156–159
Cohen, I., 21, 40, 42, 62, 159
Cohn, M., 12, 40, 42, 49, 64, 88, 91, 167, 168, 170
Cold agglutinins, 171
Coley, W., 16
Colon, 16
Columbus, 202
Compartmentalisation, 182

Competition, 14, 15, 66, 126, 137, 153, 155, 156, 182, 192
Complexity, 5, 6, 24, 37, 40–43, 46, 47, 79, 84, 86, 95, 117, 122, 124, 152
Conflict, 7, 11, 15, 16, 18, 19, 45, 88, 108, 119, 122, 125, 126, 136–138, 140, 144, 153, 159, 179–182, 185, 186, 191–197, 201
Conflict-mediating, 15, 16
Conflict-mediation, 180, 195
Conflict modifiers, 181, 192–195, 201
Constitutive defence, 118, 150, 151
Contraceptives, 183
Convergence, 123, 124
Cooperating, 15, 46
Costly defences, 147, 150
Cross-reactivity, 38, 61
Cut-off, 61, 62, 92, 112–115
Cyclophilins, 194, 195
Cyclosporine A, 194
Cytokine, 39, 47, 77, 110, 161, 162, 188, 189, 194, 198, 200, 201, 208, 209

Damasio, A., 14, 91
Danger model, 14, 91
Darwin, C., 5–7, 9, 14, 15, 61, 66, 71, 72, 84, 87, 96, 102, 121, 122, 126, 129, 136, 137, 140, 173, 174, 176, 205
Dasein, 94, 98, 99
Davies, P., 176
Dawkins, R., 7, 120, 154, 180
Decision, 27, 43, 54, 61, 62, 76, 77, 79, 88, 102–105, 109, 110, 112, 113, 141, 193, 199
Deduction, 26, 27
Degeneracy, 43, 62, 166
Dembski, W., 119
Dendritic cells, 13, 14, 64, 77–79, 185, 196, 227–229
Descartes, R., 19, 20, 22, 23, 93, 95, 125, 129, 130
Design, 5, 35–37, 39, 119, 129
Design principles, 48, 50
Determinate possible, 204, 205
Development, 16, 25, 26, 28, 45–47, 53, 54, 64, 67, 68, 72, 73, 78, 82, 84, 85, 96, 108, 118, 123, 125–128, 133–135, 143, 145, 154, 165, 177, 180, 184, 185, 191, 193, 194, 196, 198, 200
Dictyostelium discoideum, 143–145, 156
DimA, 144, 145
Discrimination, 54, 64, 88, 89, 91, 110, 113, 115, 133, 147
Distal cause, 74, 75

Dreyer-Bennet hypothesis, 11, 43
Drive, 77, 78, 123, 129, 131, 135, 145, 153, 156, 191, 192, 199
Dualism, 20, 21, 23, 24, 97
Duffy, 147
Dynamic engagement, 50

Ebert, D., 157
Eco-devo, 67
Ecological fallacy, 75
Ecological immunology, 67
Ecological validity, 43, 76, 78
Ecology, 45, 66, 67, 76, 141, 202, 205
Ecosystem, 24, 45, 53, 101, 118, 155
Ehrlich, P., 8–10, 17–20, 45, 59–61, 63, 80, 88, 129, 130
Embodied drive, 86, 129, 131, 135, 191, 192, 199
Emotion, 88, 132
Empedocles, 3
Endosymbiosis, 151, 152
Environments, 1, 5, 6, 23, 24, 40, 43, 48, 53–55, 65, 67, 68, 71, 74, 76, 77, 80–82, 84–87, 90, 92, 94, 103, 105, 106, 110, 115, 117, 118, 125, 128, 131, 133, 135, 136, 142, 146, 156, 163, 170, 176–178, 198–202
Epicurus, 129
Epidemic, 2, 4, 6, 92, 104, 160, 201–204
Epigenetic, 128, 184
Epistemology, 20, 21, 35, 42–44, 83, 89, 93, 102
Epstein-Barr virus, 169, 182, 188
Erwinia carotovora, 141, 142
Eschericia coli, 156
Essentialist, 61, 72
Estrogen, 68
Eukaryotes, 123, 135, 141, 152, 193–195
Evo-devo, 67
Evo-eco, 67
Evolutionary biology, 76, 126, 127, 174, 205
Evolutionary ecology, 45
Evolutionary theory, 71, 141, 177, 205
Evolvability, 145, 146, 184, 198
Ewald, P., 158, 159
Existentialist, 24, 25, 98
Expected importance, 113, 151
Explain, 1, 8, 10, 13, 27, 28, 31, 34, 37, 40, 42, 45, 48, 50, 59, 63, 66, 69, 75, 78, 84, 88, 105, 108, 109, 129–131, 136, 138, 140, 142, 144, 145, 152, 157, 160, 164, 167, 170, 173–175, 177, 181, 183, 184, 188, 191, 198–201, 206

Explanation, 3, 5, 9, 16–18, 21, 31, 37, 40, 45, 46, 58, 65, 68, 70, 71, 74, 75, 78, 91, 92, 94, 96, 110, 117, 129, 130, 134, 136–138, 144, 149, 160, 161, 170, 171, 173–175, 178, 188, 189, 193, 194, 203
Extraterrestrial life, 22, 23

False positive, 33, 111–115
Fc-receptor, 99, 100, 208
Fenner, F., 63, 160
Fitness, 15, 43, 46, 52, 62, 69, 70, 73, 76, 78, 81, 97, 98, 104–106, 108, 111, 112, 118, 127, 131, 135, 137, 140, 142, 144–151, 153, 155, 157–159, 161, 166, 168, 173, 175–177, 179, 191, 198, 199, 203, 204
Fleck, L., 29, 30, 82
Flexible, xiii, xviii, 34, 45, 54, 67, 71, 86, 103, 163, 171
Frank, S. A., 144, 152, 157
Function, 1, 12, 14, 15, 32, 35, 40, 45–51, 53, 56, 61, 62, 65, 66, 68, 72, 76, 85, 89, 90, 99, 102, 112, 118, 120, 121, 125, 128, 134, 135, 145, 147, 148, 149, 151, 152, 163–168, 170, 173–175, 177, 193, 195–198, 207–209
Functional, 11, 15, 30, 38, 43, 46, 55–57, 59, 62, 79, 91, 97, 109, 127, 128, 130, 141, 150, 153, 156, 167–169, 173–175, 193, 195, 199, 200, 203, 209, 210

Gene duplications, 35, 164, 167, 170
Germ, 11, 15, 16, 29, 46, 126, 144, 147, 152, 154, 168, 177, 180, 181, 197
Germain, R., 28
Gershon, R., 30
Gilman, A., xvi
Goal-directed behaviour, 119, 129, 131, 174
Golub, E., xvii, 4, 7, 8, 68, 106
Goodness-of-fit, 43, 59–61
Gould, S. J., 34, 122, 123, 146, 194
Greek medicine, 4
Group, 24, 28–30, 32, 40, 45, 46, 57, 58, 66, 67, 71, 73–76, 105, 108, 114, 124, 136–144, 146, 156, 160, 161, 179–181, 184, 186, 194, 202, 209
Grünbaum, A., 45

Hamilton's rule, 138
Hapten, 11–13
Heat shock proteins, 14, 76, 194, 195
Hedonic tone, 132
Hegel, G. W. F., 23, 83, 95, 206
Heidegger, M., xv, xvi, xviii, 20, 23, 86, 87, 93–95, 100, 102, 130, 131
Helminth parasites, 201
Hepatitis C virus, 182, 196
Herd immunity, 75, 76, 146
Hippocrates, 3
History, xiii, xvii, xviii, 1, 2, 8, 23–25, 29, 35, 36, 45, 63, 65, 66, 68, 74, 81, 83, 85–87, 95, 97, 101, 106, 107, 110, 119, 124, 125, 128, 142, 147–149, 159, 163, 173, 179, 188, 205, 206
HIV, 40, 147, 157, 158, 162, 182, 198
Hoffmeyer, J., 101
Homo sapiens, 2, 61, 65, 81
Homology, 124, 168
Honest signs, 103, 108
Horror autotoxicus, 88, 130
Hosts, xi, xii, 2, 4, 6, 14–16, 37, 45, 47, 48, 50, 54, 56–58, 61, 63–65, 67, 68, 77, 87–90, 100, 103, 108, 118, 121, 132, 141, 149, 152–161, 169, 181–183, 185, 188, 190–192, 195, 197, 200–206
Human genome, 124, 152, 154, 171, 182
Human leukocyte antigen (HLA), 158, 183, 209
Humoral theory, 3, 4, 45
Humours, 3, 4, 5,
Husserl, E., 93, 95, 102
Hygiene hypothesis, 106, 200, 201
Hypervariable, 60, 169
Hypothesis, xi, xii, xv, xvi, 4, 5, 11, 12, 16, 18, 26–29, 31–37, 40–43, 45, 49, 55, 58, 69, 71, 75, 81, 84, 91, 104, 120, 142, 145, 147, 149, 155, 157, 159, 162–165, 169, 170, 178, 183, 190, 191, 193, 194, 196, 199–201, 203

Idiotype network theory, 49
IgA, 54, 113, 199, 209, 209
IgE, 48, 113, 199, 201, 208, 209
IgG, 48, 54, 100, 113, 149, 196, 208–210
IgM, 54, 113, 171, 208–210
IL4, 198, 201
Immortality, 118, 177
Immune system, xi–xiii, xv, xvii–xix, 1, 11, 14, 16, 17, 20, 21, 24–26, 34, 35, 37, 39, 40, 42–56, 58, 59, 61–63, 65–68, 71, 73, 76–80, 82, 93, 88–94, 96–100, 102, 103, 105–109, 111–113, 117, 188, 127, 128, 133, 135, 145, 146, 148, 149, 154, 157–159, 162–165, 167, 168, 170, 173–175, 178–184, 187–193, 195–201, 203–205, 207–210
Immunobiology, xiii, 45, 66, 67
Immunocompetence handicap hypothesis, 69
Immunocompetence, xii, 67–71, 148, 198, 209
Immunodeficient, 67, 190

Immunodominance, 76, 78, 79
Immunoeducation, 25, 26
Immunoglobulin, 11, 12, 47, 51, 53, 54, 56, 59, 60, 62, 92, 99, 100, 113, 128, 154, 162–167, 169–172, 186, 197, 209
Immunoglobulin superfamily, 47, 164, 170
Immunology, xiii, xv–xviii, 1, 8, 10, 11, 14, 18, 19, 27–30, 36, 37, 39, 45, 59, 62–67, 71, 75, 83, 90, 102, 105–107, 205
Immunosemiosis, xvii, 105, 107
Immunosenescence, xiii, 173
Immunotherapy, 8, 16, 17, 26
Indeterminate possible, 204, 205
Indians, 202
Individuality, 20, 142, 144, 179, 180
Inducible defences, 150
Induction, 26, 27, 37, 38, 65, 135, 158, 169, 179, 189
Infection, 2, 4–6, 16, 25, 26, 38, 39, 47, 50, 56–58, 67, 69, 70, 75–78, 92, 101, 104, 109, 111, 112, 147–150, 155, 158, 160, 162, 163, 169, 171, 178, 181, 182, 185, 189, 190, 192, 194–196, 198, 200–202
Inflammation, xiii, 7, 15, 134, 178, 179, 194, 195
Innate, 16, 22, 28, 46–48, 68, 70, 76, 77, 79, 83–85, 89, 98, 106, 111, 112, 149–151, 162, 164, 174, 178, 191–193, 195–199, 207, 208, 210
Innateness, 85
Instructive theories, 9, 10
Instrumental realism, 42–44
Instrumentalism, 40
Instrumentalists, 40
Integrated explanations, 88, 110
Integrity, 7, 15, 17, 109, 132, 141, 180, 181, 191
Intelligent design, 35, 119
Intentional explanation, 174
Interfering RNAs, 195
Interferon-α, 169, 188, 196
Interferon-β, 209
Interpretant, 98–100, 107
Intersubjectivity, 18

Janeway, C. A., 13, 28, 31, 66, 91
Jawed vertebrates, 46, 47, 103, 146, 155, 162–165
Jenner, E., 8, 36, 38, 57
Jerne, N. K., 10, 28, 29, 49, 59, 170, 205

Kant, I., xviii, 22, 23, 83, 84, 86, 87
Kerr, B., 156
Kierkegaard, S., 95, 96, 205
Kimura, M., 122
Kin recognition, 139
Kingdoms, 46, 124, 206
Kitasato, S., 8
Klein, J., 88–90, 105, 122, 138, 161, 165, 183
Know thyself, 93, 97
Koch, R., 7
Kuhn, 29, 30

Lamarckian mechanism, 55
Landsteiner, K., 12, 13, 17, 60, 61, 63
Laws, 1, 22, 97, 98, 120, 129
Lehrman, D., 84–86
Leibniz, G. W., 125, 126
Lens metaphor, 76, 79, 80, 208
Lewontin, R. C., 24, 34, 85, 86, 123, 127, 140, 142, 160, 194
Life, xii, xii, xvi, xvii, 1–3, 5–7, 20, 22, 23, 25, 45–47, 63, 65, 68, 71, 74, 82, 84, 86, 98, 99, 101, 102, 104, 108, 117–125, 123–139, 127–130, 142, 144, 146–149, 152, 169, 173, 176–180, 182, 183, 187, 193, 200, 202, 204–207
Likelihood, 5, 32–36, 41, 42, 59, 104, 112, 115, 151
Likelihood ratio, 33, 34, 104, 112, 115, 190
Linkage, 118, 145, 151, 176, 196
Lorenz, K., 84, 85, 87
Lucky split, 162, 164
Lymnaea stagnalis, 148
Lymphocyte, xiii, 10–13, 46–58, 64, 76–79, 99, 107, 109–111, 113, 142, 149, 154, 158, 160, 163–165, 168, 172, 178, 181, 183, 186, 187, 189, 191, 192, 196, 197, 208–210

Macrophage, 13, 99, 100, 178, 196, 208–210
Major histocompatibility complex (MHC), 12, 13, 51–53, 60, 77, 78, 89, 138, 139, 161, 162, 165–169, 176, 183, 184, 187, 188, 208–210
Malaria, 147, 158, 176, 200, 202
Malfunction, 106, 173, 175–177, 181, 188, 191–193, 198–200
Malfunctional, xviii, xix, 79, 109, 173, 176, 177, 191, 193, 199, 200, 205
Malignant, xiii, 15–17, 108, 117, 178, 182, 185–187, 191, 193
Maternal antibody transmission, 53–55
Maturana, H., 132–134
Matzinger, P., 14, 65, 90, 183
Maynard Smith, J., 120, 134, 142, 152, 178
Mayr, E., 74, 126, 131, 174
Mazumdar, P. M., xvii, 8, 9, 59–61

Mechanism, xii, xii, xvii, 6, 9, 10, 12, 15, 16, 18, 28, 34, 35, 37, 39, 45–48, 50, 52, 53, 55, 56, 58, 59, 63–67, 69, 70, 73, 75, 77, 79, 84, 86, 88, 101, 113, 117, 118, 120, 122, 130, 134–136, 138–142, 144–151, 154, 157, 159–165, 168, 171, 172, 174, 177, 178, 180–182, 187–189, 191–197, 199–201, 204, 208
Medawar, P. B., 27, 28, 63
Memory, 12, 37, 49, 54–59, 62, 78, 198
Mental plasticity, 82
Metaimmunology, 19, 20
Metaphysics, 20, 61, 86, 87, 89, 98, 101, 125, 126, 173
Metchnikoff, E., 7–9, 15, 16, 45, 80, 108, 178, 179
Methylation, 184
Michod, R., 16, 140, 144, 145, 147, 179, 181
Microbes, xiii, 2, 4, 5, 9, 13–16, 24, 26, 38, 39, 47, 61, 65, 83, 92, 103, 111, 123, 124, 130, 141, 143, 152, 155–157, 159, 200, 202, 203, 205
Microchimerism, 183
Microglia, 196
Migration pathways, 56
Mitochondrion, 136, 152, 153, 181, 184, 191, 194, 195, 197
Modern synthesis, 126, 127
Module, 22, 53, 80, 127, 128, 131, 135, 141, 145, 146, 168
Morris, S. C., 123
Mosaicism, 63, 153, 181, 184, 185, 201
Multi-level selection, 140, 145
Multi-level selection theory, xviii, 16, 76, 140, 146
Multiple sclerosis, 58, 184, 186–188, 194, 196
Multiply realised, xi, 57, 188
Mutualism, 136, 152, 154
Mutualistic, 136
Mycobacterium tuberculosis, 182
Myxococcus xanthus, 143
Myxomavirus, 160, 161

Natura naturans, 117, 130, 203–205
Natura naturata, 130, 203–205
Natural killer cells, 207, 208
Natural selection, xvii, 5, 6, 9, 10, 14–16, 23, 24, 28, 34, 35, 52, 69, 96–98, 117, 120–122, 124–127, 129–113, 135–138, 140, 142, 143, 147, 151, 155
Network, xv, 35, 39, 40, 46–50, 70, 85, 98, 99, 103, 107, 108, 127, 133, 170, 200, 201, 207, 208
Neutral theory, 122, 166
Non-self, xii, xviii, 13, 14, 63, 66, 89–92, 96, 105, 117, 146, 147, 170, 171, 179–181, 188, 192, 196
Nonsynonymous changes, 166
Norm of reaction, 85
Notch, 77, 78
Nothingness, 105, 106

Ockham, William of, 72
Ockham's razor, 41
Odour, 139, 183
Ontology, 20, 44, 89, 90
Optimal defence, 149
Organismal self, 88–91, 96
Organisms, xi, xii, xvi, xix, 1, 4, 5, 10, 15, 17, 23, 24, 35, 37, 39, 43, 45–48, 50, 52–56, 58, 61–64, 67, 68, 70–73, 76, 77–82, 84–89, 92, 94–99, 101, 103–105, 108–112, 115, 117, 118, 120–156, 165–168, 170, 173–181, 184, 188, 191, 194, 197–202, 204, 207, 209
Original antigenic sin, 58, 158
Ornamentation, 69–71
Orosz, C., xvi, xvii, 50, 205, 206
Owen, R., 63, 69
Owl of Minerva, 206

Paley, W., 5, 35, 176
Parallel processing, 50
Parasites, xi–xiii, xvii, 5–7, 14, 15, 47, 50, 61, 67, 68, 88, 89, 118, 138, 147, 148, 154–159, 161, 178, 183, 186, 192, 201, 203, 205, 206
Parasitism, 144, 152, 154, 181, 182, 185, 193
Participant, 24–26, 97, 106, 107
Passer domesticus, 149
Pasteur, L., 4, 8, 9, 36
Pathogenicity, 141, 142, 155–157, 160, 185, 205
Pathology, xviii, 7, 25, 37, 147
Pattern recognition receptors, 13, 14, 76, 207
Pedagogue, 25, 26
Peirce, C. S., xviii, 19, 27, 28, 31, 35, 84, 98–100, 102, 106
Peripheral tolerance, 64, 65, 188, 193
Phagocytic cells, 7, 8, 171
Phenomenological, xiii, 93, 94, 102
Phenomenology, 93, 98, 102
Phenotypic plasticity, 81, 86, 205
Philosophy, xvii, 20, 72, 83, 87, 93, 94, 98, 102, 129, 206
Phylogenic layering, 50
Phylogenomics, 124
Pirquet, C. von, 17

Plato, 71, 72, 87, 175
Pleiotropy, 176
Pluralist, 60
Popper, K., 27, 81
Predict, 16, 37, 40–43, 47, 52, 203, 204
Predictions, xvi, 26, 27, 31, 34, 37, 41, 46, 170, 194, 203
Predictive accuracy, 37, 40–43, 151
Preunderstanding, 23
Price equation, 140
Primary immune response, 54, 56, 78, 209
Probability, 31–36, 74, 103, 104, 112, 113, 138, 143, 151, 163, 166, 190
Probiotics, 16
Programmed cell death, 134, 135, 139, 181
Prokaryotes, 123–125, 135, 141, 143, 193, 194
Protecton, 168
Proximate causes, 74, 75, 174, 178, 188, 199, 201

Quasispecies, 182
Quine, W., 31, 32
Quorum sensing, 141, 142

Rabbit, 8, 160, 161
RAG, 51, 155, 162, 164, 165
Random genetic drift, 122, 166
Re-enacting of ancient conflicts, 179, 191, 193
Realism, 40, 42–44
Realists, 40
Recombination, 10, 11, 51, 52, 167, 169
Regulatory T cells, 30, 52, 54, 179, 188, 189, 195, 200, 209
Relations, xi, xviii, 7, 16, 18, 19, 21, 23, 30, 45, 48, 56, 57, 68, 75, 76, 83, 87, 95, 96, 98–102, 107, 117, 158, 201, 203, 206
Reliable signs, 108, 151
res cogitans, 19, 22, 23
res extensa, 19, 22, 23, 93
Retroviral elements, 154, 195
Rheumatoid arthritis, 32, 33
Rheumatoid factor, 32, 33
Richet, C., 17
Rock-paper-scissors, 156, 186, 190
Rokitansky, K., 7
RRNA, 123, 124, 153
Rules, 43, 54, 97, 98, 119, 138, 142, 146, 183, 204

SARS, 202–204
Sartre, J. P., 97, 106
Schwann, T., 6
Science, xii, xv–xix, 1, 6, 14, 18–22, 24, 27–30, 36, 37, 40, 48, 61, 63, 71, 72, 81–83, 87, 105, 107, 119, 120, 203, 205
Secondary immune response, 54, 59
Segerstråle, U., 23
Selective filter, 52, 168
Self and nonself, xviii, 13, 19, 88, 89, 91, 105, 170, 188, 196
Self-organising behaviour, 49, 52
Self-preservation, 66, 129–131, 140
Self, xi, xii, xvii, xviii, 13, 15–17, 19–21, 62, 64, 66, 88–99, 104, 105, 131, 133, 146, 147, 170, 180, 181, 188, 189, 196, 205, 208
Selfish, 52, 136, 137, 140, 141, 143, 153, 154, 161, 180, 181, 191
Semiosis, 99–102, 106–109, 120, 205
Senescence, 177–179
Sensory drive, 156
Sexual ornamentation, 69
Sherlock Holmes, 31
Sickle-cell disease, 176
Side-chain, 8–10
Sign, xvii, 54, 78, 87, 88, 97, 99–105, 107, 109–112, 115, 120, 151, 156, 175, 188
Signal, 11–13, 47, 49, 52, 64, 70, 71, 76, 77, 79, 92, 97, 101, 102, 108–111, 135, 141, 142, 156, 181, 184, 193, 194
Signal detection theory, 61, 110, 111
Signs, xviii, 2, 3, 33, 54, 71, 76, 86–88, 97–103, 106, 108–111, 115, 129, 146, 151, 163, 175, 186, 187, 200
Silverstein, A. M., xvii, 1, 5, 8, 11, 59, 66, 130
Simplicity, xii, xvi, 40–44, 95
Simpson's paradox, 75, 76
Situated, 2, 73, 89, 94, 105, 107, 135
Sjögren's syndrome, 183, 190
Skjervheim, H., 23
Smallpox, 36–38, 43, 58, 61, 68, 202
Sober, E., 5, 32, 33, 41, 76, 85, 140, 144, 160, 161
Social behaviour, 109, 136, 144
Somatic, 11, 15, 47, 108, 144, 145, 147, 153–155, 162, 163, 165, 168, 171, 172, 180, 181, 184, 191
Spatio-temporal mismatch, 173, 199, 201
Species, xvii, 2, 5–7, 9, 14, 16, 17, 35, 38, 42, 46, 47, 59–62, 65, 66, 72, 73, 80, 84, 85, 87, 90, 118, 119, 123–125, 128, 129, 136–138, 141, 143, 145, 146, 148, 151, 156, 159, 162, 165, 166, 177–179, 183, 185, 186, 188, 201, 202, 205
Specificity, xvii, 8–10, 12, 28, 37, 38, 42, 43, 54, 59–62, 64, 109, 112, 113, 145, 168, 209

Spectator, 23–26, 93, 97
Spinoza, B. de, 129, 130
Spiteful, 136, 139, 140
Spontaneous generation, 6, 120
Stem cell, 51, 183
Stewart, J., 93, 170, 185
Stoic, 129
Superantigen, 169
Superfamily, 47, 164, 170
Suppressor cells, 30, 58
Symbiogenesis, 151, 152
Symbiosis, 151
Synonymous changes, 166
Synthetase 2–5A, 195, 196
System, xi–xiii, xv, xvii–xix, 1, 11, 14, 16, 17, 20, 21, 24–26, 29, 34, 35, 37, 39, 40, 42–56, 58, 59, 61–63, 65–68, 71, 73, 76–80, 82, 83, 85, 88–94, 96–113, 115, 117, 118, 120, 123, 124, 127, 128, 131–135, 139, 141, 142, 145, 146, 148–151, 154, 156–159, 162–165, 167, 168, 170–175, 178–184, 187–193, 195–201, 203–205, 207–210
Systemic lupus erythematosus, 58, 184, 186, 187, 190
Systems biology, 85
Szathmáry, E., 120, 134, 143, 152, 178

T cell receptor, 11, 13, 51–53, 60, 62, 65, 92, 128, 162–171, 179, 182, 197, 210
T helper cell, 12, 51, 77, 199, 209
Talmage, D. W., 10, 29, 30, 62
Tauber, A. I., xiii, xvii, 7, 8, 11, 15, 18, 88, 90, 92, 93, 164
Taylor, C., 20, 21, 23
Teleology, 130, 131, 173
Teleonomy, 131
Testosterone, 68–71
Th1, 39, 48, 77, 78, 161, 200, 201, 209
Th2, 39, 48, 77, 78, 161, 200, 201, 209
Thales, 3
The epistemological construal, 21–24
The immune self, xvii, 88–94, 96, 99, 205
The immunological self, 89–92, 95, 96
Theory, xii, xvii, xviii, 3–11, 14–16, 18, 23, 24, 27–30, 36, 37, 39, 42, 45, 48, 49, 61, 63, 64, 71, 73, 76, 82–84, 96, 109–111, 122, 123–135, 125, 126, 129, 137, 140, 141, 146, 149, 152, 159, 166, 177, 188, 204, 205
Thought collective, 29, 30
Three domain world, 124
Thucydides, 3
Tolerance, 12, 16, 56, 58, 63–65, 89, 90, 128, 147, 159, 183, 188, 189, 193, 200
Toll-like receptors, 13, 14, 77–79, 194, 207
Tonegawa, S., 11, 43
Trade-off, 1, 41, 65, 68, 70, 71, 117, 118, 148, 149, 177, 189, 203
Trait-group selection, 144, 160, 161
Transcendental, 22, 23, 82, 83, 119, 205
Transcendental self, 97
Transdifferentiate, 51, 183
Transmissible tumours, 185
Transplanted, xiii, 186, 193, 194
Transposable elements, 154, 155, 164
Triadic relations, 98–100
True positives, 111–115
Trypanosoma brucei, 26, 39
Tumour necrosis factor, 135, 187
Tunable activation threshold hypothesis, 109, 110
Twin, 37, 54, 63, 90, 128, 183–185, 187, 205
Two-stage selection, 168

Uexküll, J. von, 87, 102
Ultimate causes, 71, 74, 131, 178
Umwelt, 87, 94, 98, 102, 188
Unconceived alternatives, 31
Unitarian, 60, 61, 125
Unnatural immunity, 39

Vaccine, 17, 25, 36–40, 57, 159, 203, 204
Vaccinology, 1, 36, 37, 39, 205
Varela, F., 94, 132–134
Variable connectivity, 50
V-D-J segments, 171
VH4-34, 171, 172
Virchow, R., 7
Virulence, 45, 47, 68, 73, 90, 156–161
Vitalism, 119, 120, 129–132, 173
Von Bertalanffy, L., 48

Walford, R., 178, 179
War metaphor, 14, 15, 17, 25, 203
Weismann, A., 126
Williams, G. C., 137, 146, 177, 178, 188
Wilson, E., 23
Wissenschaft, 18
Witebsky, E., 17, 189

X chromosome, 185
X-ray crystallography, 60

Zoonotic diseases, 2